BIOLOGICAL AND MEDICAL PHYSICS, BIOMEDICAL ENGINEERING

Please view available titles
in *Biological And Medical Physics, Biomedical Engineering*
on series homepage http://www.springer.com/series/3740

BIOLOGICAL AND MEDICAL PHYSICS, BIOMEDICAL ENGINEERING

The fields of biological and medical physics and biomedical engineering are broad, multidisciplinary and dynamic. They lie at the crossroads of frontier research in physics, biology, chemistry, and medicine. The Biological and Medical Physics, Biomedical Engineering Series is intended to be comprehensive, covering a broad range of topics important to the study of the physical, chemical and biological sciences. Its goal is to provide scientists and engineers with textbooks, monographs, and reference works to address the growing need for information.

Books in the series emphasize established and emergent areas of science including molecular, membrane, and mathematical biophysics; photosynthetic energy harvesting and conversion; information processing; physical principles of genetics; sensory communications; automata networks, neural networks, and cellular automata. Equally important will be coverage of applied aspects of biological and medical physics and biomedical engineering such as molecular electronic components and devices, biosensors, medicine, imaging, physical principles of renewable energy production, advanced prostheses, and environmental control and engineering.

Nancy J. Woolf
Avner Priel
Jack A. Tuszynski

Nanoneuroscience

Structural and Functional Roles of the Neuronal Cytoskeleton in Health and Disease

With 49 Figures

Springer

Dr. Nancy J. Woolf
University of California
Los Angeles
Department of Psychology
Lab. NanoNeuroscience
Los Angeles CA 90095-1563
USA
nwoolf@ucla.edu

Prof. Jack A. Tuszynski
University of Alberta
Department of Physics
Edmonton AB T6G 1Z2
Canada
jtus@phys.ualberta.ca

Dr. Avner Priel
University of Alberta
Department of Physics
Edmonton AB T6G 1Z2
Canada
apriel@phys.ualberta.ca

ISSN 1618-7210
ISBN 978-3-642-03583-8 e-ISBN 978-3-642-03584-5
DOI 10.1007/978-3-642-03584-5
Springer Heidelberg Dordrecht London New York

Library of Congress Control Number: 2009942091

Cover design: SPi Publisher Services

Printed on acid-free paper

Springer is part of Springer Science+Business Media (www.springer.com)

Preface

We wrote this book to describe a new emerging discipline that integrates neuroscience and nanoscience yielding a unique perspective on the very complex organization of the material substrate for cognitive processes. Nanoneuroscience focuses on computationally relevant biomolecules found inside neurons. Because of recent technological advances at the nanometer scale, scientists have at their disposal increasingly better ways to study the brain and the biophysics of its molecules. During the past century the focus in neurobiology has been on the neuron and its synapses. Today we can expand on these basic principles to include the biomolecules that determine operations of synapses and other computationally relevant functions performed inside dendrites. By expanding our scope of knowledge of what participates in neural computation, we exponentially expand the mind-brain computer analogy through the realization that each neuron has a processing capability conceivably reaching or even exceeding that of a silicon-based multiprocessor. Put simply, our synapses feed information into a massively powerful intraneuronal matrix of cables and conduits of information – a system we refer to as Nature's nanowires.

Although we review all the internal structures inside neurons, a central theme of the book is how neurotransmitters act upon receptor molecules, triggering signal transduction molecules that affect cytoskeletal filaments inside dendrites of neurons, whereupon cytoskeletal proteins contribute to information processing and fundamental operations of neurons. Multiple findings are presented supporting the notion that naturally occurring nanowires are not only basic to intracellular transport, but also of fundamental importance to learning, memory, and possibly even higher consciousness. We present our own research, as well as research coming out of other laboratories, with particular emphasis on the most recent findings. Why do we hypothesize that the interiors of neurons, in particular the cytoskeletal filaments, play a role in higher cognitive function and how might they achieve that role? For one, the cytoskeleton undergoes structural change during learning. Second, an abnormal cytoskeleton appears to be a primary etiological factor contributing to neurological disorders, such as Alzheimer's disease, and to psychiatric disorders, such as bipolar affective disorder and schizophrenia – disorders associated with deficits in memory function or impaired mental state. Third, cytoskeletal proteins are capable of propagating signals enabling them to form intracellular circuits that compute and then transmit information form one part of the neuron to another. This capability is well suited to accommodate cognitive operations.

This book is intended as an accessible resource for those with interests in neural computation or brain mechanisms of higher cognition or disease states. Interested readers might be those with backgrounds in neuroscience, physics, physiology, psychology, biophysics, biochemistry, computer science, or bioengineering – or virtually anyone with a curiosity about the brain-mind interface and what nanotechnology has revealed and might be expected to demonstrate in the years ahead. The book is virtually self-contained but we have provided the reader with numerous references so that more in-depth studies of individual topics covered in this book can be initiated.

Many colleagues have facilitated our efforts in the writing of this book. We would like to thank all current and past collaborators who have collected data or contributed to the core ideas conveyed in this book. First and foremost, Adele Behar contributed extensively to the research performed through her generous tangible support and much appreciated encouragement. In fact, we would like to dedicate this book to her as a token of our appreciation for her steadfast support over a number of years and for her encouragement and motivation to continue against many odds. Michael Weiner is similarly thanked for his gracious tangible support and numerous fascinating ideas and suggestions. Stuart Hameroff and others at the Center for Consciousness Studies at the University of Arizona are owed a great debt for many of the original ideas reviewed and extended in this book. Horacio Cantiello and his colleagues at Harvard University are credited with those breakthroughs leading to the measurement of signal conduction along biomolecules. Travis Craddock is gratefully acknowledged for his Master's thesis work on the double-well potential in the macromolecule tubulin, a particularly critical piece in the puzzle of how cytoskeletal proteins might compute outcomes. Lastly, Michelle Hanlon is thanked for her extensive work on finalizing the manuscript.

Los Angeles, USA
Edmonton, Canada
July 2009

Nancy J. Woolf
Jack A. Tuszynski
Avner Priel

Contents

List of Acronyms

+TIPs: plus-end tracking proteins
ADAS: Alzheimer's Disease Assessment Scale
ADF: actin depolymerizing factor
ADHD: attention deficit/hyperactivity disorder
AFM: atomic force microscopy
ALS: amyotrophic lateral sclerosis
AMPA: alpha-amino-3-hydroxy-5-methyl-4-isoxazolepropionic acid
ApoE: apolipoprotein E
APP: amyloid precursor protein
Arp2/3: actin-related protein complex
ASPM: microencephaly-associated
AVV: adeno-associated virus
BDNF: brain-derived neurotrophic factor
BECON: Bioengineering Consortium
CaMK: calcium/calmodulin-dependent kinase
cGMP: cyclic guanine monophosphate
CLIP: cytoplasmic linker protein
CNS: central nervous system
CPEB: cytoplasmic polyadenylation element binding protein
CTNF: ciliary neurotrophic factor
DCX: doublecortin
DISC: disrupted in schizophrenia
DISC1: disrupted-in-schizophrenia
EEG: electroencephalogram
F-actin: filamentous actin
FAD+: flavin adenine dinucleotide
fMRI: functional magnetic resonance imaging
FMRP: fragile X mental retardation protein
FRET: fluorescence resonance energy transfer
G-actin: globular actin
GDNF: glial-derived neurotrophic factor
GDP: guanine diphosphate
GFP: green fluorescent protein
GSK: glucagon synthase kinase
GTP: guanine triphosphate
HAP: huntingtin-associated protein
HIP: huntingtin-interacting protein
IP_3: inositol 1,4,5-triphosphate
LIS: lissencephaly
LTP: long-term potentiation
MAP: microtubule associated protein
MARK: MAP/microtubule affinity-regulating kinase
MECP: methyl CpG-binding protein

MEMS: microelectromechanical systems
mGlu: metabotropic glutamate receptor
MMSE: Mini Mental Status Exam
MPTP: 1-methy-4-phenyl-1,2,3,6-tetrahydropyridine
MRI: magnetic resonance imaging
mRNA: messenger RNA
MT: microtubule
mtDNA: mitochondrial DNA
NAD+: nicotinamide adenine dinucleotide
NASA: National Aeronautics and Space Administration
nCAM: neural cell adhesion molecule
NF-H: neurofilament heavy
NF-L: neurofilament light
NF-M: neurofilament medium
NGF: nerve growth factor
NMDA: N-methyl-D-aspartate
NRTN: neurturin
NT: neurotrophin
PI-PLC: phosphoinositide-specific phospholipase-C
PK: protein kinase
PKA: cAMP-dependent protein kinase
PSD: postsynaptic density protein
PSEN: presenilin
RTK: receptor tyrosine kinases
SER: smooth endoplasmic reticular
SNPs: single nucleotide polymorphisms
SOD1: superoxide dismutase
STM: scanning tunneling microscopy
STOP: stable-tubule-only-polypeptide

1

Introducing Nanoneuroscience as a Distinct Discipline

Summary

Nanoneuroscience is a new emerging discipline that seeks to solve certain hitherto intractable problems in the neurosciences using nanoscientific perspectives and tools. These state-of-the-art methods stand to meet some of the most challenging feats in neuroscience, such as finding better means of diagnosis, treatment, and prevention for various neurological, neurodevelopmental, and neuropsychiatric disorders. Nanotechnology is arguably one of the most optimal ways currently available to address the core essence of higher cognitive functions. A nanoscale emphasis on the mechanical interactions of biomolecules is uniquely capable of demonstrating the multiple ways in which neurons communicate and transmit signals, ranging from the traditional means of interneuronal and intraneuronal communication to novel modes of biomolecular computation. Notable milestones in nanoscience include the development of instruments and techniques enabling interactions with small surfaces or individual molecules, such as scanning tunneling microscopy (STM), atomic force microscopy (ATM), and nanotweezers. These tools operate in the nanometer size range and have the potential to reveal details about molecular events and subcellular operations within neurons. Nanoscientists have also developed a wide variety of nanomaterials – carbon nanotubes, nanoparticles, nanowires, and quantum dots, among others – that can be used to probe and stimulate neurons or parts of neurons. Nanoparticle-based drug delivery systems (or gene therapy delivery systems) showing enhanced ability to cross the blood-brain barrier could potentially be used to treat a number of neurological, neurodevelopmental, and neuropsychiatric diseases. Nanomaterials, used alone and in hybrid combinations with other materials, can be used to diagnose nervous system disorders, to measure neurotransmitter levels or electrical activity in discrete brain sites, to stimulate discrete brain sites, and finally, to build potential nanoscale prosthetic devices that restore normal neural activity patterns and cognitive function.

1.1 The Definition of Nanoneuroscience

Nanoneuroscience is a science that bridges neuroscience and nanotechnology by concurrently addressing the fundamental goals of these (until recently) two separate fields. The quintessential goal of neuroscience is to understand how the nervous system operates – how it processes information to bring about various actions and mental states – and then to apply that basic knowledge to practical issues such as managing nervous system disease. Nanotechnology focuses on the study of both biological and non-biological materials at the very small end of the length scale [1]. Materials having dimensions in the range of 0.1 - 100 nm are in the domain of nanotechnology [2]. This is the size range of many biomolecules, the primary determinants of neural function.

In those cases where diseases of the nervous system are the primary focus, nanoneuroscience has obvious overlap with nanomedicine – the biomedical application of nanotechnology. The mission of nanomedicine was defined succinctly in June of 2000, when over 600 multidisciplinary scientists attending a conference organized by the U.S. government NIH Bioengineering Consortium (BECON) decided upon the following topics as research priorities for the next decade [3]:

- Methods for fabricating nanostructures.
- Therapeutic applications of nanotechnology.
- Biomimetic nanostructures.
- Biological nanostructures.
- The nanoscale electronic-biological interface.
- Nanodevices for early detection of disease.
- Nanoscale study of individual molecules.
- Nanotechnology and tissue engineering.

Other U.S. governmental agencies have devoted major effort and funding support to nanomedicine and similar efforts are ongoing around world, particularly in the most technologically advanced countries of Western Europe and Asia. The NIH Nanomedicine Roadmap Initiative has in its first phase funded several centers working on the chemical and physical properties of nanoscale biomolecules, many of which constitute the living cell [4]. The second phase of this initiative will include the development of nanoscale devices capable of sensing a chemical imbalance or disease state, or of drug-delivery. Although a significant portion of nanomedicine focuses on diseases of the nervous system, nanoneuroscience is not merely a subset of nanomedicine since it has potential applications to nanoelectronics and biological computer design. The U.S. National Aeronautics and Space Administration (NASA) has designated the design of a nanoelectronic "brain" for space exploration as one of its near-term goals; however, engineering more realistic brain-like computing is expected to extend beyond the year 2030 [5]. Nanoneuroscience research stands to affect the course of those missions.

As shown in Figure 1.1, nanoneuroscience links neuroscience to nanoscience with particular emphasis on applying new techniques to elucidate the cellular and molecular underpinnings of disorders affecting the nervous system, as well as to better understand the cellular and molecular bases of behavior and cognition. Nervous system disorders amenable to nanoneuroscience investigations include Alzheimer's disease, Parkinson's disease, Huntington's disease, major depression, bipolar disorder, schizophrenia, and a variety of neurodevelopmental disorders. In addition to the unique perspective nanoscience offers in terms

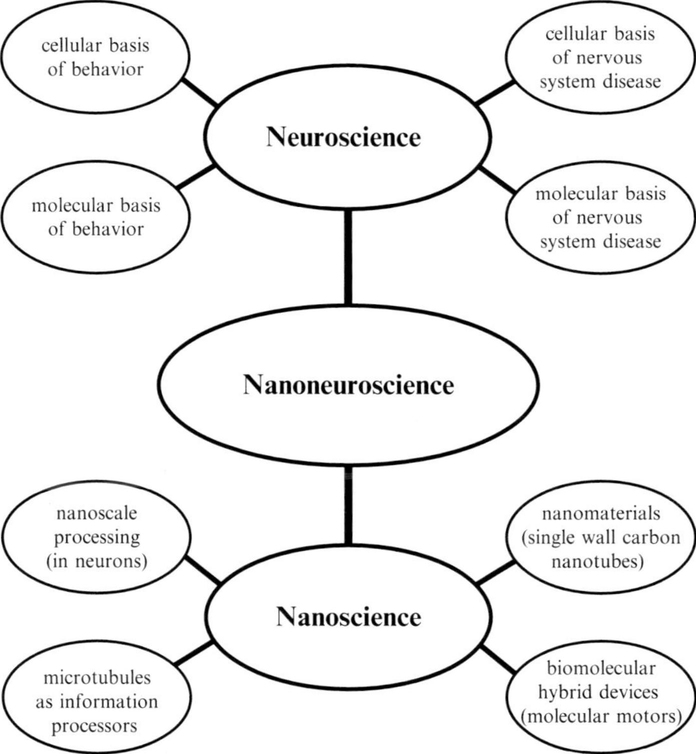

Fig. 1.1. Relationships of nanoneuroscience to other disciplines

of utilizing nanoscale level processing of information, there are novel nanomaterials, such as single-walled carbon nanotubes and biomolecular/silicon-based hybrid devices that can be used to study information processing in neurons, and ultimately to repair dysfunctional neurons [6, 7]. Nanotechnological tools operate at the appropriate scale to realistically interact with neurons – the nerve cells that are universally regarded as the computationally relevant cells of the nervous system – as well as with the smaller glial cells. These small-scale tools and devices are expected to advance molecular imaging, tissue engineering, and drug delivery well beyond what is currently possible [8].

1.2 Current Issues in Neuroscience

Cutting-edge research in the neurosciences can be broadly broken down into two categories: (1) understanding the biophysical basis of cognitive phenomena such as perception, learning, memory, language, thinking, attention, and awareness, and (2) making progress towards finding treatments or preventions for neurological, neuropsychiatric, and neurodevelopmental diseases – many of which are lacking satisfactory treatments at present. Nanoscience has great potential with respect to both these avenues, because not only do the tools of nanoscience enable greater precision in measurement, the nanoscience perspective differs from traditional approaches in neuroscience with its emphasis on biomolecular computation. Traditional neuroscience conceptualizes compu-

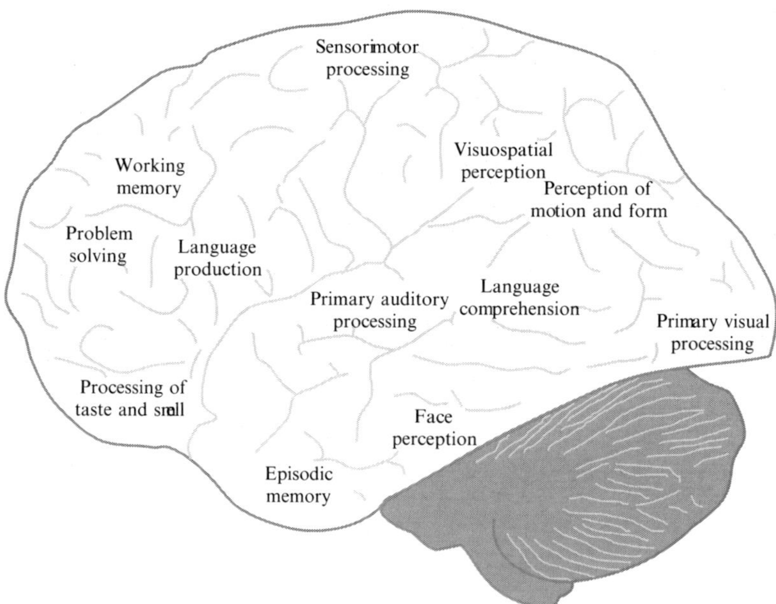

Fig. 1.2. The neocortex is the site of higher cognitive functions. The outer portion of the cerebral cortex, which is mostly neocortex, can be divided up into functionally specific areas. Although these divisions are to some degree valid, each cortical area participates in more than one function, and multiple cortical areas contribute to individual functions.

tation at the level of the neuron or neural network. Our current understanding of the most complex cognitive functions of the nervous system is primarily one of specific neurocognitive networks [9]. Individual higher cognitive functions are attributed to neural networks in specific regions of the brain (Figure 1.2); nonetheless, it is well known that this principle of localized function has its limitations [10]. For one, the restricted localization of a given cognitive function is at variance with widespread areas of brain contributing to individual cognitive functions and with specific brain areas participating in multiple cognitive functions [11]. Secondly, assigning a particular cognitive function to a circumscribed brain site does not explain how the brain executes that particular cognitive function. Although many computational models attempt to explain how the brain accomplishes various higher cognitive functions [12, 13, 14, 15], the precise mechanisms underlying many higher cognitive functions remain elusive. Nanoneuroscience research may help clarify some of these issues by including biomolecular computations into the overall scheme.

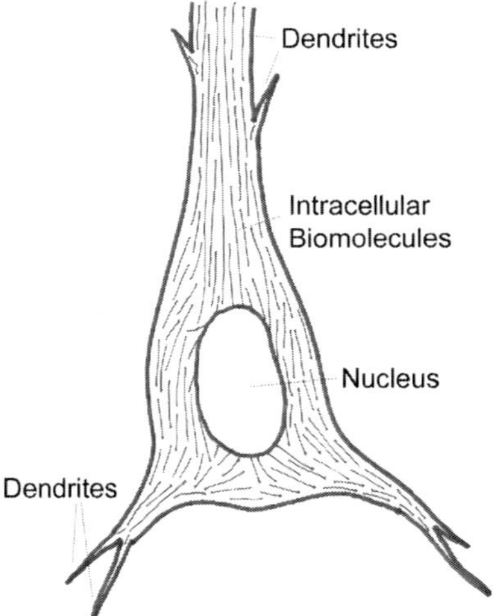

Fig. 1.3. Pyramidal cells are one of the major types of neurons in the neocortex and hippocampus. These large cells have intracellular biomolecules distributed throughout the cell body and dendrites, many of which are cytoskeletal proteins and molecules that interact with them. These cytoskeletal networks were first detected with silver stains as described by Spanish neuroanatomist Santiago Ramón y Cajal [46].

1.2.1 The Great Mysteries of Neuroscience: Higher Cognitive Functions

Although the neuroscientific community presently knows a great deal about the structure of the brain, a sufficiently detailed explanation and realistic picture of how activity in different areas of the brain comes together to bring about complex mental states is lacking. That higher cognitive function depends primarily on selective activation of interconnected circuits in the neocortex makes intuitive sense. The neocortex is the most recently acquired part of the cerebral cortex – having reached large dimensions in mammals, particularly humans (see Figure 1.2). Higher cognitive functions are as a class those mental activities that involve the act of knowing: such as perception, memory, judgment, and reasoning. Consciousness and attention are also cognitive functions; however, they might be alternately viewed as modulators of other cognitive states [16, 17]. Heightened awareness and attention possess the ability to amplify (while low levels of awareness and attention dampen) perception and memory retrieval. Thus, it is generally assumed that higher cognitive function depends on consciousness [18]. This rule is not without exception, however. Unconscious processing of information is detectable on functional magnetic resonance imaging (fMRI) even when a person is not aware of that information processing [19]. Without a doubt, consciousness remains an enigma on many levels.

Given its unique attributes among the cognitive functions, many neuroscientists view consciousness as a separate phenomenon from basic sensory perception and place the neural circuitry for global consciousness in the frontal lobe [20]. Other neuroscientists believe the neural circuitry for consciousness to be more widespread and to overlap with the neural circuitry involved with processing perceptual functions, consciousness for visual events in visual cortex, consciousness for auditory events in auditory cortex, and so on [21, 22]. Although both possibilities may well be partly correct, the nanoneuroscience approach to higher cognitive function slightly favors the latter view, since smaller scale structures (i.e., molecules within neurons) are deemed computationally relevant. According to the nanoneuroscience perspective, the electrical characteristics of molecules and macromolecular structures within the neuron stand to contribute to the neuron's computational capability [23, 24, 25]. The collective computational power of large numbers of macromolecules in neurons would be expectedly large. Moreover, since the molecular composition of neurons does not appreciably vary across cortical regions, all cortical regions should in principle possess roughly equivalent computational power. Accordingly, a large cortical pyramidal cell (see Figure 1.3) would be expected to have a similarly massive computational power regardless of whether it was located in the visual cortex or in the prefrontal cortex because its macromolecular internal structures and their energetic properties are similar. Once these computational mechanisms are more fully understood, nanoneuroscience may enable researchers to establish how the vast number of molecules in a pyramidal cell of primary visual cortex might be devoted to sensory perception, yet be able to simultaneously compute aspects of visual consciousness.

Sensation and Perception: A Case for Second-Order Transduction Based on Biomolecular Computing

Sensation and perception are fundamental cognitive functions, with much of the human cerebral cortex being devoted to processing primary sensation or giving rise to perception. Sensation and perception are often studied in tandem but are actually distinct phenomena. Nanoneuroscience provides tools or new perspectives that could help bridge the explanatory gap between sensation and perception.

Sensation is the operation of the senses or the collection of raw data. The broad stokes of sensory processes have been reasonably well understood for quite some time, and this is especially true color vision [26, 27]. Humans are sensitive to a narrow range of electromagnetic radiation called the visible spectrum. These frequencies far exceed the maximal firing rates of neurons (see Figure 1.4). The same is true for frequencies of high pitch sounds. Due to this mismatch, the nervous system must significantly lower the frequency of high frequency information. Nonetheless, resolution or detail is not lost because the nervous system converts stimuli of higher frequencies into a spatial map. Special receptor organs for each of the senses initiate the conversion process. In the case of vision, there are three kinds of cones in the retina, each of which are maximally sensitive to light having different spectral properties [28, 29] With auditory stimuli, it is the basilar membrane embedded in the cochlea that possesses a tonotopic map, wherein hair cells of the most proximal part of the basilar membrane (where sound waves enter) are sensitive to high pitch tones and hair cells in the most distal part of the basilar membrane are sensitive to low pitch tones [30].

For each of the senses – sight, hearing, touch, taste, smell, balance, and body position – transduction of some external energy occurs. Transduction occurs, for example, when a photochemical response in rods and cones of the retina results in decreased levels of second messenger cyclic guanine monophosphate (cGMP) and subsequent decreased sodium ion influx and decreased release of the neurotransmitter glutamate [32]. Transduction also occurs when waves of airborne molecules vibrate the eardrum causing the ossicles (the smallest bones in the body) in the middle ear to vibrate against the oval window, which in turn causes waves in the fluid-filled cochlea, which perturb the hair cells in the basilar membrane [33].

Unlike the operations of the senses, which are reasonably well understood, perception – which involves forming ideas in the mind that organize, interpret, and give meaning to sensory stimuli – is far from being understood. Neuroscientists have a thorough understanding of the paths that sensory information travels, not only up to the primary sensory areas of the neocortex, but beyond to what are called association areas of the cortex. Located in the higher tiers along the hierarchy of cortical processing, perception of sight, sound, and touch relies heavily on neural circuits in association cortex of the temporal-parietal region and the prefrontal cortex, whereas phylogenetically older, archicortex cortex handles taste and smell. Unification of visual and touch information, for example, occurs in the parietal lobe, which enables visual-spatial processing (such as being able to identify a set of keys based on the way they feel to the touch) [35]. The perception of taste, on the other hand, is based on a fusion of smell and taste information that occurs in the orbitofrontal cortex lying at the base of the frontal lobe [34]. Still, the coming together of separate streams of sensory information does not explain how a mental representation of that information is created. A major hurdle remains in explaining how various stimulus parameters – often deriving from different modalities – unify into distinct indivisible percepts. One possibility is that biomolecules in neurons throughout the cerebral cortex undergo a second-order transduction process (see Figure 1.4). Assuming that at least some subsets of biomolecules in neurons are capable of computation, many second-order transduction processes may exist. Moreover, multiple second-order transduction processes could occur according to different time scales thereby affording the opportunity to encode mental representations in a number of alternate ways.

To summarize, biomolecular second-order transduction processes offer potential solutions to the following difficulties in reconciling explanation gap between sensation and perception:

1. Biomolecular computing in individual neurons, which increases computational power of individual neurons, can in principle explain how perception arises from sensation. Perception is not merely a hierarchical step or two above sensation [36]. Perception is arguably a vastly more complex phenomenon requiring significantly more computational capability than sensation.
2. Perception involves being aware of the entire context within which the object of interest exists; however, it is not uncommon for major scene changes to go completely undetected [37, 38, 39]. This translates into simultaneous computations of vast amounts of previous experiences and updating that information with current levels of attention and arousal. Because of the large number of molecules per neuron, biomolecular computing in individual neurons could in principle process massive quantities of information, making each neuron (as well as the overall network) more computationally powerful.
3. Perceptual processes often fill in missing sensory stimuli. This has been shown for the blind spot and several Gestalt phenomena, such as fill-in, closure, and continuation [40, 41, 42]. Tasks of this type depend on searching massive stores of previous sensory information. When instructions are added or task complexity increases, these perceptual processes can also tap into judgment and reasoning skill. Again, biomolecular computing within individual neurons would be expected to increase computing capability.
4. A second-order transduction process linked to a biomolecular computation could produce a succinct and unique representation that is both

External Energies	**Maximal Firing Rate of Neurons**	**Higher Level Analyses**
Visible light 350 - 700 nm THz range		Molecular vibrations, hinge motion in proteins psec - nsec GHz - THz range
Audible sound waves 50 - 20,000 Hz		Unwinding of DNA μsec Enzyme-catalyzed reactions msec
Mechanical vibration to skin 1 - 200 Hz	100 - 1,000 Hz	Complex field dynamics in brain 0.5 - 100 Hz range

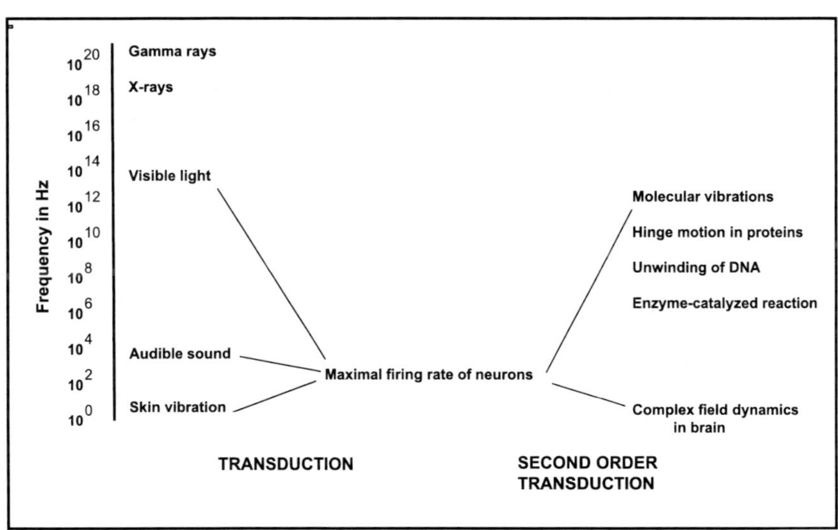

Fig. 1.4. External energies have widely discrepant frequencies. The nervous system uses transduction to convert divergent external energies into neural activity. This process is limited by the maximal firing rate of neurons, and in many cases the maximal firing rate of neurons is much lower than the frequency of the external energy. To overcome the limitations imposed by the maximal firing rate, nanoneuroscience beckons exploration of potential second-order transduction events that would enable biomolecular structures within neurons to compute information relevant to higher cognitive states.

modality-specific and transmittable from one type of cortex to another without losing its modality-specificity.

Nanoneuroscience, in particular the study of biomolecules at the nanoscale level and biomolecular computing, stands to advance our understanding of sensation and perception. Inclusion of biomolecular computing in neural network models will make such models more realistic, while greatly increasing the computational ability of single neurons. Given the vast amount of neuroscientific information presently available, there nonetheless remain key fundamental questions left unanswered. The physical basis for the great divide between sensing and perceiving is one of those uncertainties, but the way in which the nervous system stores information – the processes of learning and memory – also is not completely solved.

Learning and Memory

A nanoneuroscience perspective is needed to advance our understanding of learning and memory function at the level of biomolecular processing. There are two main ways the nervous system responds during learning and memory consolidation: (a) changes at the individual synapse level, and (b) changes within overall networks of neurons. Changes to individual synapses can be studied in cell culture, tissue slice, or living animals. Long-term potentiation (LTP), the most often used paradigm, involves measuring baseline input-output characteristics of a small set of neurons, followed by delivering an intense tetanus stimulus, and then measuring any enhancement of output following input such as was delivered to establish baseline characteristics [43, 44]. LTP experiments demonstrate that synapses do change their response characteristics following intense stimuli. This provides experimental support for notions proposed by Donald Hebb [45] and Santiago Ramón y Cajal [46] that the most likely physical substrate for learning is change in synaptic strength. To study changes in the overall neural network, recordings are made from experimental animals or computer-based programs are used to simulate neural activity in complex neural circuits. Many computer programs have been designed that can learn (which is referred to as artificial intelligence) in the sense that they generate a particular output after having been fed with particular inputs on repeated trials [47, 48, 49].

Both LTP studies and computer models of learning and memory provide valuable insights; however, both approaches focus narrowly on specific aspects of the problem. Learning in living organisms is vastly more complex than changes at individual synapses and computer-based neural network models do not realistically capture biological processes or the many nuances that affect learning. In a very realistic sense, the nervous system has the potential to take into account virtually any previously stored information when processing new information. This is illustrated by the strong influence the surrounding context has.

As is the case for perception, context is very important to associative learning, or what is also called classical or Pavlovian conditioning. An organism learns to associate one stimulus with another in a context-dependent manner. As an example, a laboratory animal will readily learn to associate a tone with foot shock delivered soon after the tone onset or offset [50, 51]. In addition to pairing the tone with the shock, the animal also learns the parameters of the chamber in which it was trained to the extent that it will respond to the training chamber with the same conditioned fear elicited by the foot shock. With human subjects, context-dependent effects are even more complex. For example, old memories are updated when subjects return to an original learning context, whereas entirely new memories are formed when subjects are placed in novel contexts [52]. It stands to reason that significant computational power is needed to execute learning and memory functions, which take into account massive stores of information relevant not only to current stimuli, but also to surrounding contexts. Nanoneuroscience perspectives that explore the likely possibility of biomolecular computations and novel means of storage of information (e.g., in structural proteins of neurons) offer new and exciting ways to expand our current understanding of learning and memory mechanisms.

The following points argue the need for biomolecular computing to greatly enhance the capacity of the nervous system for learning and memory:

1. Each memory may be more information rich than could be encoded solely by altered synaptic efficacy, even though long-term plasticity appears to increase synaptic computational ability greatly [53]. Biomolecular computing within individual neurons of a neural network would be expected to greatly increase the overall computational power of individual synapses, whole neurons, and the larger neural network, thereby enabling memory to encode more complex representations in various context-specific arrangements.
2. LTP, often thought to represent permanent changes to synaptic strength, typically decays with time constants of 2 - 3 hr, 3.5 days, or 25 days, depending on the subcellular mechanisms involved [54], whereas, human long-term memory lasts indefinitely. A link between LTP and biomolecular storage of information within neurons provides a potential solution (see Chapters 3, 4, and 6).
3. A potential problem exists with the synapse acting as both the input channel and the storage site for memory. In order to maintain fidelity, input channels need to operate more or less consistently from one occasion to another. Memory models that posit storage in the synapse itself are possibly flawed for this reason. Placing memory storage sites in biomolecules inside neurons (e.g., in the subsynaptic zone) circumvents this potential problem [55].
4. Like perceptual processes, memory function requires the assimilation of information deriving from different sensory modalities. A second-order transduction process, in which biomolecular computing in individual

neurons compacts information, would be expected to facilitate learning and memory consolidation.

To make the leap from synaptic activity to leaning and memory – and then subsequently to even higher cognitive functions such as language, thinking, problem solving, and consciousness – new approaches, such as those offered by nanoneuroscience are needed. If nanoscientists at NASA expect to assimilate a realistic brain-like computer by 2030, nanoneuroscientists might reasonably expect to have a better understanding of how the human brain operates – at its highest levels – in roughly the same time frame.

1.2.2 Neurological, Neurodevelopmental, and Neuropsychiatric Disorders: Prospects for Nanoneuroscience

Diseases of the nervous system are among the most prevalent and difficult to treat health problems across the globe. The prevalence of Alzheimer's disease in the United States for the year 2007 was estimated to be 5 million cases [56], and worldwide there was roughly five times that number. Alzheimer's disease erodes memory and intellect, and since it is an age-related disorder, the numbers of individuals affected by Alzheimer's disease are expected to triple or quadruple over the next few decades as the aged population soars. Alzheimer's disease is only one of a number of age-related neurological disorders. Parkinson's disease, a neurological disorder associated with impaired movement, also increases exponentially with age. There are currently over 1 million cases of Parkinson's disease in the United States alone [57]. Huntington's disease, another neurological disorder characterized by uncontrolled abnormal movements, affects some 30,000 in the U.S. [58].

Alzheimer's, Parkinson's, and Huntington's diseases are neurodegenerative diseases. The subsets of neurons that degenerate in each of these degenerative diseases have been linked to specific biomolecules. In Alzheimer's disease, for example, neurofibrillary tangles form because of abnormal cytoskeletal protein accumulation inside the neuron (see Figure 1.5). An increased understanding of these biomolecules at the nanoscale level should shed light on why these biomolecules are vulnerable in these diseases. This will be discussed more fully in Chapter 5.

Some of the processes that underlie neurodegeneration recapitulate or reverse primary developmental sequences. Perhaps for this reason, there are some similarities, and even some degrees of overlap between neurodegenerative diseases and neurodevelopmental disorders. Alzheimer's disease, for example, occurs at an earlier age in individuals with Down syndrome, a neurodevelopmental disorder due to trisomy 21, a failure of chromosome 21 to divide during meiosis [59].

While progress has been good in terms of understanding the degenerative processes underlying Alzheimer's, Parkinson's, and Huntington's disease, this understanding has not necessarily led to highly successful treatments or preventative measures for these diseases. More information about the mechanistic

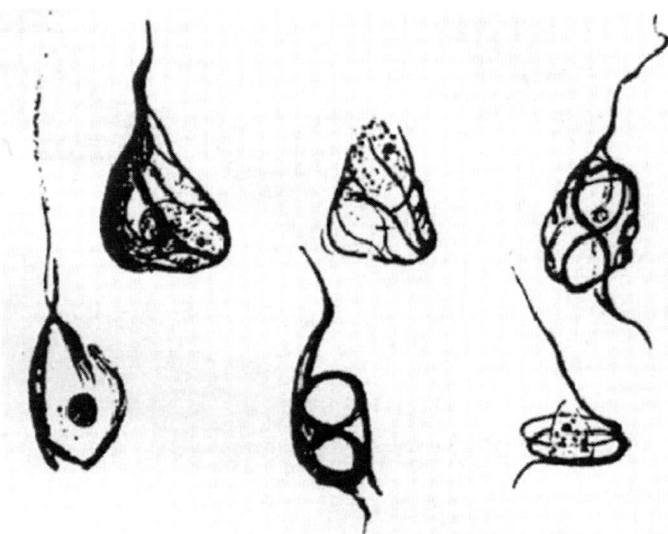

Fig. 1.5. Alois Alzheimer first described the symptoms and the neuropathology associated with Alzheimer's disease during a speech in 1906. Alzheimer's disease brains have high concentrations neurofibrillary tangles, made up of twisted cytoskeleton-based filaments that interact strongly with silver stains as shown in these original drawings [58].

interactions between biomolecules is needed to advance these fields, and this is perhaps where nanoneuroscience approaches might be able to penetrate the barriers that have existed to finding treatments or preventions for these neurological diseases. On the other hand, there is no clear understanding of the biological bases for most of the neuropsychiatric disorders [60]. Schizophrenia and bipolar disorder are two of the more serious neuropsychiatric disorders. In both disorders, there are only a small number of studies indicating changes to neuronal structures, and these reported changes are limited to particular parts of the brain [61]. Paradoxically, despite an inadequate understanding of the neurobiological underpinnings of mental disorders, there are highly effective pharmacological treatments for these disorders. For this reason, prevailing neurochemical theories of neuropsychiatric illnesses (e.g., the "dopamine hypothesis" of schizophrenia; the "monoamine" or "catecholamine hypothesis" of depression) use the reverse logic that if a pharmaceutical agent that stimulates or blocks a particularly neurotransmitter is an effective treatment for that mental disorder, a deficit or excess in that particular neurotransmitter is posited as the root cause of the disorder.

Although antipsychotic medications (i.e., neuroleptic drugs) in most cases block receptors for the neurotransmitter dopamine and effectively treat the positive symptoms of schizophrenia (e.g., the hallucinations, delusions, and disorganized thought patterns), neuroleptic drugs also act on a variety of receptors other than dopamine receptors [62]. Moreover, genetic association studies have attempted to find polymorphisms among dopamine-related genes with only limited success [63]. The complexities of schizophrenia along with the complexities of the actions of dopamine blockers make this a difficult puzzle to solve, and other reasons exist suggesting the "dopamine hypothesis" is too oversimplified to explain schizophrenia or the psychotic thinking that accompanies the disorder. For one, there are new generation antipsychotic medications that are quite effective at treating both the positive and negative symptoms of schizophrenia, the latter of which include lack of emotion and social withdrawal. Most of the new generation atypical antipsychotic medications block dopamine receptors, as do the older antipsychotic drugs, but one in particular, primarily blocks a receptor site for a different neurotransmitter called serotonin [64]. The effectiveness of this medication casts dispersions on a narrowly defined dopamine hypothesis since the drug presumably produces clinical improvements without having its prominent effect on dopamine systems.

Major depression and attention deficit/hyperactivity disorder (ADHD) are improved by antidepressant and stimulant drugs, respectively, both of which increase levels of monoamine neurotransmitters: dopamine, norepinephrine, and epinephrine, and serotonin [65, 66]. Clinicians less frequently use antidepressants to treat depressed episodes of bipolar disorder because they may cause switching to mania, a belief that has recently been challenged following careful reexamination [67]. Because the early antidepressant drugs identified in and around the 1950's enhanced actions of the catecholamines: norepinephrine, epinephrine, and dopamine, the "catecholamine hypothesis" sufficed as an early explanation for mood disorders [68]. Although there may be abnormalities in these neurotransmitter systems, such as polymorphisms among genes related to catecholaminergic and serotonergic systems in select patient populations with mood disorder [69, 70], these correlations do not prove direct causation. The genetics of mood disorders is multifactorial as it appears to involve many gene contributions that may or may not affect expression of the disorder depending on the presence or absence of key environmental triggers.

Although blocking dopamine or certain serotonin receptors ameliorates the symptoms of schizophrenia, and stimulating monoaminergic receptors combats mood disorders, it may not be the monoamine systems per se that are defective in mental disorders, at least not across all patients. Rather, the primary defect may lie in those computationally relevant biomolecules inside of neurons that are influenced by psychiatric medications. In Chapter 2, discussion will cover how receptor proteins (including receptors that are also ion channels) catalyze activity in intracellular proteins, such as signal transduction molecules, adaptor proteins, and cytoskeletal structures that connect the various parts of the neuron together. Nanoneuroscience is unique in its ability to uncover novel modes of signaling within neurons and to show how classical activation of synapses by neurotransmitters trigger these computationally relevant structures. The nanoneuroscience approach should finally enable researchers to nail down how current psychiatric drugs achieve their effects and to find ways to improve current drugs by optimizing their effects on biomolecular computing in neurons.

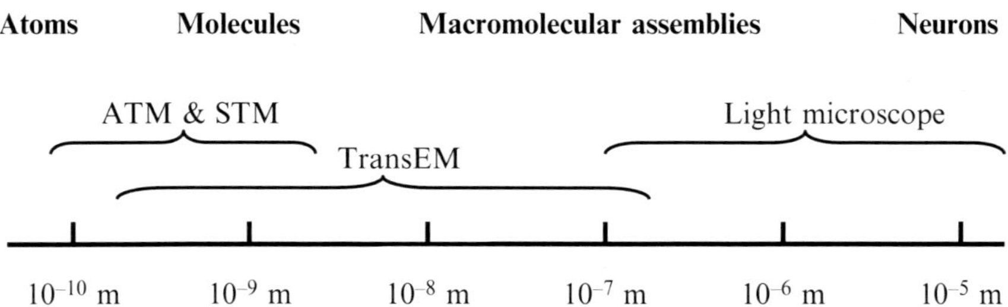

Fig. 1.6. Size ranges relevant to nanoscience and microscopes having sensitivities at those ranges. ATM: atomic force microscopy; STM; scanning tunneling microscopy: TransEM: transmission electron microscopy. Adapted from [71].

1.3 Current Issues in Nanoscience

Nanoscience and nanotechnology deal with objects at the small scale, that is 0.1 - 100 nm (1 nm is a billionth of a meter). "Nanos" derives from the Greek word meaning "dwarf". Since individual molecule sizes range between 1 - 20 nm, molecules acting singly fall squarely under the domain of nanoscience (see Figure 1.6). Also, molecules bound together into larger polymers or macromolecular assemblies may fall under the domain of nanoscience, depending on how large the final structure is, or in the case of long polymers, depending on the size of their diameters.

While there is an overwhelming enthusiasm for nanoscience and nanotechnology, there are also some public safety concerns as might be expected. One of the major causes for concern associated with nanotechnology is that individual molecules have much higher degrees of reactivity because of their increased surface area [72]. Another cause for concern is that unforeseen environmental problems might emerge after a large number of nanoparticles are already released into the environment [73]. These issues are not expected to significantly deter progress in this field.

1.3.1 The Origins of Nanoscience

Nanoscience has roots in physics, chemistry, electrical engineering, and mechanical engineering, and like that any multidisciplinary field, it appears to have many early proponents. In his famous lecture "*[There's] Plenty of Room at the Bottom*" delivered on December 29, 1959, the Nobel Prize winning physicist Richard P. Feynman of California Institute of Technology was among the first to allude to nanoscience and nanotechnology as being the inevitable progression of physics and chemistry [74]. It was in this lecture that Feynman made the startling revelation that it would someday be possible, without inventing any new laws of physics, to rewrite all the information contained in all the books worldwide onto a cube of matter the size no larger than a speck of dust. As Feynman pointed out, his prediction should not have surprised biologists, who already knew at that time that DNA strands measuring a few nm in width stored the complete blueprint for an entire organism.

Although Feynman clearly embraced the notion, the first person to actually use the term "nanotechnology" was Nori Taniguchi, who coined the term in 1974 [75]. Taniguchi's claims to fame include machining integrated circuits and nanoelectronic devices with accuracies in the 1 nm range which is equivalent to three atoms in width [76]. Some years later, K. Eric Drexler popularized nanoscience and nanotechnology in his imaginative books *Engines of Creation* and *Nanosystems: Molecular Machinery, Manufacturing, and Computation* [77, 78].

1.3.2 The Mission of Nanoscience

The overarching mission of nanoscience is to advance technology and improve the environment and public health by operating at the atomic and molecular levels. Specific nanoscale applications include nanoelectronics, nanochemistry, nanobiology, nanomedicine, and the development of commercial nanoproducts.

Nanoelectronics – the design of small electronics components, such as transistors-should be able to pick up where Moore's Law[1] predicts that manufacturers can no longer increase the power of integrated computer circuits through continued miniaturization. Physical laws at play change as the size of each transistor approaches the size of single molecules or atoms, and most scientists agree that it is going to be very difficult to manufacture transistors for computer circuits much below 10 nm in size using current methods – although there are ways to get around these problems using nanoconstruction strategies [81].

Nanochemistry and nanobiology stand to contribute greatly to nanomedicine [82]. These fields focus on the nanoscale properties of molecules and atoms working singly or in macromolecular structures that compose the living cell. Disease states can be understood in terms of these nanoscale operations, and nanoscience has already contributed new diagnostic tools and drug delivery devices.

[1] In 1965 Gordon Moore observed that the number of transistors that can be cheaply placed on an integrated circuit doubles every 2 years [79]; this remains true today and has come to be known as Moore's Law [80].

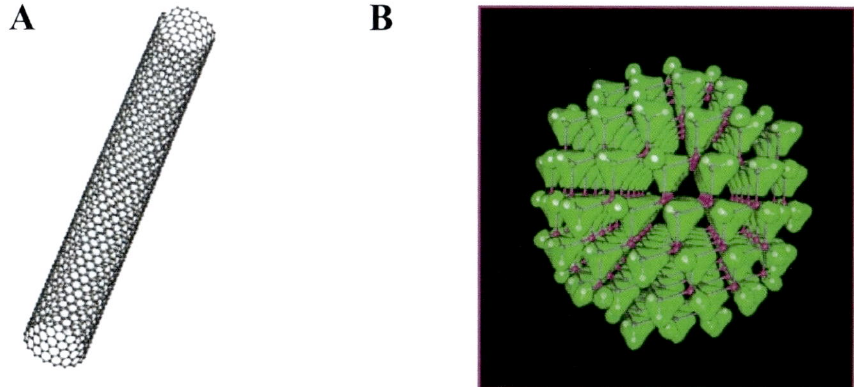

Fig. 1.7. Carbon nanotubes are cylinders of graphene (A). Quantum dots, such as the one illustrated (B), consist of a few hundred to a few thousand atoms. Reprinted with permission from Prof. Lin-Wang Wang [85].

1.3.3 Nanostructures, Nanoparticles, and Nanodevices

Carbon nanotubes, molecular motors, silicone/molecular hybrid devices, nanoparticles, and quantum dots are just a few of the products enabled by nanotechnological design (see Figure 1.7). Each of these products has different applications, and for each type of product, the number of applications has yet to be determined.

Carbon nanotubes are an example of a nanostructure having a wide range of possible applications in both scientific and in commercial settings. Sumio Iijimi discovered carbon nanotubes in 1991 [83]. Single-walled carbon nanotubes are best described as seamless rolls of one-atom thickness sheets of graphite (also called graphene), in which all carbons bond in a hexagonal lattice (see Figure 1.7). Carbon nanotubes can have armchair, zigzag, and chiral configurations, depending on the structure of the underlying lattice [84]. The widths of single-walled carbon nanotubes vary and can measure less than 1 nm. Multiwalled carbon nanotubes consist of 2 - 50 nanotubes nested within one another, attaining widths up to 1 m. Multiwalled carbon nanotubes are strong, yet flexible – a valuable combination of their physical properties. Carbon nanotubes can be quite long, over a few mm in some cases. There are also carbon nanotubes with sealed off ends called nanohorns, which are particularly suitable as drug-delivery devices [86]. Carbon nanotubes have possible applications in nanoelectronics since they are semiconductive. Since carbon is a biocompatible material, carbon nanotubes also have potential applications in nanobiology and nanomedicine. It is conceivable that carbon nanotubes could deliver drugs to very specific targets, as they are able to penetrate living cells and transport materials to specific parts of damaged cells.

Molecular motors represent yet another nanomachine of potential usefulness, such as mass transport, force and torque generation to mention but a few examples of their utility. Carlo Montemagno and George Bachand attached a single molecule of the enzyme ATPase to a metal substrate making a very tiny, nanoscale propeller [87]. A possible application for this nanodevice may lie in drug delivery. Only time will tell what other purposes this hybrid nanodevice or others like it will serve. The ATPase propeller melded together organic and inorganic materials of nanoscale proportions. Silicone is another potential nanodevice component, which may be useful in the transition from traditional transistors to novel nanoelectronic components.

Quantum dots are nanoparticles with semiconductive properties that emit light with demonstrated superiority over traditional fluorophores – giving them numerous biological applications [88, 89]. To date, quantum dots have been used to immunolabel cells, track cellular movements, and in frequency resonance energy transfer. There are a number of other nanoparticles, some of which have met with a fair degree of commercial successes. Silver nanoparticles, for example, can be added to plastic containers or woven into fabrics to retard bacterial growth [90, 91].

The making of a prototype nanoparticle or nanodevice shows what is possible, but for these nanostructures to come into widespread usage, there have to be ways to rapidly and economically manufacture them. Currently, there are top-down and bottom-up methods of constructing nanoparticles and nanodevices [92, 93]. Top-down approaches start with a larger material and use a process such as lithography to produce the smaller scale product. Bottom-up approaches literally assemble nanodevices one molecule or atom at a time. The bottom-up method of assembly is efficient because many nanostructures can self-assemble, in much the same way that macromolecular structures assemble in nature.

It is unlikely that it would be possible to construct nanodevices, were it not for nanoscale tools, which include scanning tunneling microscopy (STM) and atomic force microscopy (ATM) (see Figure 1.6). The Nobel Prize winners Gerd Binnig and Heinrich Rohrer developed STM in the early 1980's, some 50 years after the electron microscope was invented [94]. Binnig, Quate, and Gerber soon thereafter developed the AFM [95]. STM and AFM enable the viewing of individual molecules and their constituent atoms by mechanically sensing the sample. In 1985 Conrad Schneiker foresaw the potential for STM to be used to manipulate nanoscale matter at the molecular and atomic level [96]. Soon thereafter Schneiker and Stuart Hameroff located the University of Arizona Medical School set up the first STM there.

Nanotweezers can also be used to manipulate very small macromolecules or single molecules [97]. Nanotweezers can be constructed from carbon nanotubes attached to microelectrodes, and they can secure nanostructures as well as measure their electrical properties.

1.4 Applications in Nanoneuroscience

There is an increasing number of studies using nanoparticles or nanodevices to detect or stimulate neural activity, neural growth, neuronal transport, and ion channel activity (see Table 1.1 for a comprehensive list). Capabilities and advantages of these techniques include being able to more precisely probe nerve cells and exert finer control. Nanoscale materials may in some cases be less damaging to neurons due to their reduced size; however, whether residual nanoparticles and nanodevices are safe remains to be established. For some nanoscale materials there are other advantages, such as increased biocompatibility.

Table 1.1. Nanoscale methods with applications to neuroscience.

Nanoelectronics	Capabilities and advantages
Silicon nanowires and carbon nanotubes make contacts with single neurons [98]-[101]	Possible to stimulate and record from multiple sites of individual neurons. Less cell disruption than with microelectrodes.
Multiwalled carbon nanotubes used to stimulate hippocampal neurons *in vitro* [102].	Strong, yet flexible electrodes. Biocompatible.
Quantum dot- FRET imaging of action potentials [103].	High resolution.
Nanobiology	
Growing neural stem cells and primary sensory neurons on matrices of carbon nanotubes [6, 7, 104, 105].	Precise control over cell growth (e.g., dendritic branch patterns). Potential for manufacturing neural prosthetic devices.
Quantum dots used to measure retrograde transport of nerve growth factor (NGF) [106, 107].	Ability to measure single molecules of NGF.
Nano-patch-clamp used to detect activity at ion channels [108]	Enables high throughput drug screening.
Atomic force microscopy (ATM)-patch clamp techniques [109, 110]	Combines cell imaging and recording. Enables high throughput study of membrane transport.
Nanostructured silica thin films for detection of proton transport across ion channels [111]	
Quantum dot detection of actin filament "fences" limiting lateral movement of receptor clusters [112].	Highly sensitive live-cell imaging.

1.4.1 Using Nanotechnology to Study Brain Tissue Response

Nervous system physiology can be approached from the perspective of nanoelectronics using nanoscale probes. A number of nanoscale materials have proved successful at stimulating and recording from neurons including nanowires made from silicon and carbon nanotubes [98]-[101]. Charles Lieber and colleagues at Harvard University succeeded in stimulating and recording from multiple sites located along single neurons plated onto matrices of carbon nanowires [98]. In an unprecedented technical feat, these researchers were able to "connect" to neurons in a physiologically realistic manner. Each separate nanowire, measuring in the range of tens of nanometers, made an individual contact with a particular point along the process of a neuron that was about the dimensions of a synapse. Using nanowires, it was possible to make as many as 50 contacts onto individual mammalian neurons. Using standard technologies, it would be extremely difficult, if not impossible, to stimulate or record from multiple sites along a single process of a neuron. Thus, implementing nanotechnology solved the problem of scale and also resulted in a high ratio of successful recordings to failures - a ratio that was much greater than typically experienced when using traditional micron-scale electrodes [98].

Multiwalled nanotubes have also been implemented physiologically, for example, to probe the activity of neurons in hippocampal tissue slices [102]. Quantum dot- FRET is another method that has been used to study nerve cell responses in fine detail [103]. Using quantum dots as FRET acceptors from voltage-sensitive dyes enables an accurate assessment of surface potential. The method provides high resolution; however, the high degree of toxicity may limit the widespread use of this technique.

Although there is already a great deal known about the neurophysiology of neurons, fine details need to be confirmed, elaborated upon, or in some cases, revised. There is vast potential for nanowires, nanotubes, and other nanoscale devices to expand our current understanding of how neural inputs are processed and to assess any biomolecular perturbations (i.e., computations) that may occur inside neurons activated by those inputs.

Nanotechnological approaches have also shed light on neurobiological issues, such as neuronal growth, retrograde transport of growth factors, ion channel activity, and cytoskeletal protein activity (see Table 1.1). Exposing neuronal cells grown in culture to matrices of carbon nanotubes facilitates cell adhesion and the extension and branching of dendrites [6, 7, 104, 105]. These morphological changes occur along with increased electrophysiological activity in the neuronal culture networks. Being able to modulate both structure and function is essential to developing nanoscale prosthetic devices that could restore normal function and patterns of brain activity following neural damage, and it would appear that such outcomes are possible.

Nerve growth is out of balance in a number of neurological disorders, and nanoneuroscience provides new ways to study these phenomena. Certain identified chemical factors influence the behavior of the growth cone, a specialized structure at the growing tip of an axon. These factors, called neurotrophins, also influence neuron survival once growth is completed. Nerve growth factor (NGF) is often considered the prototypical neurotrophin. Discovered in 1950's by Rita Levi-Montalcini, NGF causes a massive proliferation in the number of neurites growing from sympathetic ganglion cells. In addition to sympathetic ganglion cells, neurons located in the basal forebrain are sensitive to NGF [114]. These NGF-sensitive basal forebrain neurons synthesize acetylcholine and project axons to the entire cerebral cortex and hippocampus [115, 116]. It is not the cholinergic phenotype that is responsible for CNS neuron sensitivity to NGF. Other neurons in the brain that synthesize acetylcholine are not sensitive to NGF [117]. NGF transport from the cerebral cortex back to cholinergic basal forebrain neurons is impaired in Alzheimer's disease and may contribute to the cholinergic basal forebrain cell loss typical of this disorder [118], although other plausible interpretations are possible [119]. Since it is of marked interest how NGF is transported from target tissues back to the cell body (i.e., from the cerebral cortex and hippocampus back to the basal forebrain), studies of retrograde transport of NGF abound.

One difficulty in assaying NGF is that this growth factor is present in very small quantities in brain. The sensitivity problem can be overcome by using quantum dots to study NGF transport. In rat dorsal root ganglion neurons quantum dots reveal remarkable detail of NGF transport-down to the level of a single molecule. In many cases, there appears to be only one NGF molecule per endosomal vesicle, which is transported along microtubules in the axon [106]. Each endosomal vesicle ranges 50 - 150 nm in diameter. In one study, NGF was measured to travel retrogradely along the axon at speeds of slightly above 1 μm/s, while pausing about 30% of the time. Another study using quantum dots showed bi-directional movement for NGF and slower transport rates [107]. This second study assessed NGF transport in neurites and was performed on PC12 cells rather than neural cells.

In addition to enabling new ways to probe electrical activity of neurons and study neuronal transport, nanotechnology provides a cornucopia of ways to study ion channels and their attachments to the actin cytoskeleton. The nano-patch clamp, consisting of a microstructured glass chip, is capable of making whole cell recordings with the high throughput needed for efficient drug screening [108]. AFM-patch clamp combination techniques enable simultaneous imaging and electrophysiological recording in a low vibrational noise environment [109, 110]. Nanostructured silica thin films, containing alternating hydrophilic and hydrophobic regions, have been devised that are capable of detecting proton transport across ion channels with high throughput [111]. The high sensitivity of nanotechnology applied to live cell imaging is illustrated by quantum dot detection of actin filament "fences" surrounding receptor clusters localized in cell membranes [112]. Besides facilitating imaging and drug screening, nanotechnological studies will continue to reveal biophysical characteristics of ion channels, such as those for gramicidin, sodium, potassium, calcium, chloride, and the acetylcholine receptor [120], as well as synthesized ion channels [121]. Ion channels play a vital role in neuroscience since neurotransmission and propagation of neural activity along the membrane relies on select populations of ion channels.

Nanotechnological tools and devices might also be expected to play a major role in the future of neuropharmacology. Single-walled carbon nanotubes can under appropriate circumstances act as ion channel blockers [122]. Agonists and antagonists of receptor sites on ion channels can be manipulated by optically activated nanoswitches that behave as nanotoggles, nanokeys, or nanotweezers [123]. These approaches have additionally provided valuable insights regarding the mechanisms by which ion channels open or close.

The number one reason why nanotechnology will likely continue to advance our understanding of how the nervous system operates is that it enables us to purview the operations at the level of the single molecule. Demand for high resolution imaging will only continue to increase for existing nanotechnologies – such as carbon nanotubes and quantum dot imaging – as they are increasingly used to specifically probe neurons or glial cells. In addition to a more detailed understanding of how nerve cells perform their known functions, novel modes of neural processing are likely to be revealed by nanoneuroscience methods. In the chapters to come, some novel modes of signaling that might be particularly relevant to higher cognitive function will be outlined.

1.4.2 Nanoneuroscience Approaches to Neurological, Neurodevelopmental, and Neuropsychiatric Disorders.

Nanomedicine has already begun to advance development of diagnostic measures and potential treatments for key neurological, neurodevelopmental, and neuropsychiatric disorders [124]. While these nanoscale approaches are still in the preliminary stages of development, significant breakthroughs have been achieved for actual patient populations. Nanoparticles, nanodevices, and nanoscale drug delivery devices are also being tested on animal models of various nervous system diseases or in cell culture. These initial studies, while showing promise, indicate that more neurotoxicity studies are needed before some of these nanoscale interventions are attempted on human patients.

Potentially Damaging Effects of Nanoparticles on the Blood-Brain Barrier

Nanoparticles must be able to cross the blood-brain barrier in order to be useful therapeutically in relation to nervous system diseases [125]. An issue that arises, however, is that nanoparticles may be neurotoxic or may alter the blood-brain barrier. Breakdown of the blood-brain barrier can lead to brain tissue edema or the entry of toxins and large molecules that are ordinarily excluded from the brain and spinal cord. These drawbacks must be addressed while exploring the potential of nanoparticles in diagnostics or drug delivery.

Using Nanoparticles for Diagnosing Nervous System Disorders

A number of recent studies have addressed potential nanoscale diagnostics for Alzheimer's disease [126, 127]. These nanoneuroscience approaches have focused on components of the two main neuropathologies in Alzheimer's disease-senile plaques and neurofibrillary tangles (see Figure 1.5). Senile plaques are enriched with amyloid-β protein. Although there is no direct proof that decreasing amyloid-β protein accumulation will improve the cognitive symptoms associated with Alzheimer's disease, it is currently the most popular hypothesis [128]. The popularity of the "amyloid hypothesis" took hold when a mutation in the gene coding for amyloid-β precursor protein (APP) was found to occur in some hereditary forms of early onset Alzheimer's disease. If indeed amyloid-β protein deposition is key to the etiology of the disorder, early detection of an amyloid-β diffusible ligand in the cerebral spinal fluid should be a reliable marker for Alzheimer's disease. Higher levels, compared to that of controls, have been noted in Alzheimer's disease patients using a sensitive nanoparticle-based assay that employs a DNA bio-barcode and antibodies to amyloid-β protein [129, 130]. Due to its extreme sensitivity, this method may be able to detect increasing levels of amyloid-β protein diffusible ligands before Alzheimer's disease begins, and thereby enhance the possibility of preventing the degeneration that accompanies the disorder. A potential drawback to this method is that higher than normal amyloid-β protein diffusible ligands occur in other neurological disorders, as well as in healthy aged persons without neurological disease.

The accumulation of neurofibrillary tangles inside large pyramidal neurons (see Figure 1.5) also contributes to cognitive impairment in Alzheimer's disease patients. Tangle formation begins with hyperphosphorylation of tau protein and its subsequent coiling into paired-helical filaments, which appear under the microscope as neuropil threads. Researchers recently altered mice strains having the mutant human APP gene to carry zero, one, or two copies of the mouse tau gene [134]. Surprisingly, as these mice aged, those animals with no tau showed no sign of cognitive impairment, even though the mice had the expected profile of senile plaques previously noted for this strain of mouse. This finding strongly suggests that cognitive symptoms of Alzheimer's disease cannot be completely explained by the amyloid hypothesis, and that microtubule-associated proteins, such as tau play a complex but significant role. Refocusing nanoparticle-based strategies on tau is not difficult. The nanotechnological approaches already mentioned for diagnosing Alzheimer's disease based on amyloid-β protein can also be applied directly to tau protein, as well as to the microtubules themselves. There is evidence, for example, that the microtubule matrix itself is modified as senile plaques and neurofibrillary tangles accumulate, but rather than being a cause or effect, this reduction of microtubules in Alzheimer's disease neurons appears to be an independent associated event [135]. These issues are discussed further in Chapter 5.

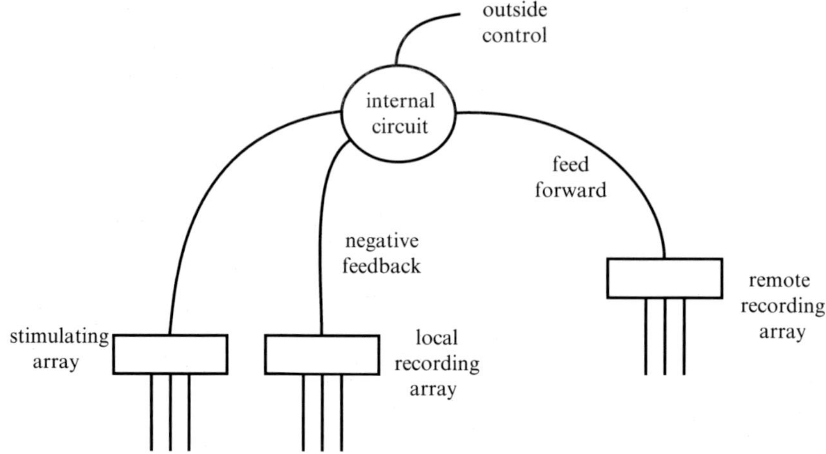

Fig. 1.8. Nanoelectrode arrays can be designed to record and stimulate multiple brain sites, or be under an external control.

Nanoparticle-Carriers and Nanoscale Drug Delivery

Nanoparticles have been considered in possible treatment strategies aimed at attacking amyloid-β protein in Alzheimer's disease brain. In one such approach, nanoparticles are used as drug delivery systems, which encapsulate the hydrophilic drugs that attack amyloid-β protein [131]. This strategy enables the drug to pass through the blood-brain-barrier [132].

Nanotechnology facilitates the ability of neuroprotective agents to enable recovery from spinal cord injury. Nanowires made from TiO2 subsequently coated with three different anti-inflammatory drugs demonstrated enhanced therapeutic efficacy compared with those drugs that were administered alone [133]. Animals treated with the drug-coated nanowires exhibited improved motor activity to a greater extent than afforded by drug application alone, presumably due to enhanced ability to cross the blood-brain barrier without disruption to nervous tissue.

Another potential use for nanomaterials is in the assembly of nanoparticle-based vectors for introducing genes into neurons of diseased brain [136]. The genetic basis of nervous system disease ranges from autosomal dominant disorders, such as Huntington's disease, to genetic linkages in limited populations of Alzheimer's or Parkinson's disease patients (discussed further in Chapter 5). Nanoparticles applied as vehicles for gene therapy might be expected to hold an important place in nanomedicine across the board.

Nanoscale Sensing and Stimulating Probes: Combining Diagnostics with Delivery

Nanotechnological approaches to Parkinson's disease include the development of highly sensitive nanoprobes to sense dopamine in degenerating brain regions and nanoelectrodes to stimulate other select populations of neurons in brains of Parkinson's patients. Parkinson's cases suffer a severe (>80%) loss of dopaminergic cells in the mesencephalon. Since dopamine levels in Parkinson's disease are in the nanomolar range, sensitive nanoprobes are ideal for detecting dopamine levels and assessing the severity or the stage of Parkinson's disease.

Nanoprobes measuring dopamine can also be designed to provide a feedback mechanism that controls delivery of electrical currents or drug particles. Although it is not well understood why dopaminergic cell loss occurs in Parkinson's disease, or why dopamine replacement only serves as a temporary solution, more sophisticated devices that pump dopamine into specific brain regions will require feedback such that too much dopamine is not infused. Excess dopamine has been found to be neurotoxic in the degenerating brain of the Parkinson's disease patient [137]. Thus, it is essential that any probe demonstrates enough sensitivity to assay the reduced dopamine levels present in Parkinson's disease and the moderate levels that are administered therapeutically; such is technically possible using current nanotechnological tools. One group of researchers recently developed an extremely sensitive sensor using polymer-doped carbon nanotubes that detects dopamine in the nanomolar range – levels typical of Parkinson's patients [138].

There is a need for nanomedicine to improve treatment strategies and to provide novel alternative treatments for Parkinson's disease, since the current pharmacological interventions for treating Parkinson's disease-L-dopa, carbidopa, dopamine agonist, or cholinergic antagonist therapies – eventually fail in controlling the motor symptoms of the disorder typically within a few years after the initiation of treatment [139]. Nanomedicine stands to advance non-pharmacological treatments for Parkinson's disease, including stem cell therapies, gene therapies, and deep brain stimulation [140]. Current nanotechnological designs have already made headway regarding improvements to deep brain stimulation. Standard stimulating electrodes are typically in the range of 1 mm. This poses increased risk of damage to stimulating brain sites and risk of infection. A recently designed trimodal (three-pronged) nanoelectrode array simultaneously records local electrical activity, assays local dopamine levels, and delivers electrical current [141]. The reduced size of such a device enables highly sensitive monitoring of local electrical activity and local dopamine levels, greater precision of stimulation, and reduced risk of neural damage and infection. Similar designs might include arrays with stimulating nanoelectrodes connected to recording electrodes that register brain activity levels at local and remote sites (see Figure 1.7).

Possible Applications of Nanoneuroscience to Neurodevelopmental and Neuropsychiatric Disorders

Neurodevelopmental and neuropsychiatric disorders are also candidates for nanotechnological methodologies being applied to their detection and treatment. Any nanotechnologies that prove effective in detection would be particularly welcome as aids in diagnosing schizophrenia and mood disorders, since there are currently no definitive clinical tests for these disorders.

Nanotechnology-based drug delivery could broaden the scope of drugs used to treat neuropsychiatric illnesses and developmental disorders such as fragile-X syndrome, Turner's syndrome, William's syndrome, velo-cardio-facial syndrome, autism, ADHD, Rett syndrome, and Down syndrome. Interventions using carbon nanotubes or nanocapsules to deliver drugs would be expected to enhance drug uptake and could facilitate localizing drug to precise regions of the brain.

Nanodevices might also prove useful. Nanoelectrode arrays containing sensor components that detect abnormal activity in one brain area and stimulating probes that provide electrical stimulation to another brain area (see Figure 1.7) could be used to self-correct activity imbalances across brain regions. Activity imbalances are implicated in mood disorders. Patients with bipolar mania, for example, present with increased activity in the left amygdala and decreased activity bilaterally in the orbitofrontal cortex [142]. A potentially successful nanodevice might be programmed to respond to increased activity in the left amygdala (or other brain region depending on the patient) by stimulating the orbitofrontal cortex bilaterally. Rectifying activity imbalances would be expected to prevent periods of abnormal activity leading to brain damage that perpetrates the disease process at the cellular level.

Nanoparticle-induced gene therapy might also have potential for treating neuropsychiatric illnesses. Dysbindin and disrupted-in-schizophrenia (DISC1) are genetic markers for subpopulations of those affected by schizophrenia, schizoaffective disorder, and bipolar disorder [143, 144]. Dysbindin and DISC1 proteins interact with cytoskeletal proteins such as microtubules, suggesting a possible fundamental problem in the cytoskeleton of those affected by schizophrenia and mood disorders. Thus, gene therapy directed at strengthening the cytoskeleton or nanoprobes that specifically interact with the cytoskeleton might be useful in diagnosing or treating these disorders [145]. The issue of applying nanotechnology to diagnose and treat neurological and neuropsychiatric disorders will be revisited in Chapter 5.

References

1. N. Taniguchi, "On the Basic Concept of 'Nano-Technology' ": Proc. Intl. Conf. Prod. Eng. Tokyo, Part II, Japan Society of Precision Engineering, 1974.
2. Franks A. Nanotechnology. J. Phys. E Sci. Instrum. 1987 **20**:1442-1451.
3. Malsch I. Biomedical applications of nanotechnology. The Industrial Physicist, 2002.
4. NIH Roadmap for Medical Research. http://nihroadmap.nih.gov/ Retrived on Sept. 21, 2008.
5. Biomimetics and bio-inspired systems. [http://ipt.arc.nasa.gov/Graphics/nano tech_nasamissions.pdf] Retrived on Sept. 21, 2008.
6. Lovat V, Pantarotto D, Lagostena L, Cacciari B, Grandolfo M, Righi M, Spalluto G, Prato M and Ballerini L (2005) Carbon nanotube substrates boost neuronal electrical signaling, Nano Letters, **5**: 1107-1110.
7. Sucapane, A., Cellot, G., Prato, M., Giugliano, M., Parpura, V., Ballerini, L. Interactions Between Cultured Neurons and Carbon Nanotubes: a Nanoneuroscience Vignette Journal of Nanoneuroscience 1:1-7, (2008).
8. Silva GA. Neuroscience nanotechnology: progress, opportunities and challenges. Nat Rev Neurosci. 2006 Jan;7(1):65-74.
9. Bressler SL, Tognoli E. Operational principles of neurocognitive networks. Int J Psychophysiol. 2006 May;60(2):139-48.
10. Love BC, Gureckis TM. Models in search of a brain. Cogn Affect Behav Neurosci. 2007 Jun;7(2):90-108.
11. Hart J Jr, Anand R, Zoccoli S, Maguire M, Gamino J, Tillman G, King R, Kraut MA. Neural substrates of semantic memory. J Int Neuropsychol Soc. 2007 Sep;13(5):865-80.
12. Just MA, Carpenter PA, Varma S. Computational modeling of high-level cognition and brain function. Hum Brain Mapp. 1999;8(2-3):128-36.
13. D'Esposito M. From cognitive to neural models of working memory. Philos Trans R Soc Lond B Biol Sci. 2007.
14. Friston KJ , Price CJ. Dynamic representations and generative models of brain function.Brain Res Bull. 2001 Feb;54(3):275-85.
15. Handy TC. Capacity theory as a model of cortical behavior. J Cogn Neurosci. 2000 Nov;12(6):1066-9.
16. Coslett HB. Consciousness and attention. Semin Neurol. 1997 Jun;17(2):137-44.

17. Sergent C, Dehaene S. Neural processes underlying conscious perception: experimental findings and a global neuronal workspace framework. J Physiol Paris. 2004 Jul-Nov;98(4-6):374-84.
18. Jack AI, Shallice T. Introspective physicalism as an approach to the science of consciousness. Cognition. 2001 Apr;79(1-2):161-96.
19. Lau HC, Passingham RE. Unconscious activation of the cognitive control system in the human prefrontal cortex. J Neurosci. 2007 May 23;27(21):5805-11.
20. Osaka N. [Active consciousness and the prefrontal cortex: a working-memory approach] Shinrigaku Kenkyu. 2007 Feb;77(6):553-66.
21. Tong F. Primary visual cortex and visual awareness. Nat Rev Neurosci. 2003 Mar;4(3):219-29.
22. Clavagnier S, Falchier A, Kennedy H. Long-distance feedback projections to area V1: implications for multisensory integration, spatial awareness, and visual consciousness. Cogn Affect Behav Neurosci. 2004 Jun;4(2):117-26.
23. Priel A, Ramos AJ, Tuszyński JA, Cantiello HF. A biopolymer transistor: electrical amplification by microtubules. Biophys J. 2006 Jun 15;90(12):4639-43.
24. Tuszyński JA, Portet S, Dixon JM, Luxford C, Cantiello HF. Ionic wave propagation along actin filaments. Biophys J. 2004 Apr;86(4):1890-903.
25. Priel A, Tuszyński JA, Cantiello HF.(2005) Electrodynamic signaling by the dendritic cytoskeleton: toward an intracellular information processing model, Electromagnetic Biology and Medicine,24:3,221 - 231
26. Young, T. (1802). On the theory of light and colours. Philosophical Transactions of the Royal Society of London, 92, 12-48.
27. von Helmholtz, H. (1850). Über die Theorie der zusammengesetzten Farben. Archiv für Anatomie, Physiologie und wissenschaftliche Medizin, 461-482.
28. Wyszecki, Günther; Stiles, W.S. (1982). Color Science: Concepts and Methods, Quantitative Data and Formulae, 2nd ed., New York: Wiley Series in Pure and Applied Optics
29. Dartnall, H. J. A., Bowmaker, J. K. & Mollon, J. D. Proc. R. Soc. B220, 115–130 (1983).
30. von Békésy, G. (1949) On the resonance curve and the decay period different points along the cochlear partition. Journal of the Acoustical Society of America, 21, 245-254
31. S. Ramon y Cajal
32. Maeda T, Imanishi Y, Palczewski K. Rhodopsin phosphorylation: 30 years later. Prog Retina Eye Res. 2003 Jul;22(4):417-34.
33. Shim K. The auditory sensory epithelium: the instrument of sound perception. Int J Biochem Cell Biol. 2006;38(11):1827-33.
34. de Araujo IE, Rolls ET, Kringelbach ML, McGlone F, Phillips N. Taste-olfactory convergence, and the representation of the pleasantness of flavour, in the human brain. Eur J Neurosci. 2003 Oct;18(7):2059-68.
35. Zacks JM, Michelon P. Transformations of visuospatial images. Behav Cogn Neurosci Rev. 2005 Jun;4(2):96-118.
36. Wang X. Neural coding strategies in auditory cortex. Hear Res. 2007 Jul;229(1-2):81-93.
37. Cole GG, Liversedge SP. Change blindness and the primacy of object appearance. Psychon Bull Rev. 2006 Aug;13(4):588-93.
38. Beck MR, Levin DT, Angelone B. Change blindness blindness: beliefs about the roles of intention and scene omplexity in change detection. onscious Cogn. 2007 Mar;16(1):31-51.

39. Simons DJ, Rensink RA. Change blindness: past, present, and future. rends Cogn Sci. 2005 Jan;9(1):16-20.
40. Awater H, Kerlin JR, Evans KK, Tong F. Cortical representation of space around the blind spot. J Neurophysiol. 2005 Nov;94(5):3314-24.
41. Meng M, Remus DA, Tong F. Filling-in of visual phantoms in the human brain. Nat Neurosci. 2005 Sep;8(9):1248-54.
42. Shuwairi SM, Curtis CE, Johnson SP. Neural substrates of dynamic object occlusion. J Cogn Neurosci. 2007 Aug;19(8):1275-85.
43. Teyler TJ, Discenna P. Long-term potentiation as a candidate mnemonic device. Brain Res. 1984 Mar;319(1):15-28.
44. Bliss TV, Collingridge GL. A synaptic model of memory: long-term potentiation in the hippocampus. Nature. 1993 Jan 7;361(6407):31-9.
45. D.O. Hebb (1949) The Organization of Behavior, Lawrence Erlbaum Associates.
46. Ramón y Cajal, S.R. (1907) Regeneración de los Nervios. [Translation and edited by J. Bresler (1908) Studien über Nervenregeneration, Johann Ambrosius Barth].
47. Fiori S. Nonlinear complex-valued extensions of Hebbian learning: an essay. Neural Comput. 2005 Apr;17(4):779-838.
48. Vogel DD. A neural network model of memory and higher cognitive functions. Int J Psychophysiol. 2005 Jan;55(1):3-21.
49. Doya K. Metalearning and neuromodulation.Neural Netw. 2002 Jun-Jul;15(4-6):495-506.
50. Woolf NJ, Zinnerman MD, Johnson GV. Hippocampal microtubule-associated protein-2 alterations with contextual memory. Brain Res. 1999 Mar 6;821(1):241-9.
51. Rogers JL, Kesner RP. Cholinergic modulation of the hippocampus during encoding and retrieval of tone/shock-induced fear conditioning. Learn Mem. 2004 Jan-Feb;11(1):102-7.
52. Hupbach A, Hardt O, Gomez R, Nadel L. The dynamics of memory: context-dependent updating. Learn Mem. 2008 Aug 6;15(8):574-9.
53. Abbott LF, Regehr WG. Synaptic computation. Nature. 2004 Oct 14;431(7010):796-803.
54. Abraham WC. How long will long-term potentiation last? Philos Trans R Soc Lond B Biol Sci. 2003 Apr 29;358(1432):735-44. 45 LTP fades in real animals
55. Woolf, N.J. Microtubules in the cerebral cortex: role in memory and consciousness. In Tuszyński, J.A. (ed.) The Emerging Physics of Consciousness, 2006, Springer Verlag.
56. Alzheimer's Association Report: 2007 Alzheimer's Disease Facts and Figures [http://www.alz.org/national/documents/Report_2007FactsAndFigures.pdf]
57. Parkinson's Disease Foundation: Fact Sheet. [http://www.pdf.org/Publications/factsheets/PDF_Fact_Sheet_1.0_Final.pdf].
58. Alzheimer A. Über eigenartige Krankheitsfälle des späteren Alters. Zeitschrift für die Gesamte Neurologie und Psychiatrie 1911; 4: 356-85.
59. Wisniewski KE, Wisniewski HM, Wen GY. Occurrence of neuropathological changes and dementia of Alzheimer's disease in Down's syndrome. Ann Neurol. 1985 Mar;17(3):278-82.
60. Iritani S. Neuropathology of schizophrenia: a mini review. Neuropathology. 2007 Dec;27(6):604-8.

61. McIntosh AM, Job DE, Moorhead WJ, Harrison LK, Whalley HC, Johnstone EC, Lawrie SM. Genetic liability to schizophrenia or bipolar disorder and its relationship to brain structure. Am J Med Genet B Neuropsychiatr Genet. 2006 Jan 5;141(1):76-83.
62. Nestler EJ, Hyman SE, Malenka RC. Molecular neuropharmacology, a foundation for clinical neuroscience. McGraw-Hill, New York, 2001.
63. Talkowski ME, Bamne M, Mansour H, Nimgaonkar VL. Dopamine genes and schizophrenia: case closed or evidence pending? Schizophr Bull. 2007 Sep;33(5):1071-81. Epub 2007 Jul 14.
64. Littrell RA, Schneiderhan M. The neurobiology of schizophrenia. Pharmacotherapy. 1996 Nov-Dec;16(6 Pt 2):143S-147S.
65. Cipriani A, Geddes JR, Furukawa TA, Barbui C. Metareview on short-term effectiveness and safety of antidepressants for depression: an evidence-based approach to inform clinical practice. Can J Psychiatry. 2007 Sep;52(9):553-62.
66. Greydanus DE, Sloane MA, Rappley MD. Psychopharmacology of ADHD in adolescents. Adolesc Med. 2002 Oct;13(3):599-624.
67. Carlson GA, Finch SJ, Fochtmann LJ, Ye Q, Wang Q, Naz B, Bromet EJ.Antidepressant-associated switches from depression to mania in severe bipolar disorder. Bipolar Disord. 2007 Dec;9(8):851-9.
68. Schildkraut JJ. The catecholamine hypothesis of affective disorders: a review of supporting evidence. Am J Psychiatry. 1965 Nov;122(5):509-22.
69. Jabbi M, Kema IP, van der Pompe G, te Meerman GJ, Ormel J, den Boer JA. Catechol-o-methyltransferase polymorphism and susceptibility to major depressive disorder modulates psychological stress response. Psychiatr Genet. 2007 Jun;17(3):183-93.
70. Lasky-Su JA, Faraone SV, Glatt SJ, Tsuang MT. Meta-analysis of the association between two polymorphisms in the serotonin transporter gene and affective disorders. Am J Med Genet B Neuropsychiatr Genet. 2005 Feb 5;133(1):110-5.
71. L. Stryer. Biochemistry, 2nd Ed.1981, W.H. Freeman and Co. San Fransisco, CA.
72. Tetley TD. Health effects of nanomaterials. Biochem Soc Trans. 2007 Jun;35(Pt 3):527-31.
73. Donaldson K. Resolving the nanoparticles paradox. Nanomed. 2006 Aug;1(2):229-34.
74. Feynman, R.P. Plenty of Room at the Bottom, December 1959, The Archives of California Institute of Technology. [http://www.its.caltech.edu/~feynman/plenty.html].
75. Taniguchi, N. "On the Basic Concept of 'Nano-Technology'," Proc. Intl. Conf. Prod. Eng. Tokyo, Part II, Japan Society of Precision Engineering, 1974.
76. Taniguchi N. (Ed.) Nanotechnology: Integrated Processing Systems for Ultra-precision and Ultra-fine products 1996, Oxford University Press, USA.
77. Drexler KE. Engines of Creation: The Coming Era of Nanotechnology, 1986, Random House, New York.
78. Drexler KE. Nanosystems: Molecular Machinery, Manufacturing, and Computation, 1992, John Wiley
79. Moore GE. Electronics, Volume 38, Number 8, April 19, 1965
80. Lundstrom M. Applied physics. Moore's law forever? Science. 2003 Jan 10;299(5604):210-1.

81. Likharev K, Mayr A, Muckra I, Türel O. CrossNets: high-performance neuromorphic architectures for CMOL circuits. Ann N Y Acad Sci. 2003 Dec;1006:146-63.
82. Jain KK. Applications of nanobiotechnology in clinical diagnostics. Clin Chem. 2007 Nov;53(11):2002-9.
83. Ajayan, PM and Iijima, S. Nature, 358, 23-23 (1992).
84. Graham AP, Duesberg GS, Seidel RV, Liebau M, Unger E, Pamler W, Kreupl F, Hoenlein W. Carbon nanotubes for microelectronics? Small. 2005 Apr;1(4):382-90.
85. A New Model of Quantum Dots: Rethinking the Electronics Scientific. Research News Berkeley Lab. June 15, 2005, contact: Lin-Wang Wang, (510) 486-5571, lwwang@lbl.gov [http://www.lbl.gov/Science-Articles/Archive/Quantum-Dot-Electronics.html]
86. Venkatesan N, Yoshimitsu J, Ito Y, Shibata N, Takada K. Liquid filled nanoparticles as a drug delivery tool for protein therapeutics. Biomaterials. 2005 Dec;26(34):7154-63.
87. Montemagno CD, Bachand G. Nanotechnology (Vol. 10, No.3).
88. Alivisatos AP, Gu W, Larabell C. Quantum dots as cellular probes. Annu Rev Biomed Eng. 2005;7:55-76.
89. Doty RC, Fernig DG, Lévy R. Nanoscale science: a big step towards the Holy Grail of single molecule biochemistry and molecular biology. Cell Mol Life Sci. 2004 Aug;61(15):1843-9.
90. Vigneshwaran N, Kathe AA, Varadarajan PV, Nachane RP, Balasubramanya RH. Functional finishing of cotton fabrics using silver nanoparticles. J Nanosci Nanotechnol. 2007 Jun;7(6):1893-7.
91. Marini M, De Niederhausern S, Iseppi R, Bondi M, Sabia C, Toselli M, Pilati F. Antibacterial activity of plastics coated with silver-doped organic-inorganic hybrid coatings prepared by sol-gel processes. Biomacromolecules. 2007 Apr;8(4):1246-54.
92. Liu Y, Cui T. Silica nanowires fabricated with layer-by-layer self-assembled nanoparticles. J Nanosci Nanotechnol. 2006 Apr;6(4):1019-23.
93. Zhang S. Fabrication of novel biomaterials through molecular self-assembly. Nat Biotechnol. 2003 Oct;21(10):1171-8.
94. Robinson AL. Electron Microscope Inventors Share Nobel Physics Prize: Ernst Ruska built the first electron microscope in 1931; Gerd Binnig and Heinrich Rohrer developed the scanning tunneling microscope 50 years later. Science. 1986 Nov 14;234(4778):821-822.
95. Binnig G, Quate CF, Gerber C. Atomic force microscope. Phys Rev Lett. 1986 Mar 3;56(9):930-933.
96. Tourney, C. The man who understood the Feynman machine. Nature Nanotechnology, 2007, 2: 9-10.
97. Kim P, Lieber CM. Nanotube nanotweezers Science. 1999 Dec 10;286(5447):2148-50.
98. Patolsky F, Timko BP, Yu G, Fang Y, Greytak AB, Zheng G, Lieber CM. Detection, stimulation, and inhibition of neuronal signals with high-density nanowire transistor arrays. Science. 2006 Aug 25;313(5790):1100-4.
99. Mazzatenta A, Giugliano M, Campidelli S, Gambazzi L, Businaro L, Markram H, Prato M, Ballerini L. Interfacing neurons with carbon nanotubes: electrical signal transfer and synaptic stimulation in cultured brain circuits. J Neurosci. 2007 Jun 27;27(26):6931-6.

100. Park I, Li Z, Li X, Pisano AP, Williams RS. Towards the silicon nanowire-based sensor for intracellular biochemical detection. Biosens Bioelectron. 2007 Apr 15;22(9-10):2065-70.
101. Wallace R. Neural membrane microdomains as computational systems: Toward molecular modeling in the study of neural disease. Biosystems. 2007 Jan;87(1):20-30.
102. Wang K, Fishman HA, Dai H, Harris JS. Neural stimulation with a carbon nanotube microelectrode array. Nano Lett. 2006 Sep;6(9):2043-8.
103. Nadeau JL, Clarke SJ, Hollmann CA, Bahcheli DM. Quantum dot-FRET systems for imaging of neuronal action potentials. Conf Proc IEEE Eng Med Biol Soc. 2006;1:855-8.
104. Jan E, Kotov NA. Successful differentiation of mouse neural stem cells on layer-by-layer assembled single-walled carbon nanotube composite. Nano Lett. 2007 May;7(5):1123-8.
105. Meng J, Song L, Meng J, Kong H, Zhu G, Wang C, Xu L, Xie S, Xu H. Using single-walled carbon nanotubes nonwoven films as scaffolds to enhance long-term cell proliferation in vitro. J Biomed Mater Res A. 2006 Nov;79(2):298-306.
106. Cui B, Wu C, Chen L, Ramirez A, Bearer EL, Li WP, Mobley WC, Chu S. One at a time, live tracking of NGF axonal transport using quantum dots. Proc Natl Acad Sci U S A. 2007 Aug 21;104(34):13666-71.
107. Echarte MM, Bruno L, Arndt-Jovin DJ, Jovin TM, Pietrasanta LI. Quantitative single particle tracking of NGF-receptor complexes: transport is bidirectional but biased by longer retrograde run lengths. FEBS Lett. 2007 Jun 26;581(16):2905-13.
108. Brueggemann A, George M, Klau M, Beckler M, Steindl J, Behrends JC, Fertig N. Ion channel drug discovery and research: the automated Nano-Patch-Clamp technology. Curr Drug Discov Technol. 2004 Jan;1(1):91-6.
109. Lehmann-Horn F, Jurkat-Rott K. Nanotechnology for neuronal ion channels.
110. Pamir E, George M, Fertig N, Benoit M. Planar patch-clamp force microscopy on living cells. Ultramicroscopy. 2008 May;108(6):552-7.
111. Yang TH, Yee CK, Amweg ML, Singh S, Kendall EL, Dattelbaum AM, Shreve AP, Brinker CJ, Parikh AN. Optical detection of ion-channel-induced proton transport in supported phospholipid bilayers. Nano Lett. 2007 Aug;7(8):2446-51.
112. Tamkun MM, O'connell KM, Rolig AS. A cytoskeletal-based perimeter fence selectively corrals a sub-population of cell surface Kv2.1 channels. J Cell Sci. 2007 Jul 15;120(Pt 14):2413-23.
113. Dr. Horacio Cantiello, Harvard University, personal communication.
114. Levi-Montalcini R, Dal Toso R, della Valle F, Skaper SD, Leon A. Update of the NGF saga. J Neurol Sci. 1995 Jun;130(2):119-27.
115. Woolf NJ. Cholinergic systems in mammalian brain and spinal cord. Prog Neurobiol. 1991;37(6):475-524.
116. Mufson EJ, Ginsberg SD, Ikonomovic MD, DeKosky ST. Human cholinergic basal forebrain: chemoanatomy and neurologic dysfunction. J Chem Neuroanat. 2003 Dec;26(4):233-42.
117. Woolf NJ, Jacobs RW, Butcher LL. The pontomesencephalotegmental cholinergic system does not degenerate in Alzheimer's disease. Neurosci Lett. 1989 Jan 30;96(3):277-82.

118. Mufson EJ, Conner JM, Kordower JH. Nerve growth factor in Alzheimer's disease: defective retrograde transport to nucleus basalis. Neuroreport. 1995 May 9;6(7):1063-6.
119. Siegel GJ, Chauhan NB. Neurotrophic factors in Alzheimer's and Parkinson's disease brain. Brain Res Brain Res Rev. 2000 Sep;33(2-3):199-227.
120. Jordan PC. Fifty years of progress in ion channel research. IEEE Trans Nanobioscience. 2005 Mar;4(1):3-9.
121. Sakai N, Mareda J, Matile S. Ion channels and pores, made from scratch. Mol Biosyst. 2007 Oct;3(10):658-66.
122. Park KH, Chhowalla M, Iqbal Z, Sesti F. Single-walled carbon nanotubes are a new class of ion channel blockers. J Biol Chem. 2003 Dec 12;278(50):50212-6.
123. Gorostiza P, Isacoff E. Optical switches and triggers for the manipulation of ion channels and pores. Mol Biosyst. 2007 Oct;3(10):686-704.
124. Azzazy HM, Mansour MM, Kazmierczak SC. Nanodiagnostics: a new frontier for clinical laboratory medicine. Clin Chem. 2006 Jul;52(7):1238-46.
125. Sharma SH, Nanoneuroscience: emerging concepts on nanoneurotoxicity and nanoneuroprotection. Nanomedicine, 2007, 2:753-758.
126. Fradinger EA, Bitan G. En route to early diagnosis of Alzheimer's disease–are we there yet? Trends Biotechnol. 2005 Nov;23(11):531-3
127. Keating CD. Nanoscience enables ultrasensitive detection of Alzheimer's biomarker. Proc Natl Acad Sci U S A. 2005 Feb 15;102(7):2263-4.
128. Marx J. Alzheimer's disease. A new take on tau. Science. 2007 Jun 8;316(5830):1416-7.
129. Georganopoulou DG, Chang L, Nam JM, Thaxton CS, Mufson EJ, Klein WL, Mirkin CA. Nanoparticle-based detection in cerebral spinal fluid of a soluble pathogenic biomarker for Alzheimer's disease. Proc Natl Acad Sci U S A. 2005 Feb 15;102(7):2273-6.
130. Nam J-M, Thaxton, C. S. Mirkin, C. A. (2003) Science 301, 1884-1886.
131. Härtig W, Paulke BR, Varga C, Seeger J, Harkany T, Kacza J. Electron microscopic analysis of nanoparticles delivering thioflavin-T after intrahippocampal injection in mouse: implications for targeting beta-amyloid in Alzheimer's disease. Neurosci Lett. 2003 Feb 27;338(2):174-6.
132. Roney C, Kulkarni P, Arora V, Antich P, Bonte F, Wu A, Mallikarjuana NN, Manohar S, Liang HF, Kulkarni AR, Sung HW, Sairam M, Aminabhavi TM. Targeted nanoparticles for drug delivery through the blood-brain barrier for Alzheimer's disease. J Control Release. 2005 Nov 28;108(2-3):193-214.
133. Sharma HS, Ali SF, Dong W, Tian ZR, Patnaik R, Patnaik S, Sharma A, Boman A, Lek P, Seifert E, Lundstedt T. Drug delivery to the spinal cord tagged with nanowire enhances neuroprotective efficacy and functional recovery following trauma to the rat spinal cord. Ann N Y Acad Sci. 2007 Dec;1122:197-218.
134. Roberson ED, Scearce-Levie K, Palop JJ, Yan F, Cheng IH, Wu T, Gerstein H, Yu GQ, Mucke L. Reducing endogenous tau ameliorates amyloid beta-induced deficits in an Alzheimer's disease mouse model. Science. 2007 May 4;316(5825):750-4.
135. Cash AD, Aliev G, Siedlak SL, Nunomura A, Fujioka H, Zhu X, Raina AK, Vinters HV, Tabaton M, Johnson AB, Paula-Barbosa M, Avíla J, Jones PK, Castellani RJ, Smith MA, Perry G. Microtubule reduction in Alzheimer's disease and aging is independent of tau filament formation. Am J Pathol. 2003 May;162(5):1623-7.

136. Singh, R., Pantarotto, D., McCarthy, D., Chaloin, O., Hoebeke, J., Partidos, C.D., Briand, J.P., Prato, M., Bianco, A., Kostarelos, K., 2005. Binding and condensation of plasmid DNA onto functionalized carbon nanotubes: toward the construction of nanotube-based gene delivery vectors. J. Am. Chem. Soc. 127 (12), 4388-4396.
137. Fahn S; and the Parkinson Study Group. Does levodopa slow or hasten the rate of progression of Parkinson's disease? J Neurol. 2005 Oct;252 Suppl 4:IV37-IV42.
138. Ali SR, Ma Y, Parajuli RR, Balogun Y, Lai WY, He H. A nonoxidative sensor based on a self-doped polyaniline/carbon nanotube composite for sensitive and selective detection of the neurotransmitter dopamine. Anal Chem. 2007 Mar 15;79(6):2583-7.
139. Kurth MC. Using liquid levodopa in the treatment of Parkinson's disease. A practical guide. Drugs Aging. 1997 May;10(5):332-40.
140. Tan AK. Current and emerging treatments in Parkinson's disease. Ann Acad Med Singapore. 2001 Mar;30(2):128-33.
141. Li J, Andrews RJ. Trimodal nanoelectrode array for precise deep brain stimulation: prospects of a new technology based on carbon nanofiber arrays. Acta Neurochir Suppl. 2007;97(Pt 2):537-45.
142. Altshuler L, Bookheimer S, Proenza MA, Townsend J, Sabb F, Firestine A, Bartzokis G, Mintz J, Mazziotta J, Cohen MS. Increased amygdala activation during mania: a functional magnetic resonance imaging study. Am J Psychiatry. 2005 Jun;162(6):1211-3.
143. Ishizuka K, Paek M, Kamiya A, Sawa A. A review of Disrupted-In-Schizophrenia-1 (DISC1): neurodevelopment, cognition, and mental conditions. Biol Psychiatry. 2006 Jun 15;59(12):1189-97.
144. Camargo LM, Collura V, Rain JC, Mizuguchi K, Hermjakob H, Kerrien S, Bonnert TP, Whiting PJ, Brandon NJ. Disrupted in Schizophrenia 1 Interactome: evidence for the close connectivity of risk genes and a potential synaptic basis for schizophrenia. Mol Psychiatry. 2007 Jan;12(1):74-86.
145. Woolf NJ. Bionic microtubules: potential applications to multiple neurological and neuropsychiatric diseases. Journal of Nanoneuroscience 2009, 1:85-94.

2

Nanoscale Components of Neurons: From Biomolecules to Nanodevices

Summary

Neurons contain many structurally diverse nanoscale components, which individually carry out a well-defined function, or as is increasingly found to be the case, multiple functions. Nanoscale proteins are organized as systems. The neuronal membrane – embedded with multiple ion channels and receptors connected to scaffolding and effector proteins – represents a key information processing system in the neuron. In addition to receptors that mediate electrophysiological responses, there exist distinct membrane receptor populations that respond to neurotrophins and play critical roles in neural growth during development and in neural plasticity during adulthood. Despite their being touted as the main neuronal information processing system, membrane – embedded receptor systems operate relatively slowly, on the order of milliseconds to seconds. This has led researchers to probe other neuronal components in search of faster information processing speeds. DNA strands, which are well known to be the physical substrate of genes, act as semi-conductive wires when isolated outside the cell and are capable of transmitting and processing information analogously to the way a computer circuit might. Yet there is no evidence that DNA strands act as anything other than genes *in situ*. Cytoskeletal proteins form long strands that fill the entire interiors of neurons. Cytoskeletal proteins include neurofilaments, actin filaments, and microtubules. Traditional roles for the cytoskeletal proteins are mediating cell division, providing cell structure, and serving as a matrix for intracellular transport. Like DNA, microtubules are semiconductive and may transmit and process information, not only when isolated outside the cell, but also *in situ*. Nanotechnology provides new methods to investigate individual neuronal compartments and to manufacture small products ranging from mimetic molecules that interact with receptors to neural prosthetics that restore function following neural degeneration. Both recent breakthroughs and challenges relevant to creating effective interfaces between neurons and nanodevices are outlined.

2.1 Intracellular Components of Neurons

Generally speaking, a neuron has all the same intracellular components of any living cell. Neurons are immediately distinguishable from liver or kidney cells due to their elaborate shapes. In addition to the soma (the cell body), many neurons possess several dendrites and a single axon[1]. These processes contribute to the neuron's overall specific shape. An excitable neuronal membrane forms the surface of the neuron. As shown in Figure 2.1, the neuronal membrane encases the entire neuron including the soma, dendrites, and axon.

In the center of the soma lies the nucleus, which contains the genetic material in the form of DNA strands wound into double-helical configurations. Messenger RNA (mRNA) is transcribed in the nucleus of the neuron, much the same as it is in any cell. Nonetheless, the particular proteins that are expressed in neurons are specific not only to brain (or nervous system), but also to a particular neuronal type. The type of neuron includes its location in the brain (e.g., a cortical neuron versus a thalamic neuron), its morphology (e.g., a pyramidal cell versus a "star-shaped" stellate cell), and the neurotransmitter that the neuron synthesizes and releases (e.g., a GABAergic cell versus a glutamatergic cell).

Other organelles are variously distributed throughout the neuron. The Golgi apparatus and the rough endoplasmic reticulum are typically found in the soma. Synaptic vesicles can be tracked from the soma, down along the axon, to the axon terminal. Smooth endoplasmic reticulum is found at various locations in neurons, and mitochondria are particularly concentrated at active sites, such as in the nerve terminal.

Unlike other neuronal components, which are localized to limited portions of the neuron, cytoskeletal proteins extend into virtually all parts of the neuron with the exception of the nucleus. Moreover, cytoskeletal proteins have recently been shown to possess novel signaling capabilities (discussed more fully in Chapter 3). Nanotechnology has been partly responsible for uncovering these non-traditional functions.

2.1.1 The Neuronal Membrane and Protein Complexes Related to Neurotransmission

The neuronal membrane, like other plasma membrane variants, is a phospholipid bilayer measuring approximately 3 nm across that encapsulates all components of the neuron [1]. Each phospholipid is vertically aligned within the membrane and each has a polar head and a hydrophobic carbon tail. The polar heads interact with one another and with water to form the inside and outside surfaces of the membrane, respectively. The hydrophobic hydrocarbon

[1] Multipolar neurons possess several dendrites and an axon, whereas unipolar neurons have a single process that bifurcates into two, and bipolar neurons have two processes.

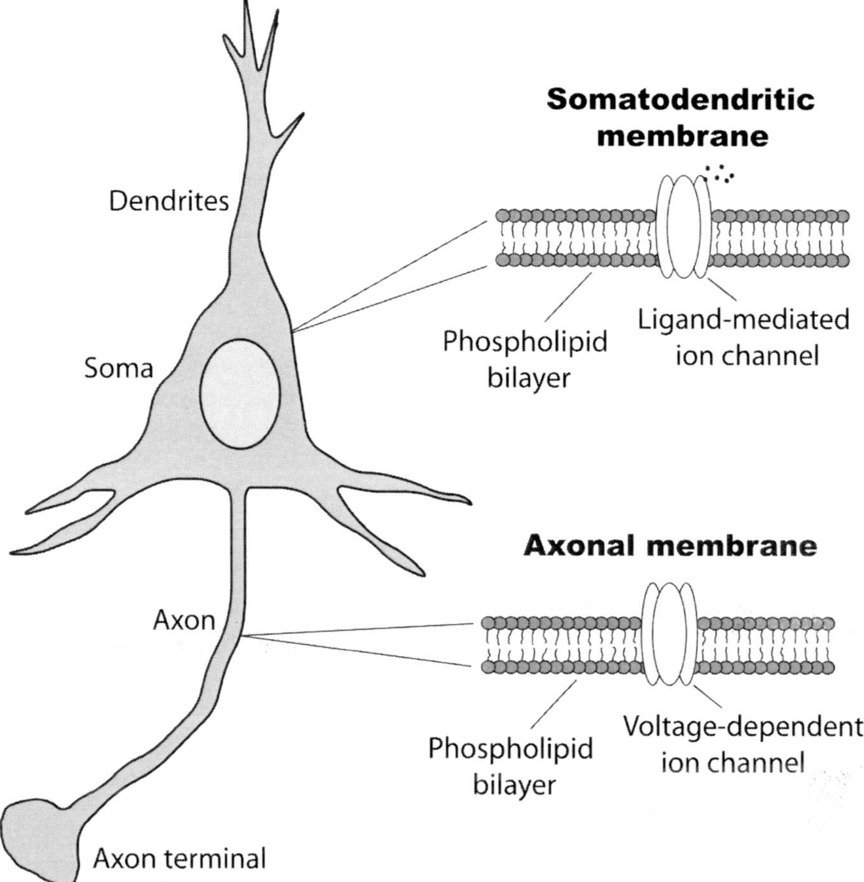

Fig. 2.1. The neuronal membrane surrounds all parts of the neuron. The somatodendritic membrane and the axonal membrane are distinguished on the basis of the types of ion channels embedded within their boundaries.

tails, which are made of fatty acid chains, aggregate and form the inner layers of the membrane (see Figure 2.1). Essential fatty acids (i.e., those required from the diet) interact with membrane phospholipids and modify the nature of the neuronal membrane, especially during early development and in aging.

According to one specific hypothesis, the fluidity of the membrane (i.e., the ease of lateral movement within the membrane) relies on essential fatty acids, such as linoleic acid, an omega-6 fatty acid, and alpha-linolenic acid, an omega-3 fatty acid [2]. Dietary increases in omega-3 fatty acids have been attempted in efforts to improve cognitive functions and as adjunct therapies for a variety of psychiatric and neurological disorders ranging from attention deficit/hyperactivity disorder (ADHD) to bipolar depression to Huntington's disease [3]. Some studies show improvement in mental status or physical health following omega-3 fatty acid regimens, while other studies fail to show improvements. The incomplete success of these fatty acid treatments warrants possible development of nanoscale approaches aimed at restoring full function to the neuronal membrane. Nanoparticle or nanodevice drug delivery systems may be able to improve on the outcome of fatty acid regimens by more directly delivering those compounds to brain.

Other nanotechnological innovations may be able to address deficiencies in membrane function associated with certain disease states as those technologies are further developed. Commercially available 10-nm "nanodiscs" composed of phospholipids surrounded by amphipathic protein scaffolds presently enable researchers to study single protein molecules that are embedded in the neuronal membrane [4]. Nanodiscs containing receptors deficient in a particular disease might soon be constructed so as to enable inserting those receptors into neuronal membranes of patients afflicted with such deficiencies.

Table 2.1. Characteristics of ions that determine neuronal excitability.

Ion	Abbreviation	Radius	Hydration energy
Sodium	Na^+	0.95Å	-105 kcal/mol
Chloride	Cl^-	1.81Å	-82 kcal/mol
Potassium	K^+	1.33Å	-85 kcal/mol
Calcium	Ca^{2+}	0.99Å	-397 kcal/mol

Adapted from [5].

2.1.2 Ion Channels and Ligand-Binding Receptor Proteins

The highly excitable neuronal membrane is a depository for a host of functional proteins – multiple selective ion channels and diverse families of receptors that bind selectively and specifically to individual neurotransmitters, neuromodulators, or neurotrophins. These proteins underlie the ability of the neuronal membrane to respond to inputs and send signals from one part of the neuron to another.

Ion Channels

Ion channels directly contribute to the excitability of the neuronal membrane and are essentially pores that enable positively or negatively charged ions to flow across the membrane. The main ions that flow across the neuronal membrane are sodium (Na^+), chloride (Cl^-), potassium (K^+), and calcium (Ca^{2+}) ions, and each has a unique size and hydration energy (see Table 2.1).

Based on identified genes, there are over 400 ion channels in existence [6]. Ion channels are generally selective to a particular ion. Any given ion channel will demonstrate a certain permeability, or range of permeabilities, that correspond with the somewhat oversimplified notion of the pore being "open" or "closed". A limited number of channels remain open at all times, whereas many ion channels open (or close) under particular circumstances. Changes in ion channel permeability can be triggered by voltage changes in the neuron, by mechanical pressure (as in special sensory receptors located in skin), and by ligand binding. Ligands controlling ion channels in the brain include neurotransmitters, neuromodulators, neuropeptides, neurohormones, neurotrophins, and drugs. Nanoparticles are also capable of directly interacting with ion channels, in some cases because of their comparable size to ligands [7].

Voltage-dependent (or gated) ion channels are found in both the somatodendritic and axonal membranes; however, it is the voltage-dependent ion channels found in the axonal membrane that are responsible for the action potential, which is also known as the spike or nerve impulse [8]. Dendritic spikes have also been recorded, and depending on the extent to which synapses cluster along dendrites, may greatly increase the computational power of individual neurons [9].

In an idealized axon, an action potential occurs when the neuronal membrane is depolarized from the normal resting potential to reach a threshold value. The Goldman-Hodgkin-Katz equation [10, 11] predicts the membrane potential at rest V_r:

$$V_r = \frac{RT}{F} \log \left(\frac{P_K[K]_{out} + P_{Na}[Na]_{out} + P_{Cl}[Cl]_{in}}{P_K[K]_{in} + P_{Na}[Na]_{in} + P_{Cl}[Cl]_{out}} \right) \tag{2.1}$$

where R is the gas constant; T is the absolute temperature; F is Faraday's constant; P_{ion} is the permeability for potassium, sodium, and chloride ion, respectively; and $[K]$, $[Na]$, and $[Cl]$ stand for the concentrations of the respective ions inside or outside of the cell.

The Goldman-Hodgkin-Katz equation yields results near $-70mV$ (originally reported as $-80mV$), which is approximately the same as the experimentally measured potential difference across the membrane with the inside of the neuron negative with respect to the outside of the neuron. Although not addressed by the earlier versions of this equation, negative surface charges of proteins exposed on the intracellular side of the neuronal membrane are currently known to contribute significantly to the resting potential.

The action potential represents a marked depolarization from the resting potential and it is the result of rapid and brief opening and closing of voltage-dependent Na^+ and voltage-dependent K^+ channels along the length of the axon. During propagation along the axon, the action potential travels in one direction only because the membrane is temporarily refractory afterwards. Na^+ channels open first, allowing Na^+ to rush into the axon briefly raising the inside potential from $-70mV$ to as high as $50mV$. As the voltage rises,

K^+ channels open which allow K^+ ions to rush out of the axon bringing the membrane potential down to slightly undershoot the resting potential for a brief period. Although the basic ionic theory of membrane currents at first appears straightforward, several questions remained unanswered for many years concerning what accounts for channel selectively, how voltage is detected, and what the hinge mechanism is for opening and closing the channel. Nanotechnology has enabled higher precision study of these issues [12, 13], some of which still require further elucidation.

Something that especially puzzled researchers was how a channel could permit the passage of larger ions while excluding smaller ones. Using nanotechnology, it has recently been discovered how K^+ channels selectively permit the flow of K^+ ions, having a radius of 1.33 Å, while excluding the smaller Na^+ ion having a radius of 0.95 Å. By constructing a type of semi-synthetic K^+ channel, MacKinnon and colleagues at the Rockefeller University and Howard Hughes Medical Institute determined that multiple (two or more) K^+ ions are needed to induce protein conformational changes to the filter portion of these K^+ channel such that passage of Na^+ is blocked [14]. The structure of the channel is customized to prefer multiple K^+ ions at their naturally occurring spacing intervals over Na^+ ions. Even though Na^+ ions could flow across K^+ channels if no K^+ ions were present; that kind of ionic imbalance occurs at a very low rate in situ.

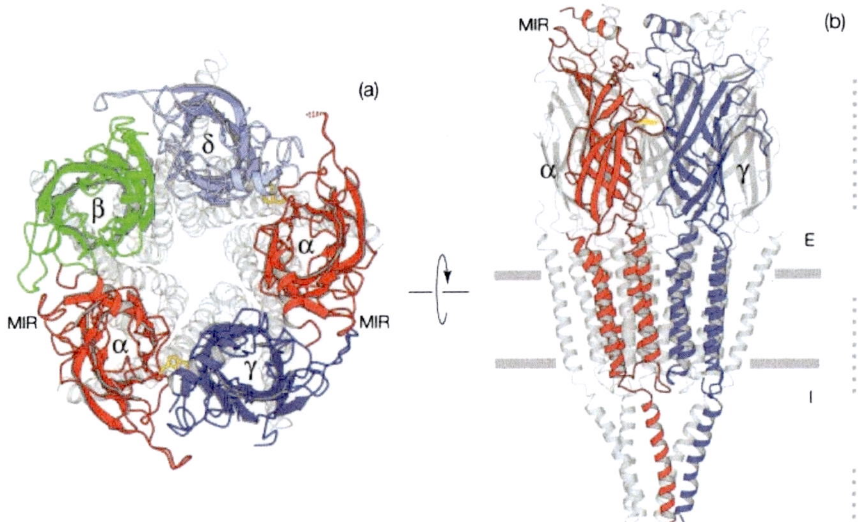

Fig. 2.2. Ribbon diagram of the nicotinic acetylcholine receptor (nAChR). The nAChR is an ionotropic receptor having a pentameric configuration with two α subunits, and β, γ, and δ subunits, as shown from above (a) and from a cross-section view through the neuronal membrane (b). Reprinted with permission; Unwin N. Journal of Molecular Biology 346:967-89 [19].

High-resolution atomic structure determination followed by molecular dynamics modeling has also shed light on the mechanisms that enable voltage-dependent ion channels to detect changes in voltage and to open or close, but the different methods have not always yielded the same result [15, 16]. Upon crystallizing K^+ channels in the open state, researchers suspected that a hinge motion in a glycine-rich conserved portion of the channel mediated opening and closing, while molecular dynamics simulations supported the hinge lying in a proline-rich region [15, 16]. In both schemes, bending at the hinge physically occludes passage through the pore; however, the overall conformation of the closed channel differs. Proposed voltage-sensing mechanisms include models in which the protein conformational rearrangement accounting for voltage-dependent response resembles a transporter, a helical screw, or a paddle [15, 16, 17]. Taking yet a different approach, quantum mechanical calculations applied to the amino acids near the gating mechanism and to the surrounding water molecules suggest that four water molecules may form a "basket" blocking the channel in the closed position [18]. This basket of water, which is confined to a nanometer space, becomes ice-like or glue-like and otherwise very stable due to the physical restrictions imposed by its nanoscale environment.

Propagation of the action potential is one the most rapid electrical events occurring in neurons. The fastest conducting axons, which are those coated with fatty myelin substance, conduct at speeds of $100m/s$. Electric currents propagate through copper wire at about 2/3 the speed of light or $2 \times 10^8 m/s$. Judging from the huge discrepancy between the speed of nerve conduction and that of electricity, either speed is not exceedingly important to higher neural function or novel modes of information processing are available to neurons that are separate from those operating at the neuronal membrane, and conceivable faster. In either case, receptors that respond to the binding of ligands such as neurotransmitters, neuromodulators, and neurotrophins are critically involved as triggering events.

Ligand-Binding Receptors

Receptors that bind neurotransmitters are essential for most instances of interneuronal communication – namely the exchange of information between two neurons. Changes in ionic conductance produced by ligand binding, which are called graded potentials, are smaller in amplitude and propagate more slowly than those produced during action potentials. Nonetheless, these graded potentials summate over time and spatial distribution giving them marked integrative capabilities. Table 2.2 lists the classic neurotransmitters and neuromodulators and their primary actions, which vary depending on whether the receptor is ionotropic or metabotropic (i.e., G-protein-coupled).

An ionotropic receptor is essentially an ion channel with a binding site for a neurotransmitter located on one or more of its subunits. The binding site is located on the extracellular surface of the recipient neuron (or muscle

Table 2.2. Receptors categorized according to the neurotransmitter-specific families, along with the ion channel fluxes and second messengers they activate.

Neurotransmitter	Receptor	Action	
		Ionotropic - Major (Minor) Ion Fluxes	Metabotropic - Effect on Second Messenger
Acetylcholine	Nicotinic	Na^+ influx	—
	M1, M3, M5	—	stimulate PI-PLC
	M2, M4	—	inhibit adenylyl cyclase
Norepinephrine	α-1	—	stimulate PI-PLC
Epinephrine	α-2	—	inhibit adenylyl cyclase
	β	—	stimulate adenylyl cyclase
Dopamine	D1, D5	—	stimulate adenylyl cyclase
	D2, D3, D4	—	inhibit adenylyl cyclase
Serotonin	5-HT1	—	inhibit adenylyl cyclase
	5-HT2	—	stimulate PI-PLC
	5-HT3	Na^+ (Ca^{2+}) influx	—
	5-HT4	—	stimulate adenylyl cyclase
Glutamate	AMPA	Na^+ (Ca^{2+}) influx	—
	Kainate	Na^+ (Ca^{2+}) influx	—
	NMDA	Ca^{2+} (Na^+) influx	—
	mGlu Class I	—	stimulate PI-PLC
	mGlu Class II	—	inhibit adenylyl cyclase
	mGlu Class III	—	inhibit adenylyl cyclase
GABA	$GABA_A$	Cl^- influx	—
	$GABA_B$	—	inhibit adenylyl cyclase
	$GABA_C$	Cl^- influx	—

(Summarized from [1]. Abbreviations: 5-TH: 5-hydroxytryptamine (serotonin); AMPA: α-amino-3-hydroxy-5-methylisoxazole-4- propionic acid; D1 - D5: dopamine receptors; GABA: γ-amino-butyric acid; M1 - M5: muscarinic acetylcholine receptors; mGlu: metabotropic glutamate receptors; NMDA: N-methyl-D-aspartic acid; PLC: phosphoinositide-specific phospholipase C)

cell) so that the neurotransmitter, when released from the input neuron, can bind to the exposed receptor. The nicotinic acetylcholine receptor is a classic example of an ionotropic receptor (see Figure 2.2). The nicotinic receptor is a pentamer, meaning it contains five subunits (α, β, γ, and δ), of which two are α-subunits that possess the binding site for acetylcholine [19]. When both sites on the extracellular domain are bound to molecules of acetylcholine (or a ligand such as nicotine), the receptor undergoes tertiary and quaternary conformational changes described as a clockwise rotation with outward motions in the transmembrane domains that in turn open the pore region of the receptor, enabling ions to enter to postsynaptic cell [20]-[21]. There appear

Table 2.3. Neurotrophins in the CNS

Neurotrophin	Location of sensitive neurons	Receptors
Nerve growth factor (NGF)	Cholinergic basal forebrain; sensory and sympathetic ganglia	TrkA, p75
Brain-derived neurotrophic factor (BDNF)	Hippocampus, cerebral cortex	TrkB, p75
Neurotrophin 3 (NT3)		TrkA, TrkB, TrkC, p75
Neurotrophin 4 (NT4)		TrkB, p75
Glial-cell-line-derived neurotrophic factor (GDNF)	Striatum, substantia nigra, sensory neurons, sympathetic neurons, and motor neurons	GFRα1, RET-receptor tyrosine kinase
Neurturin (NRTN)	Striatum, substantia nigra, sensory, sympathetic, and motor neurons	GFRα2, RET-receptor tyrosine kinase
Ciliary neurotrophic factor	Striatum, motor cortex	CNTFRα, GP130, and LIFbR

Based on [23]-[26]

to be general similarities in the way in which nicotinic, glycine, and $GABA_A$ receptors operate on a biomechanical level [22].

Metabotropic receptors are quite distinct from ionotropic receptors, with each consisting of a long polypeptide chain of approximately 300 - 1200 amino acids. These chains typically have seven transmembrane segments, as well as extracellular and intracellular domains. The N-terminus is one of the extracellular domains, and it contains the neurotransmitter-binding site. The C-terminus is the part of the intracellular domain that couples and uncouples to a G-protein complex consisting of α, β, and γ subunits [1]. In the resting state, the G-protein binds a guanine diphosphate (GDP). When neurotransmitter molecules bind the receptor, the α-subunit (or the β or γ subunit) that is coupled to the receptor loses a GDP and gains a guanine triphosphate (GTP). This leads to the dissociation of G-protein subunits, which in turn stimulate (or inhibit) second messengers that are capable of activating signal transduction cascades that can, as a consequence, lead to the opening or closing many ion channels. The two main second messengers that are triggered by receptors for classical neurotransmitters and neuromodulators are phosphoinositide-specific phospholipase C (PI-PLC) and adenylyl cyclase (see Table 2.2).

As is more fully discussed in the chapters to come, signal transduction cascades triggered by metabotropic receptors activate chemical reactions inside

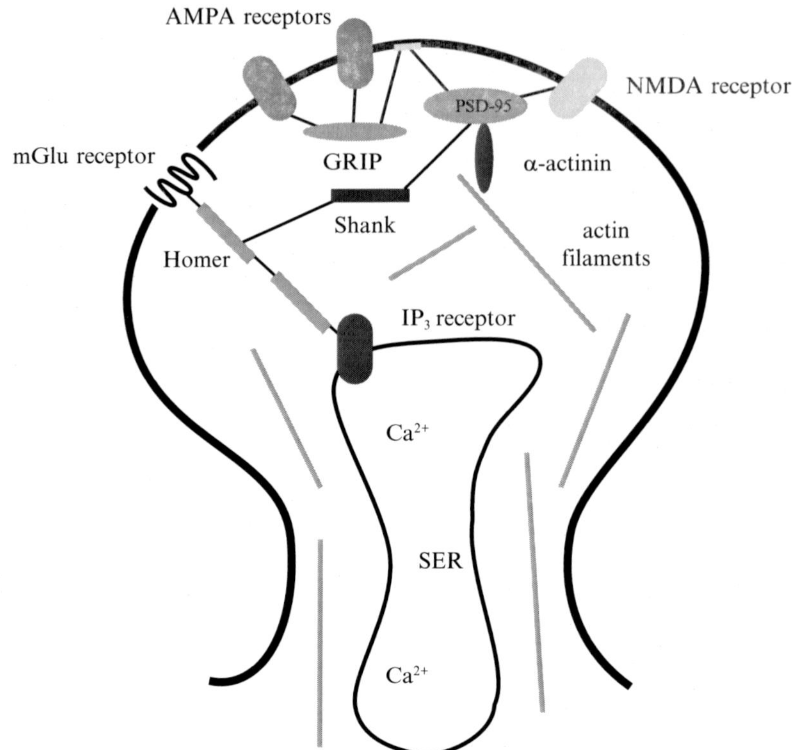

Fig. 2.3. Scaffolding proteins found in a dendritic spine postsynaptic to a glutamatergic input. Adapted from [1, 32, 33].

of neurons, with some of these cascades triggering novel communication modes within neurons [27]. Nanoscience methods have already enabled precision study of metabotropic receptor actions. Dendrimers, which are complexly branching nanostructures, can be conjugated to G-protein-coupled receptors paving the way for nanoscale pharmacology aimed at this type of receptor [28, 29].

The neuronal membrane also contains receptors for neurotrophins-molecules that sustain neuronal growth and survival. There are three families of neurotrophins and examples of each are as listed Table 2.3. Nerve growth factor (NGF), brain-derived growth factor (BDNF), neurotrophin 3 (NT3), and neurotrophin 4 (NT4) represent the first family and these bind to one or more of tyrosine kinases: TrkA, TrkB and TrkC, as well as to the p75 receptor, a member of the necrotic death receptor family [23]. One of the downstream actions of NGF is upon the neuronal cytoskeleton, which contributes to NGF's influence on neurite growth. Glial-derived neurotrophic factor (GDNF) and neurturin (NTNF) are members of another family and these bind to receptor tyrosine kinases (RTK) that are rearranged during transfection [24, 25].

GDNF can also act through the neural cell adhesion molecule (nCAM) [30]. Finally, ciliary neurotrophic factor (CTNF) is a member of yet another family of neurotrophins [26]. Neurotrophins require sensitive assays since they are found in very low concentrations in brain. Nanoparticles, such as quantum dots, have been useful and enable the detection of neurotrophins down to the level of single molecules [31].

2.1.3 Scaffolding Proteins, Signal Transduction Cascades, and Cell Adhesion Molecules

The components of a synapse consist of the presynaptic membrane, the synaptic gap, and the postsynaptic membrane. Scaffolding proteins attached to the postsynaptic membrane are critical to the postsynaptic cell's response because they link receptor molecules in the membrane to effector proteins and to the cytoskeleton. Effector proteins are responsible for signal transduction cascades inside the cell. Finally, cell adhesion molecules are responsible for keeping neurons close enough to neighboring neurons and surrounding glia such that the low levels of neurotransmitters, neuromodulators, and neurotrophins released are able to elicit significant effects.

The site of synaptic contact from a glutamatergic axon terminal onto a spine of a large pyramidal cell of the hippocampus or cerebral cortex has been well studied and many of the scaffolding proteins identified are shown in Figure 2.3. This synapse contains a large number of AMPA receptors and relatively fewer NMDA receptors and mGlu receptors linked to the inside components of the neuron by scaffolding proteins, such as postsynaptic density protein-95 (PSD95), Homer, Shank, and GRIP [1, 32, 33]. The role of these scaffolding proteins is to secure the receptors and to tether them to (a) intercellular kinases, including protein kinase C (PKC), cAMP-dependent protein kinase (PKA), and calcium/calmodulin-dependent kinase II (CaMK II), (b) the inositol 1,4,5-triphosphate (IP_3) receptor embedded in the smooth endoplasmic reticular (SER) of sacs containing Ca^{2+}, and (c) filamentous actin (F-actin) through α-actinin. Clusters of CaMK II associate with lipid rafts in the membrane, which in turn associate with PSD95 [34]. A-kinase anchoring protein tethers PKA to the membrane, to NMDA receptors, and to the cytoskeleton, and may be responsible for stimulating local polymerization of the cytoskeleton protein actin in response to receptor-mediated activity [35].

GABAergic and glycinergic synapses contain different scaffolding proteins than do glutamatergic postsynaptic sites. Gephyrin, rather than PSD95, is found in association with these inhibitory GABAergic and glycinergic synapses, and the presence of this scaffolding protein has been shown to direct synaptogenesis and growth of synapses [36]. These scaffolding proteins also ensure a proper match between presynaptic and postsynaptic elements during neurodevelopment. If a mismatch occurs between the presynaptic axon terminal and the scaffolding proteins in the postsynaptic membrane, the presynaptic element will retract.

Signal transduction cascades operate within the framework of the specific scaffolding proteins. There are (a) the messengers (i.e., the neurotransmitters), (b) the second messengers (i.e., the products of adenylyl cyclase, cAMP, and PI-PLC, IP_3, and diacylglycerol), and (c) higher order signal transduction cascades (such as the protein kinases mediating diverse biological responses). Protein kinases, such as PKC, PKA, and CaMK II, function to phosphorylate proteins of numerous varieties. Neurotransmitter receptors, ion channels, second messengers, cytoskeletal proteins, synaptic vesicle proteins, transcription factors, and the kinases themselves are regulated by phosphorylation – all in a manner that is dependent on the surrounding matrix. Nanoscience provides tools, materials and techniques with which to study and manipulate the matrix of scaffolding proteins and signal transduction in the neuron. Nanotubes and nanostructures have already been used to build scaffolds having applications to biomedicine and electronics [37, 38].

Cell adhesion molecules also contribute to the framework within which signal transduction molecules operate. Cell adhesion molecules and adhesion complexes include nCAM, synCAM, cadherin, neuroligin/neurexin, and Ephrin/EphB; these adhesion proteins bind each other or their partner proteins to link the presynaptic membrane to the postsynaptic membrane [39, 40]. Neuroligin binds to the postsynaptic membrane, specifically to PSD95, and by this means couples NMDA receptor action and signal transduction cascades induced by Ca^{2+} influx. Its partner neurexin binds to presynaptic membrane proteins. Cadherin molecules link the cytoskeletal protein actin localized to the postsynpatic membrane to actin filaments in the presynaptic membrane via an intermediary protein α-catenin. Although individual adhesion molecules operate via distinct mechanisms, those localized to the synapse share the major function of regulating synapse formation during early neural development, adult synaptogenesis, and alignment of pre- and postsynaptic elements. In addition to mediating synaptic plasticity during development, these cell adhesion molecules play critical roles in adaptive neural responses to stress and in synaptic reorganization with learning and memory.

Nanotechnology should facilitate further study of cell adhesion molecules and their particular roles in nervous system development and adult neural plasticity. Studies have already shown that nanoparticles are capable of targeting cells that have upregulated levels of cell adhesion molecules [41]. Nanocarriers functionalized with cell adhesion molecules have also been used to provide a matrix that controls cell morphology and growth [42].

2.1.4 DNA, mRNA, and the Golgi Apparatus in Neurons: Transcription, Translation, and Packaging in Synaptic Vesicles

Determining the double-helical structure of deoxyribonucleic acid (DNA) was unquestionably one of the greatest discoveries of the 20th century science; nonetheless, there were a number of equally brilliant discoveries leading up to it. Were it not for the seminal deduction of Avery, MacLeod, and McCarthy in 1944 that DNA was the substance of heredity [43], James Watson and Francis Crick would not have been likely to ever publish their famous 1953 report characterizing the 3-D structure of DNA [44]. By the early 1950's, the initial skepticism surrounding the simple nucleotide-based DNA as the genetic material had subsided and what some had originally thought to be a "stupid molecule" was ready to occupy center stage[2]. Paradoxically, an early argument for nanotechnology realizing great heights of success was how DNA, a molecule of nanoscale dimensions, is nonetheless capable of storing a vast amount of information [45]. DNA has also proved to be a useful nanoscale material, with DNA strands being used to make nanodevices, such as gears, walkers, and translation devices [46].

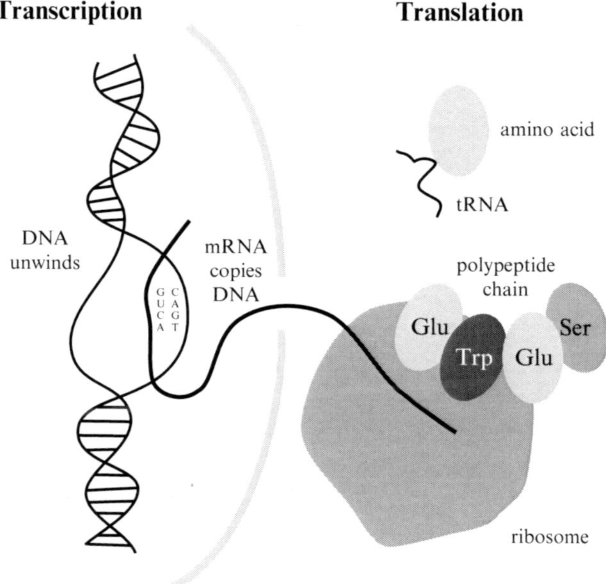

Fig. 2.4. Protein synthesis in neurons is similar, but not identical, to that in other cells. Transcription occurs in the nucleus as with other cells; however, translation occurs in both the cell body and in dendrites.

DNA and the rest of the machinery involved with protein synthesis in neurons is similar to that of most living cells with the key distinction of there being more diverse expression of select proteins due to so many different neuronal types. Neuronal differentiation, which is based on the expression

[2] As stated in a preface preceding the reprinting of the 1944 article by Avery, MacLeod, and McCarthy, University of Rockefeller president, Törsten Wiesel notes that in the 1940's many believed that DNA was too simple to encode genetic material.

of neuron-specific proteins, is a process regulated by DNA-binding proteins, promoters, and inhibitors. Neuronal type depends on selective synthesis of particular structural proteins (e.g., specific microtubule and actin binding proteins) and neurotransmitter-related proteins (e.g., specific receptors and synthetic enzymes).

As shown in Figure 2.4, the first step in protein synthesis, called transcription, occurs when a segment of DNA unwinds, and nucleotides adenine (A), cytosine (C), guanine (G) or thymine (T) of the DNA segment pair respectively with nucleotides uracil (U), G, C, and A of mRNA to form an mRNA strand. The mRNA strand then exits the nucleus through a nuclear pore enabling the second step in protein synthesis, called translation. Once in the cytoplasm, mRNA becomes associated with free ribosomes or with ribosomes associated with the endoplasmic reticulum. Transfer RNA (tRNA) binds to individual amino acids and carries them to the site of protein translation. This process consists of aligning amino acids in close proximity so that the C-terminus of one amino acid can covalently bind to the N-terminus of the next amino acid resulting in a long polypeptide chain. Every amino acid has a code signified by three unique consecutive nucleotides, called a codon, and there are also stop codons that signal the protein sequence is completed [47].

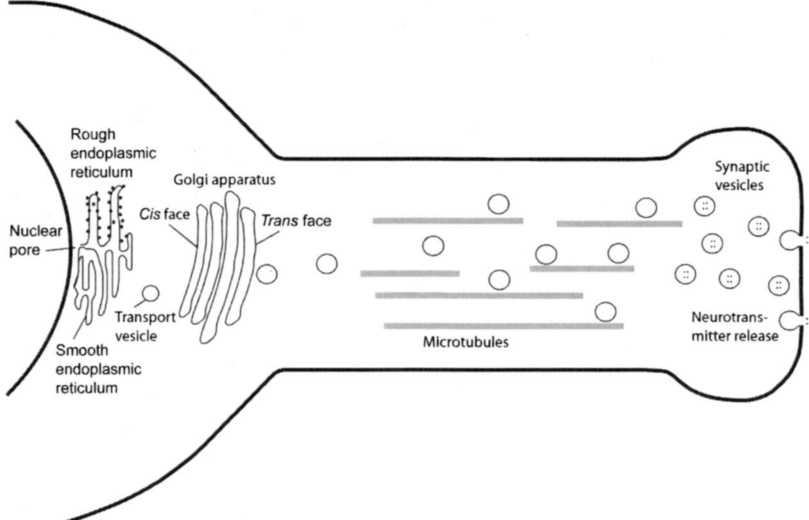

Fig. 2.5. The Golgi apparatus in neurons packages proteins synthesized by ribosomes. Ribosomes traverse membranes (called the rough endoplasmic reticulum) extending from nuclear pores to the Golgi apparatus. In neurons, many of these secretory vesicles are synaptic vesicles, which carry biosynthetic enzymes down the axon to the presynaptic terminal where large numbers of neurotransmitter molecules are synthesized and stored for release.

Protein synthesis in neurons also differs from that in other cells in that a select number of proteins are translated from mRNA and ribosomes located in the dendrite [48, 49, 50]. Among the mRNA species found in dendrites are those encoding for the α-subunit of CaMK II, microtubule-associated protein-2 (MAP2), activity-related cytoskeletal protein (Arc), β-actin, NMDA receptor subunits: NR1 and NRB2, and AMPA receptor subunits: GluR1 and GluR2. It has been proposed that synaptic activity stimulates transcription of dendritic mRNA in the neuron's nucleus, which is encapsulated in a granule and then transported down the microtubules of dendrites, and finally along actin filaments into the dendrite spine where the mRNA escapes from the granule and encodes a protein responsible for stabilizing or consolidating the spine [49, 50]. Neuronal activity or chemical stimulation is not, however, always associated with stimulated local translation. Applying NMDA down-regulates local translation of GluR1 and GluR2 AMPA receptor subunits [51]. It is likely that many factors contribute to activity-related changes in local translation, and that these changes are more robust during early development of the nervous system.

Although the mRNA species found in dendrites are functionally relevant to synaptic operations, each spine contains little more than a handful of ribosomes, and the individual strands of mRNA have lengths that in some cases exceed the width of spine heads [52]. Accordingly, nanoscale tools and materials are uniquely appropriate for future studies investigating mRNA translation in dendrites and spines.

Following protein translation in the cell body region, certain neuron-specific packaging processes occur. After a newly synthesized polypeptide chain is released from a ribosome into the cell body, it moves to a membranous structure called the Golgi apparatus (see Figure 2.5). The Golgi apparatus looks like a stack of folded membranes that thicken from cis to trans surface. This thickening is due to an increase in concentration of cholesterol and sphingolipids manufactured by the Golgi apparatus [53]. In some cases, the Golgi apparatus extends into the dendrite of the neuron [54].

Once proteins are incorporated in the Golgi apparatus some are modified post-translationally[3] and then packaged in membranous sacs called synaptic vesicles. This latter step is particularly important in neurons as synthetic enzymes that manufacture neurotransmitter molecules are packaged into synaptic vesicles in the Golgi apparatus, then sorted and directed to the microtubule tracks of the axon. Once in the axon, synaptic vesicles are transported along the microtubule tracks as cargo, carried by a member of a large family of kinesins [57, 58]. In the axon terminal, synaptic vesicles accumulate until depolarization of the axon terminal membrane resulting from the arrival of an action potential triggers a significant influx of Ca^{2+} ions. This in turn

[3] Post-translational modifications occurring in the Golgi apparatus include glycosylation of precursor proteins and membrane-bound proteins, such as cell adhesion molecules and receptors [55, 56].

activates the docking of synaptic vesicles to the presynaptic membrane and the subsequent release of neurotransmitters (and other contents of the synaptic vesicles such as ATP) into the synaptic gap [59, 60].

This sequence of protein synthesis, packaging, and synaptic vesicular transport provides the underpinnings for the traditional mechanisms of neural communication. The "astonishing hypothesis" proposed by Francis Crick – the same Nobel laureate responsible for the discovery of the structure of DNA-posits neurophysiological mechanisms operating at the level of the neuronal membrane as adequate in accounting for higher cognition, with consciousness emerging from the firing patterns among large groups or assemblies of neurons [61]. A growing number of nanoneuroscientists, however, suggest that multiple molecules inside neurons – DNA, mRNA, signal transduction molecules, scaffolding proteins, and cytoskeletal proteins – may perform biomolecular computations that complement and vastly expand upon the capacities of such neural networks.

2.1.5 The Neuronal Cytoskeleton

The cytoskeleton is a major component of all eukaryotic cells. Typical roles played by the cytoskeleton in simple to complex organisms include organizing chromosomes during cell division, enabling motility (as in forming cilia or flagella), transporting materials to specific sites in the cell, and giving a cell its characteristic structure and mechanical stability. These nanoscale structures play diverse roles that vary according to cell type. Neurons, at least adult neurons, do not appear to rely on cytoskeletal proteins for certain functions to the same extent that other cells do.

A salient difference between neurons and other cells of complex organisms is that a majority of neurons do not divide after birth. Adult neural stem cells are a prominent exception to this rule and are a topic of intense research given their potential in treating disease [62]. Stem cell research has been facilitated by recent nanotechnological developments, such as precision nanoscale imaging methods and nanomaterials suitable as growth matrices [63, 64].

Other than during development, neurons are not usually motile, so that particular function is not generally served by the cytoskeleton in adult neurons either. The cytoskeleton does, however, play two notable roles in neurons, both during adulthood and during development, and those are transporting materials from one site to another and giving the neuron its structure [65, 66]. While these roles may appear unrelated to neural communication, the way in which a neuron responds to inputs depends on receptor levels and their clustering in the membrane, properties that are determined by transport and anchoring functions of the cytoskeleton. Axonal transport of synaptic vesicles also affects the output of a neuron. Thus, to the extent that the cytoskeleton is an "intelligent" intracellular structure, capable of self-regulating its transport and growth, it possesses the unique capability to control the strength of individual inputs to the neuron and its output.

Since the cytoskeleton determines neuronal structure, a molecular definition of structural-functional relationships is achievable with certain cytoskeletal proteins being associated with certain cell shapes, neuronal compartments, and neural functions [67, 68]. Neurons can be classified on the basis of size, shape, and how many dendrites or neuritic processes they possess (if any). The neuronal shape – pyramidal or stellate – frequently correlates with a physiological response pattern [69, 70]. Moreover, the environment continuously alters neuronal structure, such that overall neuronal shape reflects previous cell activity and experience. Neurons that receive sufficient inputs, expand their dendritic arbors, whereas those that do not retract their dendritic arbors or die altogether [71].

Much of neuronal structure, as determined by the underlying cytoskeleton, is adapted to relaying specific sensory information. From the first sensory neuron to the final cortical processing unit, the majority of neurons in the nervous system play a primary role in relaying sensory information. Unipolar sensory neurons, for example, carry messages from specialized touch receptors in the skin to the spinal cord, then to the medulla, the thalamus, and finally to the cerebral cortex. A remarkable feature of cortical pyramidal cells – implicated as playing pivotal roles in higher cognitive functions, such as perception, learning, memory, and consciousness – is their development of massive dendritic trees and a vast number of inputs. The structure of these dendritic trees and arrangement of synaptic connections is determined by cytoskeletal proteins. Biomolecular computing within these large neurons would be expected to greatly expand the computational power of the neural networks to which these neurons belong. Cytoskeletal proteins, by virtue of their high concentrations in the large pyramidal neurons of the cerebral cortex are in a position to contribute substantially to such biomolecular computations [27]. Recently developed nanotechnologies that enabled precision study and fine-tuned alterations of dendrite morphology include nanosurgical techniques, nanoscale gene-delivery methods, and nanostructured scaffolds [73, 74, 75]. These and other nanotechnological methods will advance our understanding of biomolecular computing among the different cytoskeletal proteins, including how they contribute to pyramidal cell integrative capacities.

Table 2.4. The protein composition of the squid giant axon.

Protein	Fraction of axoplasmic protein (%)	Concentration (mg/ml)
α- and β-Tubulin	22	5.6
Neurofilament subunits	13	3.3
Actin	6	1.4
Total	41	10.3/24.1

Adapted from [72].

Neurons have three different kinds of cytoskeletal proteins: neurofilaments, microfilaments, and microtubules, and all these cytoskeletal proteins are developmentally regulated. Each type of cytoskeletal protein is uniquely compartmentalized within a neuron, contributing in a particular way to neuron structure, and in the case of microtubules and microfilaments, transport of materials. Cytoskeletal proteins form a major portion of the protein found in the neuronal cytoplasm. Core cytoskeletal components alone account for 41% of total protein in the axonal cytoplasm (see Table 2.4), and numerous cytoskeletal-associated proteins would be expected to increase that percentage significantly.

Neurofilaments

Neurofilaments measure 10 nm in diameter and represent members of the intermediate filament group that are exclusive to neurons [71]. Neurofilaments consist of subunits having three sizes: neurofilament light (NF-L; 68kD), neurofilament medium (NF-M; 145kD), and neurofilament heavy (NF-H; 200kD). Each neurofilament protein has a central α-helical rod domain that forms the core of the filament, an N-terminus, and a C-terminus (see Figure 2.6). The N-termini are partly responsible for binding, in a head-to-tail fashion, the neurofilament subunits together into homodimers of two NF-L subunits or, more typically, heterodimers of one NF-L with either an NF-M or NF-H subunit. Dimers are then assembled into tetramers and then into filaments containing 32 subunits in cross-section. The C-termini of NF-M and NF-H extend laterally to form side-arms that can make crossbridges with other cytoskeletal proteins. Multiple sites of phosphorylation on the side-arms of neurofilament proteins regulate their functions, many of which depend on interactions with other cytoskeletal proteins [77].

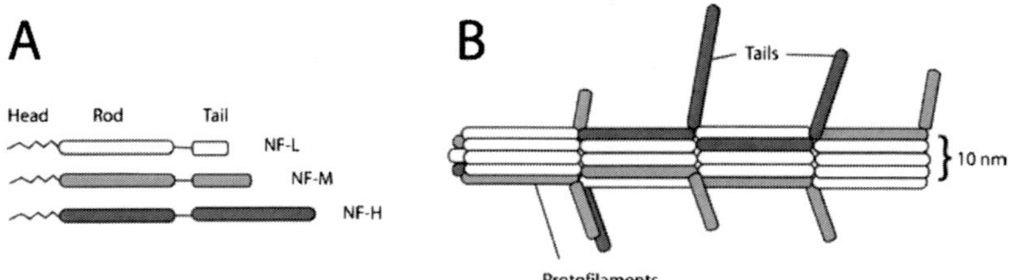

Fig. 2.6. Neurofilaments. There are three neurofilament subunits: NF-L, NF-M, and NF-H. Each has a similar head and rod region, but they differ markedly in their tail regions (A). Neurofilament subunits assemble into filaments with tail regions forming side branches (B). After [76].

Of all the cytoskeletal proteins, neurofilaments are most preferentially concentrated in axons. As observed in electron micrographs, the ratio of neurofilaments to microtubules is far greater in the axon than in dendrites or the soma of neurons [71]. An important function of neurofilaments in the axon is to provide stability. Neurofilament side-arms bind other cytoskeletal proteins, such as the microtubules, as well as linker proteins that in turn bind with microtubules thereby facilitating microtubular transport [78]. AFM studies have also revealed that neurofilaments act as nanoscale shock absorbers inside cells [79]. Axonal shock absorbers would be useful given all the transport and trafficking that occurs in the axon.

The side-arms of neurofilaments also act as spacers separating them from other neurofilaments that run parallel to the major axis of the axon, as well as from neighboring microtubules. This spacing of neurofilaments directly affects the diameter of the axon. Moreover, certain neurofilament subunits are particularly critical in determining the diameter of the axon. Genetic deletions of NF-L and of both NF-M and NF-H, but not deletions of NF-H alone, cause significant reductions in axonal caliber of murine axons [80], and deletions of NF-L impair axon regeneration following injury [81]. Even though lower concentrations of neurofilaments occur in dendrites as compared with axons, normal dendritic arborization depends on these stabilizing proteins [82]. In knockout mice having the gene for NF-L deleted, motor neurons normally displaying many dendrites showed marked decreases in dendrite number, whereas smaller neurons, which ordinarily have fewer dendrites, were unaffected. Thus, the relationship between the cytoskeleton and neuron structure is complex and depends on the specific cytoskeletal protein, the neuron type, and physiological properties of the neuron.

As illustrated in Figure 2.7, high concentrations of neurofilament proteins have been observed in pyramidal cells of the neocortex and limbic cortex. Neurofilament proteins also distinguish two broad categories of pyramidal cells: one with tufted dendrites (and projections to the spinal cord, pons, tectum, and striatum) and another having slender apical dendrites and projections to the contralateral side of the cerebral cortex [84]. These differences reflect different communication styles: the tufted-dendrite pyramidal neurons conduct action potentials more rapidly in their longer and larger diameter axons than do the slender-dendrite pyramidal neurons, which have axons that travel lesser distances. Tufted-dendrite pyramidal neurons also have different physiological response patterns (bursting type) from those of slender-dendrite pyramidal neurons (non-bursting type). Although other proteins also distinguish between these two types of pyramidal cells, it is likely that neurofilaments proteins contribute significantly to physiological responses patterns since they determine axonal diameter, and as a direct consequence, rate of propagating action potentials. Nanotechnologies that enable precise manipulation of axon morphologies will likely further delineate these types of structure-function relationships.

While neurofilaments stabilize neuronal structure during both development and regeneration, the developmental sequence of neurofilament subunit proteins suggests their individual functions. Neurofilament subunit proteins NF-L and NF-M first appear in embryonic tissue, whereas NF-H appears

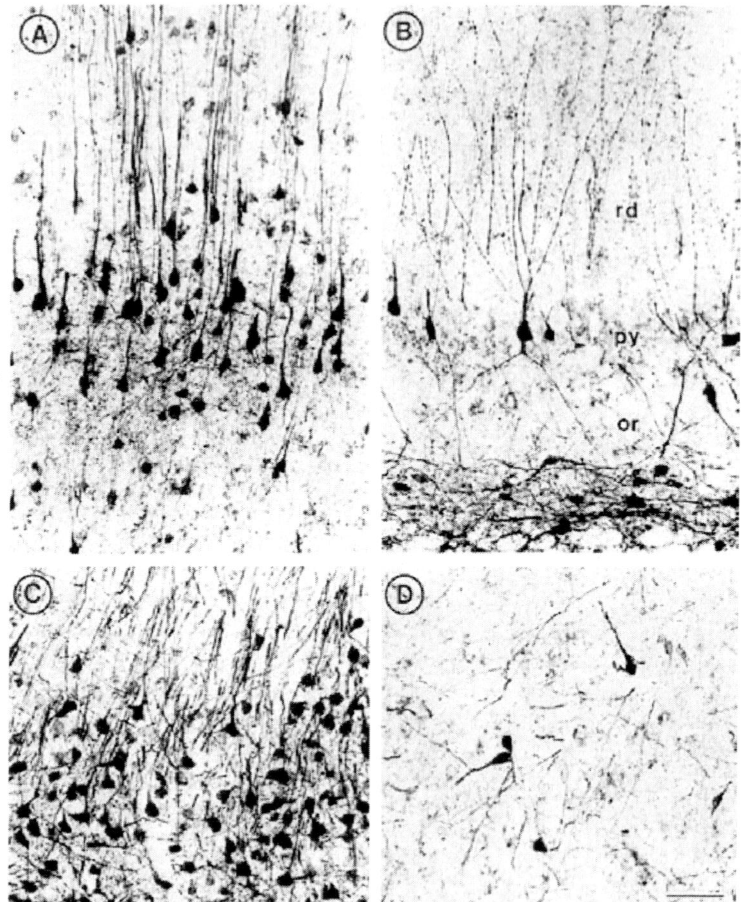

Fig. 2.7. The distribution of NF-L. Neurofilament 68kD protein immunostained pyramidal cells in (A) temporal cortex, (B) hippocampus, (C) subiculum, and (D) entorhinal cortex. From [83].

for the first time postnatally [85, 86]. This time course parallels NF-L being responsible for initiating filament formation, NF-M being responsible for filament elongation, and NF-H being responsible for interactions with other filaments through cross-bridges.

The role of neurofilaments in axonal regeneration is highly relevant to several motor neuron diseases in which nerve regeneration fails and degeneration prevails. Neurofilament protein accumulations have been detected in neurons or in cerebrospinal fluid of persons affected by amyotrophic lateral sclerosis (ALS; also known as Lou Gehrig's disease), spinal muscular atrophy, multiple sclerosis, and Charcot-Marie-Tooth disease (also known as hereditary motor and sensory neuropathy) [87]-[90]. The possibility that neurofilament protein

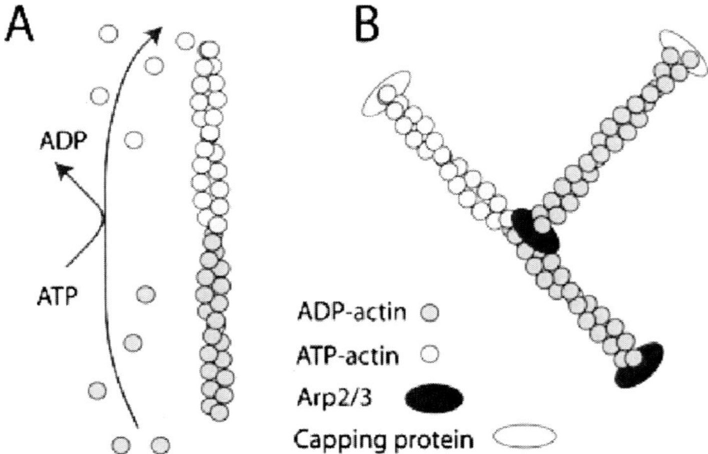

Fig. 2.8. Actin formation is ATP-dependent. ATP-actin assembles into filaments, whereas ADP-actin depolymerizes. Arp2/3 serves as a nucleating protein as well as a branching protein, in each case promoting assembly of actin. Capping proteins, on the other hand, halt actin filament assembly.

expression abnormalities may be a factor in the etiology of at least some of these cases has led researchers to employ genetic mutations that increase or decrease various neurofilament proteins in an attempt to recapitulate symptoms of specific motor neuron diseases, as well as those of neurodegenerative diseases such as frontotemporal dementia, Alzheimer's disease, and Parkinson's disease [91]-[94]. It remains to be determined to what extent normal neurofilament function can be reinforced or replaced by nanomaterials or nanodevices. Nanostructured scaffolds and nanomaterials are currently being studied in the context of potential biomedical uses such as in facilitating neural regeneration and combating degenerative disease [95].

Microfilaments

Microfilaments are single-stranded filaments, each having diameters of approximately 5 nm and variable lengths [96]. Individual microfilament strands consists of actin monomers bound into a left-handed helix, with a plus end (barbed) and a minus end (pointed) (see Figure 2.8). In the cell, soluble actin protein called globular actin (G-actin) is free to bind and form filamentous actin (F-actin). Assembly of G-actin into F-actin filaments is an ATP-dependent process [97]. Net assembly of actin filaments occurs at its plus ends, where ATP-actin is added.

There are over 100 different actin-binding proteins responsible for actin associating with the membrane, with membrane-bound receptors, and with ion channels, as well as for promoting assembly causing the depolymerization of those filaments (see Table 2.5). Actin-binding proteins, such as Arp2/3 and

Table 2.5. Actin-binding proteins and their functions.

Binding protein	Functions	References
α-Actinin	Cross-linking protein (links to NMDA receptor)	[98, 99]
ADF/cofilin	Stimulates actin filament treadmilling; depolymerization; and removal of ADP-actin from pointed end	[97, 99, 100]
Ankyrin	Anchors actin to membrane and to membrane-bound receptors and channels	[101]
Arp2/3	Facilitates nucleation and assembly	[98, 100, 102]
CapZ	Capping protein	[100]
Cordon bleu	Nucleation factor	[103]
Cortactin	Enables actin to modulate ion channel activity	[104]
Drebrin	Morphogenesis and maintenance of dendritic spines	[98]
Formin	nucleation factor	[103]
Gelsolin	Severing protein; capping protein; nucleation protein	[98, 100]
Integrin	Cell adhesion	[101]
Myosin	ATP-driven motor; transport along microfilaments	[98]
Profilin	Catalyzes exchange of ADP for ATP; adds ATP-actin to barbed end	[98, 100]
Spectrin	Cross-linking protein	[101]

profilin, regulate microfilament assembly and determine how much ATP-actin is added to the barbed end of the filament [98]. Actin filaments disassemble by losing ADP-actin from the pointed ends of the filaments. Actin depolymerizing factor (ADF)/cofilin stimulates disassembly [97, 98]. Treadmilling of actin filaments is said to occur when the rate of polymerization at barbed ends is roughly equal to the rate of depolymerization at pointed ends, resulting in a steady-state flux of subunits that preserves a constant filament length.

In neurons, the functions of actin filaments are specialized to match basic neuronal functions. Working in conjunction with the various binding proteins and other cytoskeletal proteins, actin filaments play prevalent roles in dendrite spine formation, initiation and elongation of the axonal growth cone, anchoring and controlling ion channels, and axonal transport [98]-[105].

Dendritic spines are highly specialized neuronal compartments where shape determines functional status. Spine shapes range from that of thin extensions, to stubby protrusions, to mushroom heads arising from a slender stalk [98]. Such spine shapes typically develop from very thin filopodia, which initially contain little else besides actin. Once a less developed protrusion makes contact with a post-synaptic membrane, the dendritic spine matures.

Mature spines contain high concentrations of actin filaments, actin-binding proteins, scaffolding proteins, and receptors (see Figure 2.9). Researchers have attempted to determine which proteins are key to functional spine maturation. Actin protein and drebrin (in particular drebrin A, which is a neuron-specific actin-binding protein) appear to be largely responsible for spine initiation [98].

Spines typically receive strong glutamatergic inputs and for this reason exhibit significant concentrations of NMDA and AMPA receptors. α-Actinin is a cross-linker protein responsible for binding actin to NMDA receptors and clustering these receptors to a particular region of the membrane [106, 107]. Although F-actin interacts with both NMDA and AMPA receptors, α-actinin preferentially interacts with NMDA receptors and also with spectrin. Actin provides a matrix in which NMDA receptors couple to intracellular signal transduction cascades. NMDA receptors are Ca^{2+} channels, and actin and its associated proteins tether signal transduction molecules activated by Ca^{2+} to the vicinity of the NMDA receptor. When NMDA channels open to allow Ca^{2+} entry, this triggers the calcium-activated protease, calpain, to cleave actin-associated spectrin, and spectrin in turn further modulates NMDA receptors [108].

Modulation of spine morphology by actin and its associated proteins illustrates how a collective of molecules or "molecular factory" can regulate synaptic plasticity and possibly account for the learning-related phenomenon, long-term potentiation (LTP), believed to underlie certain cognitive functions [109]. That actin filaments participating in spine plasticity are pivotal to higher cognition is consistent with the finding that many forms of mental retardation and cognitive impairment are associated with spine abnormalities and deformations. Without actin filaments, there are no spines.

In addition to anchoring receptors, such as the NMDA and AMPA receptors, actin plays a pivotal role in anchoring and clustering ion channels. Not only does the actin cytoskeleton anchor ion channels to scaffolding and signal transduction molecules in the vicinity of the postsynaptic site, actin filaments also appear to regulate the excitability of certain ion channels. Actin has been shown to bind to the membrane, via spectrin and ankyrin, and to various ion exchangers, such as the voltage-dependent Na^+ channel [101], the L-type voltage-dependent Ca^{2+} channel [110], and the voltage-dependent K^+ channel (Kv1.2) [104]. Moreover, via an association with cortactin, actin appears necessary for maintaining the excitability of voltage-dependent K^+ channels (Kv1.2). Actin filaments in the trigger zone or initial segment of the axon are responsible for concentrating voltage-dependent Na^+ channels in a manner that is essential for initiating the action potential [112]. Without actin filaments, initiation of neural firing is markedly impaired.

Actin filaments perform critical functions during neural development. Axonal growth is particularly responsible for wiring the brain and it is the final pattern of connections that at least partly underlies higher cognitive functions. After nerve cells divide, migrate, and differentiate into their specific types, they grow neurites that later become either dendrites or axons. In

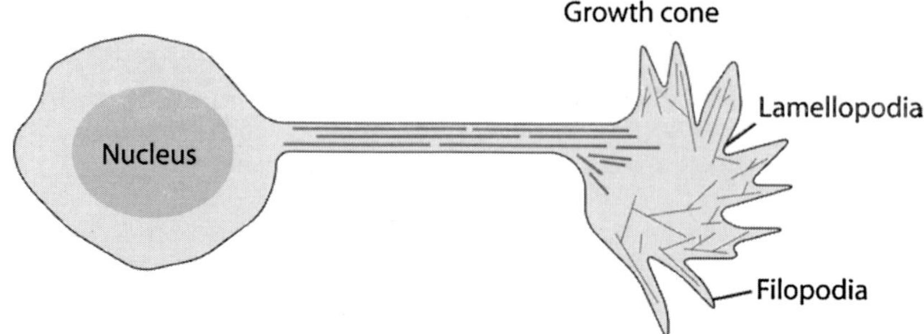

Fig. 2.9. The growth cone of the developing neuron. Actin filaments fill the growth cone lamellopodia and provide the structural basis of the extending filopodia. This contrasts with the main axon shaft, which is filled with microtubules.

many cases, the growth of an axon involves traveling a great distance from the cell body, often taking a tortuous course. In order for an axon to reach its proper target, the axonal growth cone exhibits highly motile, advancing, retracting, turning, and branching behavior – all enabled by actin [113]. As shown in Figure 2.9, the axonal growth cone has two main regions: the central region and the peripheral region out of which actin-rich filopodia extend. Many of the actin-binding proteins mediate various behaviors of actin in the growth cone. For example, Arp2/3 is responsible for the nucleation and subsequent branching of actin filaments in the filopodia into different directions, whereas ADF/cofilin proteins are responsible for the retraction of actin-rich filopodia [98, 114].

Nanotechnology has revealed more about the properties of actin and how the local environment affects the arrangements of actin filaments. Surfaces coated with nanoparticles were shown to determine the organization of actin filaments [115]. Researchers were also able to activate mechanoreceptors by stimulating actin stress fibers with optical nanotweezers, showing that actin filaments are capable of transmitting mechanical force to receptors [111]. These and other kinds of nanotechnological studies should provide a much better understanding of roles played by actin filaments in establishing brain connectivity.

Microtubules and Microtubule-Associated Proteins

Microtubules are long hollow cylindrical filaments made from heterodimers consisting of bound α-tubulin and β-tubulin monomers (see Figure 2.10). Each of these 55 kD tubulin subunits is highly conserved in eukaryotes, and not present in prokaryotes. Based on 3.5 - 3.7 $\mathring{A}$ resolution X-ray crystallographic images, the $\alpha\beta$-tubulin dimer has been computationally reconstructed as having a β-sheet core surrounded by α-helices [116]. Assembled microtubules have outer diameters of 25 - 26 nm and inner diameters of 15 nm and typically contain 13 protofilaments when assembled in living cells.

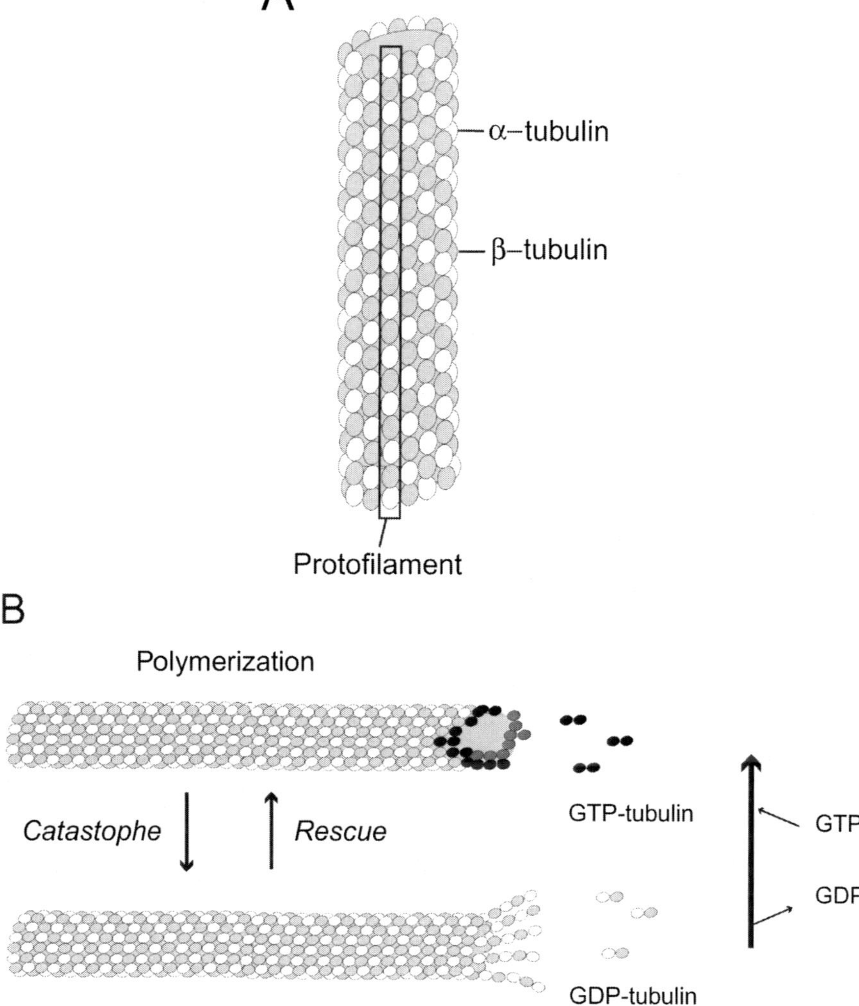

Fig. 2.10. Microtubules. Each longitudinal protofilament of a microtubule is made up of α and β tubulin subunit heterodimers (A). Assembly of microtubules is a GTP-dependent process (B). GTP-tubulin dimers polymerize to lengthen (or rescue) microtubules. Microtubules depolymerize by losing GDP-tubulin in a process called catastrophe.

Microtubule dimers assemble in a GTP-dependent process, by GTP-tubulin dimers adding to the growing ends [117]. Under certain experimental conditions individual protofilaments can form first and then co-assemble into mature microtubules. Microtubules depolymerize by losing GDP-tubulin dimers after protofilaments separate and curve away from one another forming structures resembling "ram's horns". Like actin filaments, microtubules have plus and minus ends. Polymerization or elongation typically occurs at plus ends and depolymerization or shortening occurs at minus ends. Under very specific conditions of tubulin concentration, pH and temperature, the net result may be no appreciable change in microtubule length – a process called treadmilling, which as described earlier also occurs in actin filaments.

The distribution of microtubules within neurons differs from that of either neurofilaments or actin filaments. Microtubules fill the interiors of both dendrites and axons roughly equally, whereas they are absent or sparse in dendritic spines [118]. The spacing of microtubules varies depending on their location in the neuron. Typically, microtubules in axons are spaced much closer together than those in dendrites, largely due to microtubule-associated proteins [119] that interconnect neighboring microtubules.

As listed in Table 2.6, the major microtubule-associated proteins found in brain include various isoforms of the stabilizing proteins MAP1, MAP2, and tau, proteins identified on the basis of their involvement with various disorders, and the motor proteins [120]-[130]. MAPs are developmentally regulated and compartmentalized in accordance with their specialized roles. These binding proteins determine, to a large degree, the arrangement of microtubules in a given part of the neuron and the function played by the microtubules in that neuronal compartment.

MAP1B is expressed early on during embryological development (having been detected in human and rodent embryological brain tissue), consistent with its role in axogenesis [122]. At birth, MAP2A is virtually undetectable in rodent brain, whereas, MAP2B and MAP2C are present [131]. MAP2A is readily detectable by postnatal day 15 in rodent brain, suggesting that MAP2 plays a more prominent role once neurites have been established. An exception to this may be the low molecular weight MAP2C, which is present at birth and declines to negligible levels in adult brain.

Microtubule binding proteins determine the architecture of microtubules and are also dysfunctional in many neurodegenerative and neurodevelopmental disorders. Microtubules in adult dendrites bind MAP2 preferentially, whereas microtubules in axons prefer tau [124, 132]. The process by which this adult pattern is achieved is gradual and follows a distal-to-proximal gradient. Growing axons contain both MAP2 and tau, but the amount of MAP2 begins to recede from the distal tip of the axon to the proximal part, finally disappearing altogether. The exclusive compartmentalization of MAP2 to the somatodendritic part of the neuron and tau to the axon raises interesting issues relevant to Alzheimer's disease-related brain pathology. Hyperphosphorylated tau, which forms insoluble paired-helical filament and finally neurofibrillary

Table 2.6. Microtubule-binding proteins and their functions.

Binding protein	Functions	References
Stabilizing MAPs		
MAP1A	Neural development; stabilizing microtubules in axons and dendrites	[120, 121]
MAP1B	Neural development; stabilizing microtubules in axons and dendrites	[120, 121, 122, 123]
MAP2A	Neural development; stabilizing microtubules in dendrites; signal transduction	[120, 124, 125]
MAP2B	Neural development; stabilizing microtubules in dendrites, signal transduction	[120, 124, 125]
MAP2C	Early neural development	[120, 124, 125, 126]
Tau	Neural development; stabilizing microtubules in axons; axonal transport	[120, 124, 126]
Proteins related to specific neurodevelopmental disorders		
ASPM	Affects cell division; responsible for brain enlargement and possibly the explosion in human intelligence	[127]
DCX	Neural development; cortical neuron migration	[127]
LIS1	Neural development; cortical neuron migration	[127]
CLIP-115	Regulates microtubule dynamics by binding to tips of growing microtubules	[128]
Motor proteins		
Dynein	Retrograde transport in the axon; transport to minus ends of microtubules in dendrites	[129, 130]
Kinesin	Anterograde transport in the axon; transport to plus ends of microtubules in dendrites	[129, 130]

Abbreviations: ASPM: abnormal spindle-like protein, microencephaly-associated ; DCX: doublecortin; LIS1: lissencephaly-1.

tangle material, is highly concentrated to the somatodendritic part of the neuron instead of to axons. Thus, tangles, which are mostly found in the soma and dendrites, do not concentrate where tau normally accumulates. There is evidence that tangles may be most debilitating to neuronal function because they bind healthy MAP2 and tau, and thereby impair normal microtubule function [133]. Thus, it would appear that microtubule function in the soma and dendrites of neurons is most compromised in Alzheimer's disease. This will be further discussed in Chapter 5.

Deficits in other microtubule-binding proteins have been linked to a number of neurodevelopmental disorders in which the cerebral cortex fails to develop properly. The majority of cases of lissencephaly (i.e., smooth brain) result from deletions in the microtubule-binding proteins, doublecortin (DCX) and lissencephaly-1 (LIS1), whereas microcephaly has been connected to genetic mutations in abnormal spindle-like protein, microencephaly-associated (ASPM) [127]. Williams syndrome is associated with genetic deletions of the 115kD cytoplasmic linker protein (CLIP-115), which binds microtubules [128], and Rett syndrome may be associated with decreased levels of MAP2 [134]. It is unclear what these associations suggest. Are microtubules and their interactions with linker proteins critical for cognitive function per se, or is it merely the case that deletions disrupting the cytoskeleton impair proper neuron maturation in a general way. Much depends on the proven functions of microtubules, all of which are not yet known.

The functions of microtubules in neurons differ markedly from those in other cells in the organism. Key to their function is how microtubules link many neuronal compartments together. Only microtubules, for example, carry synaptic vesicles manufactured in the Golgi apparatus of the cell body, all the way down the axon, to the actin filaments in the axon terminal or developing growth cone [135, 136]. Similarly, only microtubules provide a direct conduit from the cell body carrying protein and mRNA granules to dendritic spines and other non-spinous postsynaptic sites along the dendritic membrane [137]. It is conceivable that microtubule transport lies at the fundamental core of higher cognitive function. Transport is one of the main functions of microtubule tracks, and given its centrality to the overall function of the neuron, this nanoscale mechanism and the nanotechnological approaches available to study it will be discussed separately in Chapter 4.

Microtubules also serve as excellent models for inspiring new nanomaterials, with their unique biomechanical properties – such as high resilience and stiffness – making them of great interest to nanoengineers. Microtubules may be nature's closest version of carbon nanotubes, possessing architectures that enable biomimetic designs of carbon nanotube-based structures with similarly useful properties, including the capacity for self-assembly [138]. Nanotechnological tools like nanotweezers have also been used to study the mechanical properties of microtubules and the underlying principles of dynamic self-assembly [139]. Microtubules can also be targeted by nanoscale drug-delivery

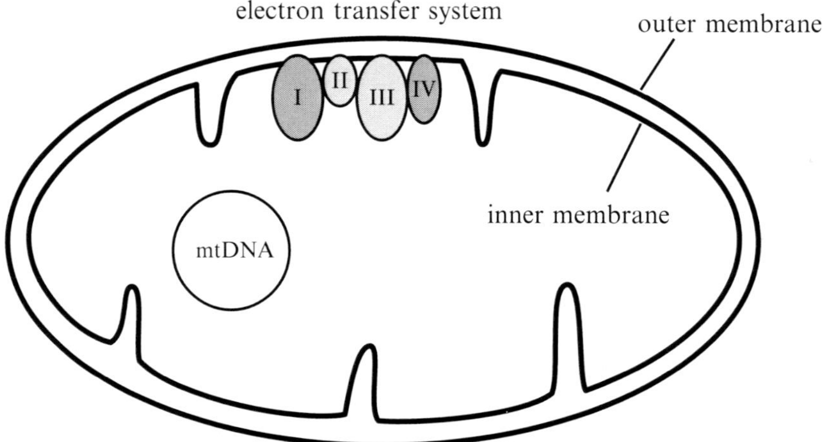

Fig. 2.11. Mitochondria in neurons operate the same as those in other cells.

methods. Nanoparticles that increase cellular uptake have been used, for example, to deliver the microtubule-stabilizing drug paclitaxel [140].

2.1.6 Mitochondria in Neurons

Due to the high energy requirements of neurons, mitochondria are found in abundant supply; this is especially the case in the initial part of the axon, nodes of Ranvier, and in the axon terminals [1]. Small mitochondria have diameters measuring 200 - 500 nm placing them at the large end of nanostructures; however, their essential function in energy metabolism establishes them as playing pivotal roles in the operations of neural nanostructures. Individual mitochondria in neurons are shaped much the same as those in other cells. As shown in Figure 2.11, each mitochondrion has an outer membrane, separating it from the cytosol of the neuron, and an inner membrane containing the essential electron transport system. Mitochondria also have their own DNA (mtDNA) responsible for encoding the proteins of the electron transport system.

The brain preferentially utilizes glucose, as opposed to fatty acids or ketones, consuming approximately 20% of the body's supply of glucose and oxygen. Much, but not all the glucose taken into neurons ends up as energy in the form of ATP; some glucose is used to manufacture the neurotransmitters glutamate and GABA [141]. In the neuron, glucose is broken down to form pyruvate, among other products. It is pyruvate that enters the mitochondria, and following oxidative phosphorylation via the tricarboxylic acid cycle, reduces nicotinamide adenine dinucleotide (NAD+) to NADH and flavin adenine dinucleotide (FAD+) to FADH2. NADH and FADH2 serve as proton donors to the electron transport system consisting of complexes I - IV, coenzyme Q, and cytochrome C. The proton gradient drives ATP synthesis in complex IV.

Mitochondria in neurons undergo cycles of fusion and fission [142]. Fusion allows mitochondria to share lipids and proteins, and fission allows mitochondria to multiple and increase the provision of energy to different parts of the neuron, Mitochondrial fission can also signal that a neuron is undergoing degeneration. Mitochondrial fission or fragmentation frequently precedes apoptosis (i.e., programmed cell death).

Transport of mitochondria in neurons occurs along microtubule tracks [143]. Although it appears that mitochondria are transported bi-directionally, transport away from the cell body is the typical direction for newly produced mitochondria, whereas transport towards the cell body is usually reserved for damaged mitochondria. Kinesins (Kif1B and Kif5B) are the prominent motor proteins responsible for transporting mitochondria away from the cell body, whereas dynein is the predominant motor protein transporting toward the cell body [144]. There are also adaptor proteins (e.g., Milton and syntubulin) that bind mitochondria to the motor protein, which will be discussed further in Chapter 4. Although microtubules are critical for transporting mitochondria in neurons, once mitochondria reach the region of the axon, dendrite, or soma (where they were recruited because of increased energy demands), they remain stationary [145].

Dysfunctional mitochondria have been noted in a number of neurodegenerative conditions including Alzheimer's, Parkinson's, and Huntington's diseases, as well as following ischemic insult and trauma [146]-[153]. The nature of this dysfunction typically involves oxidative damage due to free radicals in affected mitochondria. This kind of damage increases with aging and is therefore prominent in age-related disorders. The electron transport system in mitochondria ordinarily produces highly reactive free radical that can do damage to cell lipids, carbohydrates, nucleic acids, and proteins. A healthy young cell has sufficient antioxidants to counteract any ill effect; however, cellular changes due to aging and disease render mitochondria less well protected. In Alzheimer's disease, for example, mitochondria become increasingly vulnerable to toxic effects of amyloid-β peptide [146]. One proposal is that amyloid-β peptide interferes with fission/fusion cycles and transport of mitochondria thereby interfering with multiple neuronal functions [150].

Nanotechnology has been applied to the study of mitochondria in an effort to reveal the underlying causes of mitochondrial dysfunction and cell death. Gold nanoparticles measuring 3 nm in diameter (but not those measuring 6 nm) were able to penetrate pores in the outer mitochondrial membrane of isolated mitochondria from cardiac tissue [147]. Based on measurable permeability transitions, researchers concluded that breach of the outer membrane of mitochondria is a component of apoptotic cell death. Nanodevices are also being constructed that contain mitochondria-based biomolecular power supplies. In one example, isolated mitochondria attached to microfluidic devices supply ATP needed to drive molecular motors [148]. Many complex nanodevices of the future will need their own source of biomolecular power.

2.2 Nanoengineering and Neurons

Biomimetics, biomimcry, and bionics are all terms that refer to the efforts of bioengineers to copy living organs, cells, or cell compartments for a variety of purposes. In many cases that purpose is to seek a novel prosthetic device or treatment strategy for a biomedical disorder. Nowhere in the biomedical field is such an approach more challenging than in the neurosciences, due in part to the highly integrative function of neurons and complexities of neural networks. Despite these hurdles, nanotechnology is currently advancing diagnostic techniques and treatments for numerous nervous system disorders, ranging from enhanced precision neurosurgery for neurodegenerative disorders to modulation of psychiatric disorders [149].

As mentioned earlier in this and the previous chapter, neurons in the cerebral cortex, especially the large pyramidal neurons of the neocortex and hippocampus, are pivotally involved in higher cognitive functions, including but not limited to remembering, perceiving, problem solving, and voluntary movements of the body. Neurodevelopmental and neurodegenerative disorders have the potential to strip the afflicted individual of one or more of these most essential capabilities resulting in mental retardation, dementia, or movement disorder. Unfortunately, for many of these disorders there is no viable treatment, let alone cure. This poor prognosis extends to cases of spinal cord damage and neuromuscular diseases for which there is often little if any recovery. Novel approaches involving nanotechnology are clearly needed to restore function to the damaged nervous system.

Nanoscale or microscale biomimetics can attempt to restore function following neural dysfunction due to injury or disease at one of three levels.

- Mimic the function of an entire brain region or assembly of neurons in a part of the nervous system. Examples include microelectromechanical systems (MEMS) mimicking the functions of the sensorimotor cortex, hippocampus, retina, cochlea, or olfactory system.
- Mimic the function of individual neurons. Examples include robotic neurons and combinations of stem cell technology with nanotechnology.
- Mimic the function of individual subcellular organelles. Examples include introducing bionic microtubules into neurons in an attempt to recover lost function due to diseases associated with microtubule dysfunction.

Experiments done at Duke University and the Japan Science and Technology Agency on a monkey named Belle provide a good example of the first of these three approaches; these experiments showed that it is possible to mimic the function of an entire brain region, in this case that of the sensorimotor cortex in moving robotic limbs [154]. As Belle walked on a treadmill, researchers decoded her brain activity patterns from microelectrodes hooked up to her sensorimotor cortex and then sent these decoded messages to a robot in Japan. Watching the robot on a computer terminal, Belle was able to "think" the appropriate instructions to keep the robot in action. The long-term expectation is that this kind of technology will enable paralyzed patients to generate brain activity linked to (possibly nano-coated) microelectrodes that in turn stimulate their own muscles, a prosthetic device, or a computer cursor.

Another experimental model having the goal of replacing an entire brain region, or at least a significant portion of it, is the prosthetic hippocampal circuit being developed by Theodore Berger and colleagues at the University of Southern California [155]. These fabricated hippocampal cell circuits are modeled as realistically as possible, and the hope is they will be able to restore lost memory function in patients suffering from impairments due to disease or injury.

Yet other examples of bionic cell assemblies include biosensor devices that help those suffering from inherited and acquired visual impairments, such as retinitis pigmentosa and macular degeneration, to partially regain vision. One such device consists of a miniaturized camera, a MEMS device that receives the images and converts them into an electrical output, and an ultra thin wire that transmits that output to the retina [156]. Since their initial conception, significant advances have been made in refining the design of such devices, such as enhancing the neural interface with the retinal prosthesis [157]. Similar bionic products are in various stages of development or refinement for artificially transducing auditory or olfactory stimuli [158, 159, 160].

These models, experiments, and prototypes provide proof of principle that prosthetic devices presently offer (or will offer in the not-too-distant future) dramatic improvements for those suffering from certain types of blindness, deafness, spinal cord injury, neuromuscular disease, neurodegenerative movement disorders, and neurodegenerative memory impairment. Futuristic prosthetic devices should be able to send and receive information through wires or through wireless routes of communication. Mimetic biosensor devices will sense external stimuli, even in the absence of a fully functional sensory system, and the willful actions of the brain or what might be called "mind" will be able to control the actions of the human body or make direct interfaces with machines.

Nanotechnology is likely to make significant contributions when it comes to perfecting the neural-computer interface. At present, there is significant damage to the brain when introducing permanently implanted electrodes. Nanoparticle coatings of electrodes circumvent some of these problems by reducing tissue damage [161]. Carbon nanotubes also have been used to coat electrodes, resulting in highly biocompatible sensitive probes capable of recording electrophysiological and neurochemical responses of neurons [162]-[164]. Moreover, because of their nanoscale properties, certain materials are particularly suitable for wireless communication with prosthetic devices

[165, 166]. Last but not least, nanoneuroscience will undoubtedly make major advances towards understanding biomolecular computations, thereby making transmission of information between nervous system and prosthetic devices more seamless.

The second approach mentioned above, assembly and delivery of bionic neurons, has its distinct advantages and potential pitfalls. Bionic cells are not new; it was during the 1950's that artificial red blood cells were first made [167]. Assembling artificial or bionic neurons has proved a more difficult task, and as a further confound, there are a large number of different kinds of neurons. The diversity of neuronal types may prove to be an advantage rather an obstacle, however, only under some specific circumstances. With certain types of neurological disorders, Parkinson's and Alzheimer's disease, for example, select populations of neurons are destroyed. This raises the possibility of exclusively replacing (or aiding in the survival of) those neuronal populations (i.e., the dopamine neurons of the substantia nigra for Parkinson's disease and the cholinergic neurons of the basal forebrain for Alzheimer's disease). Such approaches have been undertaken using stem cell strategies [168, 169].

Nanotechnology has also led to advances in stem cell transplantation; examples include using nanoparticles and nanomaterials for a variety of purposes and implementing nanodelivery devices. Nanoparticles have been used to facilitate delivery of genes to stem cells [170]. Superparamagnetic nanoparticles, in particular, have been used to track the migration of stem cells implanted into the nervous system [171]. Stem cells have also been coaxed into growing on templates of nanomaterials, such as carbon nanotubes, which provide an external control of axonal morphology [172]. A wide range of additional experiments further demonstrate that nanoparticles can be incorporated into stem cells in order to image those cells, monitor their trafficking, and control their migration [173]-[175].

Nanotechnological techniques might prove useful for targeting specific neuronal populations and restoring greater functionality. A current limitation placed on transplant strategies for Parkinson's disease is that dopamine-secreting cells transplanted into the substantia nigra would have an extremely difficult time growing axons all the way to their targets in the striatum, so typically the cells are placed directly in the striatum [176]. Nanotechnological approaches, such as superparamagnetic nanoparticles guided by externally applied magnetic fields, could result in coaxing these growing axons to grow over long distances, enabling dopamine cells to be transplanted into the substantia nigra, where these neurons receive their inputs. In this suggested scenario, nanotechnology would provide techniques for proper reconstruction of neural circuits to extents not presently possible.

The third approach mentioned above – introducing bionic neuronal components into neurons – falls squarely in the domain of nanoscience. Replacing specific neuronal compartments may prove more advantageous than replacing entire neurons since connections with other neurons would be maintained. Using nanoscale materials to replace or reinforce individual proteins, such as receptors, ion channels, signal transduction molecules, and cytoskeletal proteins, appears achievable in the relatively near future. Bionic microtubules may be particularly useful in treating nervous system disorders [177].

2.2.1 Nanoparticles and Their Interactions with Receptors and Signal Transduction Molecules

The fundamental principles of neurotransmission are based on neurotransmitter molecules binding to receptor proteins embedded in the neuronal membrane, which in turn activate ion channels or signal transduction cascades. To what extent these principles can be amplified, dampened, or otherwise modified by nanoparticles or other nanoscale technologies remains to be fully elucidated, as is likely to occur over the next few decades. Nanotechnology also offers the opportunity to study ion channels and receptor actions from different perspectives to those typically taken by neurochemists and neuropharmacologists.

A number of studies have already used nanotechnology to decipher ion channel and receptor actions with great sensitivity. Individual K^+ channels embedded in a plasma membrane have been detected by using quantum dot technology [178]. Nanoparticles can also be used to regulate ion channels and receptors. In one such study, single-walled carbon nanotubes were used to block K^+ channels in a dose-dependent manner by directly occluding the pore [179]. In another study, nano-magnetic particles coated with monovalent ligands were bound to receptors located on mast cells [180]. Once in place these nanomagnetic particles were manipulated by externally applied electromagnetic fields, producing a rapid and robust clustering of ion channels, which in turn led to a rapid influx of Ca^{2+}. These types of experiments show that the nanoscale physical environment, which plays a critical role in receptor response, can be controlled by nanoparticles.

Nanotechnological tools like AFM also help researchers peer down to the activity of single receptors. In one study, AMPA receptors were shown to aggregate in stiff nanodomains of the membrane, demonstrating that a high elasticity modulus is essential to their placement and function [181]. The membrane stiffness was reversed by NMDA receptor activation, which simultaneously resulted in a disappearance of a significant portion of AMPA receptors from the membrane. This study illustrates that the nanoscale properties of the environment surrounding receptor populations are critical to regulating receptor-mediated actions in neurons.

Novel ways to stimulate receptors in the nervous system have resulted from advances in nanotechnology-based drug delivery systems [182]-[184]. Nanotubes, nanoparticles, nanocapsules, and dendrimers can be devised with special surface properties enabling them to pass the blood-brain barrier and then the neuronal membrane. There has also been significant progress in microfluidic lab-on-a-chip technologies, with applications to nervous system disorders that include guiding axonal growth with precision release of neurotrophic factors or drugs [185]. Since nanotechnological tools operate at the small scale, assays can approach the level of individual proteins or molecules. Potassium channels, for example, have been reconstituted with proteoliposome fused with lipid bilayers giving rise to efficient drug screening capable of measuring effects at the level of single channels [186].

2.2.2 DNA Nanodevices

Single or double strands of DNA are suitable biological materials for making biosensors, biomolecular actuators, and information processing devices [187]. One particularly useful nanodevice thus far constructed called the "DNA-tweezers" uses DNA hybridization as a nanomechanical switch [188, 189]. This device is able to grab onto small molecules and physically rotate, stretch, and pull them for further examination or manipulation. It is also possible to incorporate instructions into a DNA gene for DNA-tweezers to follow; hybridization between the strands will transmit the instructions [190]. Electromagnetic fields can even be used to manipulate DNA-based nanodevices. One research group has demonstrated that electromagnetic fields can be used to dehybridize DNA molecules attached to gold nanocrystals [191].

Nanoparticles and nanocapsules are presently being considered as alternatives to viral vector carriers for gene therapy, and it is believed that these approaches may solve some long-standing associated problems, such as immunological response and mutagenicity [192]-[194]. DNA-nanoparticles have proved safe when NASAlly administering genes to correct for cystic fibrosis [195]. Neurons may in some ways prove particularly responsive to gene therapies, especially those delivered via nanocarriers. Genes introduced into a living cell by a viral vector survive only as long as that living cell survives, and cellular turnover is rapid in most tissue types. Neurons on the other hand, especially the large pyramidal cells of the cerebral cortex and hippocampus, survive seemingly indefinite periods – perhaps entire lifetimes. Moreover, neurons have more compartments (i.e., dendrites and axons), and nanotechnology offers ways to introduce DNA into cells and then move those DNA strands to particular locations in the neuron. Microtubules are responsible for transporting DNA in cells, as discussed next.

2.2.3 Microtubule and Actin Filament Interactions with Nanomaterials

A number of laboratories have demonstrated that it is possible to load DNA cargo onto microtubules that in turn glide along kinesin grids [196]-[198]. In one experiment, one end of the DNA strand was attached to the microtubule using a biotin sandwich technique, and the other end of the DNA strand was thiolated and then attached to a gold nanoparticle adhered to a surface substrate [198]. The ATP-powered movement of the microtubule over a kinesin template resulted in the stretching out of the folded DNA, indicating that a microtubule-based type of nanomachinery could in principle be used to build DNA networks for use in nanoelectronics. There is also direct evidence that microtubules serve as natural gene nanocarriers in living cells [199]. As such, plain or modified microtubules may be useful as carriers of gene vectors introduced into neurons. The advantages are the inherent ability of microtubules to travel along the entire lengths of axons and dendrites, and the option to externally control their movement.

There is now solid experimental evidence that both electric and megnetic fields can control microtubule movement. Electric fields are especially effective in controlling the movement at the leading tip of the microtubule [200, 201]. Magnetic nanoparticles can be used to manipulate ion channel activity and subsequently alter cytoskeletal protein dynamics [202, 203]. Biophysical properties of microtubules contribute to their movements and interactions with nanoparticles. Microtubule mobility is independent of length and is anisotropic [204]. Movement in the axial direction predominates. Microtubules functionalized by superparamagnetic nanoparticles such as cobalt ferrite align in a direction opposite to the magnetic field lines and change orientation as the direction of the magnetic field is altered [205, 206].

Other possible uses of both microtubules and actin filaments include using them as biosensors and in nanofluidic lab-on-a-chip devices that allow the measurement of the chemical composition of single cellular compartments [207]-[209]. As mentioned above, a distinct advantage of using microtubules as nanomaterials is that these biological structures self-assemble in the presence of ATP [210].

2.3 Future Directions of Nanodevice-Cell Hybrid Designs

The challenges of building even simple nanodevices are significant but not insurmountable. As nanodevices become increasingly complex, more components of nanodevices will successfully interface with living cells. Research done thus far supports the conclusion that bioartificial organs will be able to rely on nanotechnological materials to accommodate multiple functions, such as sensing, feedback, control, and drug delivery [211].

When it comes to applications to nervous system disorders of higher cognition, nanodevices of the future are likely to concentrate on what it is that makes a neuron "intelligent". One of the themes of this book is elucidating novel ways in which biomolecules in neurons compute and transmit information – in concert with electrophysiological impulses being propagated along the membranes of dendrites, the soma, and axons. Biomolecular computations are likely to occur in a number of neuronal proteins, including microtubules and actin filaments. These filamentous structures that lie within our neurons are the most probable candidates for that kind of cellular sentience – that which decides what to do based on previous experience or history. Microtubules are strong and resilient biomaterials that share many mechanical characteristics with carbon nanotubes [138]. In the next chapter, the unique physical properties of microtubules and actin filaments that endow them with the ability to receive, propagate, and amplify signals that originate at the membrane are discussed in greater detail.

References

1. Siegel, G.J., Albers R.W., Brady, S.T. & Price, D.L. (2006) Basic Neurochemistry: Molecular, Cellular, and Medical Aspects, 7th Ed. Academic Press, New York.
2. Yehuda S, Rabinovitz S, Mostofsky DI. Neurobiol Aging. 2005 Dec;26 Suppl 1:98-102.
3. Kidd PM. Altern Med Rev. 2007 Sep;12(3):207-27.
4. Boldog T, Li M, Hazelbauer GL. Methods Enzymol. 2007;423:317-35.
5. Hille, B. Ion Channels of Excitable Membranes, 3rd Ed. 2001, Sinauer Associates, Sunderland, MA, USA
6. Zheng W, Spencer RH, Kiss L. High throughput assay technologies for ion channel drug discovery. Assay Drug Dev Technol. 2004 Oct;2(5):543-52.
7. Debbage P, Jaschke W. Molecular imaging with nanoparticles: giant roles for dwarf actors. Histochem Cell Biol. 2008 Sep 30.
8. Koch, C. Biophysics of Computation: Information Processing in Single Neurons. Oxford University Press: New York, New York, 1999.
9. Larkum ME, Nevian T. Synaptic clustering by dendritic signalling mechanisms. Curr Opin Neurobiol. 2008 Sep 16.
10. Goldman, D. E., Potential, impedance, and rectification in membranes J. Gen. Physiol. 27, (1943) 37-60
11. Hodgkin, A. L. and Huxley, A. F., The quantitative description of membrane current and its application to conductance and excitation in nerve. J. Physiol. (London) 117, (1952) 500-544.
12. Lehmann-Horn F, Jurkat-Rott K. Nanotechnology for neuronal ion channels. J Neurol Neurosurg Psychiatry. 2003 Nov;74(11):1466-75.
13. Siwy ZS, Powell MR, Petrov A, Kalman E, Trautmann C, Eisenberg RS. Calcium-induced voltage gating in single conical nanopores. Nano Lett. 2006 Aug;6(8):1729-34.
14. Valiyaveetil FI, Leonetti M, Muir TW, Mackinnon R. Science. 2006, 314:1004-7.
15. Jiang Y, Lee A, Chen J, Ruta V, Cadene M, et al. 2003. X-ray structure of a voltage-dependent K+ channel. Nature 423:33-41
16. Elinder F, Nilsson J, Arhem P. On the opening of voltage-gated ion channels. Physiol Behav. 2007 Sep 10;92(1-2):1-7.
17. Tombola F, Pathak MM, Isacoff EY. How does voltage open an ion channel? Annu Rev Cell Dev Biol. 2006;22:23-52.

18. Kariev AM, Znamenskiy VS, Green ME.Quantum mechanical calculations of charge effects on gating the KcsA channel. Biochim Biophys Acta. 2007 May;1768(5):1218-29.
19. Unwin N. Refined structure of the of the nicotinic acetylcholine receptor at 4A resolution. Journal of Molecular Biology 346:967-89
20. Law RJ, Henchman RH, McCammon JA. A gating mechanism proposed from a simulation of a human alpha7 nicotinic acetylcholine receptor. Proc Natl Acad Sci U S A. 2005 May 10;102(19):6813-8.
21. Grutter T, de Carvalho LP, Dufresne V, Taly A, Edelstein SJ, Changeux JP. Molecular tuning of fast gating in pentameric ligand-gated ion channels. Proc Natl Acad Sci U S A. 2005 Dec 13;102(50):18207-12.
22. Xiu X, Hanek AP, Wang J, Lester HA, Dougherty DA. A unified view of the role of electrostatic interactions in modulating the gating of Cys loop receptors. J Biol Chem. 2005 Dec 16;280(50):41655-66.
23. Maness LM, Kastin AJ, Weber JT, Banks WA, Beckman BS, Zadina JE. Neurosci Biobehav Rev. 1994,18:143-59.
24. Bespalov MM, Saarma M. GDNF family receptor complexes are emerging drug targets. Trends Pharmacol Sci. 2007 Feb;28(2):68-74.
25. Alberch J, Pérez-Navarro E, Canals JM. Neurotrophic factors in Huntington's disease. Prog Brain Res. 2004;146:195-229.
26. Dutta R, McDonough J, Chang A, Swamy L, Siu A, Kidd GJ, Rudick R, Mirnics K, Trapp BD. Brain. 2007 Oct;130(Pt 10):2566-76.
27. Woolf NJ. Microtubules in the cerebral cortex: role in memory and consciousness. In Tuszyński, J.A. (ed.) The Emerging Physics of Consciousness, 2006, Springer Verlag.
28. Ivanov AA, Jacobson KA. Molecular modeling of a PAMAM-CGS21680 dendrimer bound to an A2A adenosine receptor homodimer. Bioorg Med Chem Lett. 2008 Aug 1;18(15):4312-5.
29. Kim Y, Hechler B, Klutz AM, Gachet C, Jacobson KA. Toward multivalent signaling across G protein-coupled receptors from poly(amidoamine) dendrimers. Bioconjug Chem. 2008 Feb;19(2):406-11.
30. Sariola H, Saarma M. Novel functions and signalling pathways for GDNF. J Cell Sci. 2003 Oct 1;116(Pt 19):3855-62.
31. Cui B, Wu C, Chen L, Ramirez A, Bearer EL, Li WP, Mobley WC, Chu S. One at a time, live tracking of NGF axonal transport using quantum dots. Proc Natl Acad Sci U S A. 2007 Aug 21;104(34):13666-71.
32. Sala C. Molecular regulation of dendritic spine shape and function. Neurosignals. 2002 Jul-Aug;11(4):213-23
33. Calabrese B, Wilson MS, Halpain S. Development and regulation of dendritic spine synapses. Physiology (Bethesda). 2006 Feb;21:38-47.
34. Suzuki T, Du F, Tian QB, Zhang J, Endo S. J Neurochem. 2008 Feb;104(3):596-610.
35. Beene DL, Scott JD. Curr Opin Cell Biol. 2007 Apr;19(2):192-8.
36. Lardi-Studler B, Fritschy JM. Matching of pre- and postsynaptic specializations during synaptogenesis. Neuroscientist. 2007 Apr;13(2):115-26.
37. Spólnik P, Stopa B, Piekarska B, Jagusiak A, Konieczny L, Rybarska J, Król M, Roterman I, Urbanowicz B, Zieba-Palus J. The use of rigid, fibrillar Congo red nanostructures for scaffolding protein assemblies and inducing the formation of amyloid-like arrangement of molecules. Chem Biol Drug Des. 2007 Dec;70(6):491-501.

38. Guo X, Whalley A, Klare JE, Huang L, O'brien S, Steigerwald M, Nuckolls C. Single-molecule devices as scaffolding for multicomponent nanostructure assembly. Nano Lett. 2007 May;7(5):1119-22.
39. Yamada S, Nelson WJ. Synapses: sites of cell recognition, adhesion, and functional specification. Annu Rev Biochem. 2007;76:267-94.
40. Lisé MF, El-Husseini A. The neuroligin and neurexin families: from structure to function at the synapse. Cell Mol Life Sci. 2006 Aug;63(16):1833-49.
41. Zhang N, Chittasupho C, Duangrat C, Siahaan TJ, Berkland C. PLGA nanoparticle–peptide conjugate effectively targets intercellular cell-adhesion molecule-1. Bioconjug Chem. 2008 Jan;19(1):145-52.
42. Sharma RI, Shreiber DI, Moghe PV. Nanoscale variation of bioadhesive substrates as a tool for engineering of cell matrix assembly. Tissue Eng Part A. 2008 Jul;14(7):1237-50.
43. Avery OT, MacLeod CM, McCarty M. 1944 Journal of Experimental Medicine 79: 137-158.
44. Watson JD and Crick FHC. Nature 171, 737-738 (1953).
45. Feynman RP. In Miniaturization, ed. by H.D. Gilbert (Reinhold, New York 1961) p. 282
46. Seeman NC. From genes to machines: DNA nanomechanical devices. Trends Biochem Sci. 2005 Mar;30(3):119-25.
47. Lodish H, Berk A, Zipursky LS, Matsudaira P, Baltimore D, Darnell J, 2000 Molecular Cell Biology, Freeman Press, New York.
48. Kleiman R, Banker G, Steward O. J Neurosci. 1994 Mar;14(3 Pt 1):1130-40.
49. Bramham CR, Wells DG. Dendritic mRNA: transport, translation and function. Nat Rev Neurosci. 2007 37.
50. Hirokawa N. mRNA transport in dendrites: RNA granules, motors, and tracks. J Neurosci. 2006, 26(27):7139-42
51. Grooms SY, Noh KM, Regis R, Bassell GJ, Bryan MK, Carroll RC, Zukin RS. Activity bi-directionally regulates AMPA receptor mRNA abundance in dendrites of hippocampal neurons. J Neurosci. 2006 Aug 9;26(32):8339-51.
52. Schuman EM, Dynes JL, Steward O. Synaptic regulation of translation of dendritic mRNAs. J Neurosci. 2006
53. van Meer G. Lipids of the Golgi membrane. Trends Cell Biol. 1998 Jan;8(1):29-33.
54. Gardiol A, Racca C, Triller A. Dendritic and postsynaptic protein synthetic machinery. J Neurosci. 1999 Jan 1;19(1):168-79.
55. Mowla SJ, Farhadi HF, Pareek S, Atwal JK, Morris SJ, Seidah NG, Murphy RA. Biosynthesis and post-translational processing of the precursor to brain-derived neurotrophic factor. J Biol Chem. 2001 Apr 20;276(16):12660-6.
56. Duvernay MT, Filipeanu CM, Wu G.The regulatory mechanisms of export trafficking of G protein-coupled receptors. Cell Signal. 2005 Dec;17(12):1457-65.
57. Nakata T, Hirokawa N. Neuronal polarity and the kinesin superfamily proteins. Sci STKE. 2007 Feb 6;2007(372):pe6.
58. Muresan V. One axon, many kinesins: What's the logic? J Neurocytol. 2000 Nov-Dec;29(11-12):799-818.
59. Becherer U, Rettig J. Vesicle pools, docking, priming, and release. Cell Tissue Res. 2006 Nov;326(2):393-407. Epub 2006 Jul 4
60. Sudhof TC. The synaptic vesicle cycle. Annu Rev Neurosci. 2004;27:509-47.

61. Crick F. The Astonishing Hypothesis: The Scientific Search For The Soul. Charles Scribner's Sons, New York, 1994.
62. Taupin P, Gage F. Adult neurogenesis and neural stem cells of the central nervous system in mammals. J Neurosci Res. 2002 Sep 15;69(6):745-9.
63. Solanki A, Kim JD, Lee KB. Nanotechnology for regenerative medicine: nanomaterials for stem cell imaging. Nanomed. 2008 Aug;3(4):567-78.
64. Hunt JA. Regenerative medicine: Materials in a cellular world. Nat Mater. 2008 Aug;7(8):617-8.
65. Nixon RA. Dynamic behavior and organization of cytoskeletal proteins in neurons: reconciling old and new findings. Bioessays. 1998 Oct;20(10):798-807.
66. Portier MM. [Neuronal cytoskeleton: structural, functional and dynamic aspects] Rev Neurol (Paris). 1992;148(1):1-19.
67. Kasai H, Matsuzaki M, Noguchi J, Yasumatsu N, Nakahara H. Structure-stability-function relationships of dendritic spines. Trends Neurosci. 2003 Jul;26(7):360-8.
68. Touri F, Welker E, Riederer BM. Differential distribution of MAP1A isoforms in the adult mouse barrel cortex and comparison with the serotonin 5-HT2A receptor. J Chem Neuroanat. 2004 May;27(2):99-108.
69. DeFelipe J, Alonso-Nanclares L, Arellano JI. Microstructure of the neocortex: comparative aspects. J Neurocytol. 2002 Mar-Jun;31(3-5):299-316.
70. Schubert D, Kötter R, Staiger JF. Mapping functional connectivity in barrel-related columns reveals layer- and cell type-specific microcircuits. Brain Struct Funct. 2007 Sep;212(2):107-19.
71. Hirokawa N, Glicksman MA, Willard MB. Organization of mammalian neurofilament polypeptides within the neuronal cytoskeleton. J Cell Biol. 1984 Apr;98(4):1523-36.
72. Morris JR, Lasek RJ. Monomer-polymer equilibria in the axon: direct measurement of tubulin and actin as polymer and monomer in axoplasm. J Cell Biol. 1984 Jun;98(6):2064-76.
73. Sacconi L, O'Connor RP, Jasaitis A, Masi A, Buffelli M, Pavone FS. In vivo multiphoton nanosurgery on cortical neurons. J Biomed Opt. 2007 Sep-Oct;12(5):050502.
74. Houchin-Ray T, Whittlesey KJ, Shea LD. Spatially patterned gene delivery for localized neuron survival and neurite extension. Mol Ther. 2007 Apr;15(4):705-12.
75. Koh HS, Yong T, Chan CK, Ramakrishna S. Enhancement of neurite outgrowth using nano-structured scaffolds coupled with laminin. Biomaterials. 2008 Sep;29(26):3574-82.
76. Perrone Capano C, Pernas-Alonso R, di Porzio U. Neurofilament homeostasis and motoneurone degeneration. Bioessays. 2001 Jan;23(1):24-33.
77. Julien JP, Mushynski WE. Multiple phosphorylation sites in mammalian neurofilament polypeptides. J Biol Chem. 1982 Sep 10;257(17):10467-70.
78. Yang Y, Bauer C, Strasser G, Wollman R, Julien JP, Fuchs E. Integrators of the cytoskeleton that stabilize microtubules. Cell. 1999 Jul 23;98(2):229-38.
79. Kreplak L, Bär H, Leterrier JF, Herrmann H, Aebi U. Exploring the mechanical behavior of single intermediate filaments. J Mol Biol. 2005 Dec 2;354(3):569-77.
80. Lariviere RC, Julien JP. Functions of intermediate filaments in neuronal development and disease. J Neurobiol. 2004 Jan;58(1):131-48.

81. Zhu Q, Couillard-Després S, Julien JP. Delayed maturation of regenerating myelinated axons in mice lacking neurofilaments. Exp Neurol. 1997 Nov;148(1):299-316.
82. Zhang Z, Casey DM, Julien JP, Xu Z. Normal dendritic arborization in spinal motoneurons requires neurofilament subunit L. J Comp Neurol. 2002 Aug 19;450(2):144-52.
83. Woolf NJ. Cholinoceptive cells in rat cerebral cortex: somatodendritic immunoreactivity for muscarinic receptor and cytoskeletal proteins. J Chem Neuroanat. 1993 Nov-Dec;6(6):375-90.
84. Molnár Z, Cheung AF. Towards the classification of subpopulations of layer V pyramidal projection neurons. Neurosci Res. 2006 Jun;55(2):105-15.
85. Kost SA, Chacko K, Oblinger MM. Developmental patterns of intermediate filament gene expression in the normal hamster brain. Brain Res. 1992 Nov 13;595(2):270-80.
86. Julien JP, Meyer D, Flavell D, Hurst J, Grosveld F. Cloning and developmental expression of the murine neurofilament gene family. Brain Res. 1986 Dec;387(3):243-50.
87. Julien JP. Neurofilaments and motor neuron disease. Trends Cell Biol. 1997 Jun;7(6):243-9.
88. Perrone Capano C, Pernas-Alonso R, di Porzio U. Neurofilament homeostasis and motoneurone degeneration. Bioessays. 2001 Jan;23(1):24-33.
89. Luque FA, Jaffe SL. Cerebrospinal fluid analysis in multiple sclerosis. Int Rev Neurobiol. 2007;79:341-56.
90. Miltenberger-Miltenyi G, Janecke AR, Wanschitz JV, Timmerman V, Windpassinger C, Auer-Grumbach M, Löscher WN. Clinical and electrophysiological features in Charcot-Marie-Tooth disease with mutations in the NEFL gene. Arch Neurol. 2007 Jul;64(7):966-70.
91. Pérez-Ollé R, López-Toledano MA, Goryunov D, Cabrera-Poch N, Stefanis L, Brown K, Liem RK. Mutations in the neurofilament light gene linked to Charcot-Marie-Tooth disease cause defects in transport. J Neurochem. 2005 May;93(4):861-74.
92. Gotow T. Neurofilaments in health and disease. Med Electron Microsc. 2000;33(4):173-99.
93. Liu Q, Xie F, Siedlak SL, Nunomura A, Honda K, Moreira PI, Zhua X, Smith MA, Perry G. Neurofilament proteins in neurodegenerative diseases. Cell Mol Life Sci. 2004 Dec;61(24):3057-75.
94. Bugiani O. The many ways to frontotemporal degeneration and beyond. Neurol Sci. 2007 Oct;28(5):241-4.
95. Seidlits SK, Lee JY, Schmidt CE. Nanostructured scaffolds for neural applications. Nanomed. 2008 Apr;3(2):183-99.
96. Bremer A, Aebi U. The structure of the F-actin filament and the actin molecule. Curr Opin Cell Biol. 1992 Feb;4(1):20-6.
97. Chen H, Bernstein BW, Bamburg JR. Regulating actin-filament dynamics in vivo. Trends Biochem Sci. 2000 Jan;25(1):19-23.
98. Sekino Y, Kojima N, Shirao T. Role of actin cytoskeleton in dendritic spine morphogenesis. Neurochem Int. 2007 Jul-Sep;51(2-4):92-104.
99. Nakagawa T, Engler JA, Sheng M. The dynamic turnover and functional roles of alpha-actinin in dendritic spines. Neuropharmacology. 2004 Oct;47(5):734-45.

100. dos Remedios CG, Chhabra D, Kekic M, Dedova IV, Tsubakihara M, Berry DA, Nosworthy NJ. Actin binding proteins: regulation of cytoskeletal microfilaments. Physiol Rev. 2003 Apr;83(2):433-73.
101. Denker SP, Barber DL. Ion transport proteins anchor and regulate the cytoskeleton. Curr Opin Cell Biol. 2002 Apr;14(2):214-20.
102. Ishikawa R, Kohama K. Actin-binding proteins in nerve cell growth cones. J Pharmacol Sci. 2007 Sep;105(1):6-11.
103. Ahuja R, Pinyol R, Reichenbach N, Custer L, Klingensmith J, Kessels MM, Qualmann B. Cordon-bleu is an actin nucleation factor and controls neuronal morphology.Cell. 2007 Oct 19;131(2):337-50.
104. Williams MR, Markey JC, Doczi MA, Morielli AD. An essential role for cortactin in the modulation of the potassium channel Kv1.2. Proc Natl Acad Sci U S A. 2007 Oct 30;104(44):17412-7.
105. Myers KA, He Y, Hasaka TP, Baas PW. Microtubule transport in the axon: Re-thinking a potential role for the actin cytoskeleton. Neuroscientist. 2006 Apr;12(2):107-18.
106. Shirao T, Sekino Y. Clustering and anchoring mechanisms of molecular constituents of postsynaptic scaffolds in dendritic spines. Neurosci Res. 2001 May;40(1):1-7.
107. Allison DW, Gelfand VI, Spector I, Craig AM. Role of actin in anchoring postsynaptic receptors in cultured hippocampal neurons: differential attachment of NMDA versus AMPA receptors. J Neurosci. 1998 Apr 1;18(7):2423-36.
108. Wechsler A, Teichberg VI. Brain spectrin binding to the NMDA receptor is regulated by phosphorylation, calcium and calmodulin. EMBO J. 1998 Jul 15;17(14):3931-9.
109. Carlisle HJ, Kennedy MB. Spine architecture and synaptic plasticity. Trends Neurosci. 2005 Apr;28(4):182-7.
110. Cristofanilli M, Mizuno F, Akopian A. Disruption of actin cytoskeleton causes internalization of Ca(v)1.3 (alpha 1D)L-type calcium channels in salamander retinal neurons. Mol Vis. 2007 Aug 28;13:1496-507.
111. Hayakawa K, Tatsumi H, Sokabe M. Actin stress fibers transmit and focus force to activate mechanosensitive channels. J Cell Sci. 2008 Jan 29
112. Kole MH, Ilschner SU, Kampa BM, Williams SR, Ruben PC, Stuart GJ. Action potential generation requires a high sodium channel density in the axon initial segment. Nat Neurosci. 2008 Feb;11(2):178-86.
113. Kalil K, Dent EW. Touch and go: guidance cues signal to the growth cone cytoskeleton. Curr Opin Neurobiol. 2005 Oct;15(5):521-6.
114. Sarmiere PD, Bamburg JR. Regulation of the neuronal actin cytoskeleton by ADF/cofilin. J Neurobiol. 2004 Jan;58(1):103-17.
115. Lipski AM, Pino CJ, Haselton FR, Chen IW, Shastri VP. The effect of silica nanoparticle-modified surfaces on cell morphology, cytoskeletal organization and function. Biomaterials. 2008 Oct;29(28):3836-46.
116. Löwe J, Li H, Downing KH, Nogales E. Refined structure of alpha beta-tubulin at 3.5 A resolution. J Mol Biol. 2001 Nov 9;313(5):1045-57.
117. Schek HT 3rd, Gardner MK, Cheng J, Odde DJ, Hunt AJ. Microtubule assembly dynamics at the nanoscale. Curr Biol. 2007 Sep 4;17(17):1445-55.
118. Caceres A, Binder LI, Payne MR, Bender P, Rebhun L, Steward O. Differential subcellular localization of tubulin and the microtubule-associated protein MAP2 in brain tissue as revealed by immunocytochemistry with monoclonal hybridoma antibodies. J Neurosci. 1984 Feb;4(2):394-410.

119. Chen J, Kanai Y, Cowan NJ, Hirokawa N. Projection domains of MAP2 and tau determine spacings between microtubules in dendrites and axons. Nature. 1992 Dec 17;360(6405):674-7.
120. Nunez J, Fischer I. Microtubule-associated proteins (MAPs) in the peripheral nervous system during development and regeneration. J Mol Neurosci. 1997 Jun;8(3):207-22.
121. Halpain S, Dehmelt L. The MAP1 family of microtubule-associated proteins. Genome Biol. 2006;7(6):224.
122. Gonzalez-Billault C, Jimenez-Mateos EM, Caceres A, Diaz-Nido J, Wandosell F, Avila J. Microtubule-associated protein 1B function during normal development, regeneration, and pathological conditions in the nervous system. J Neurobiol. 2004 Jan;58(1):48-59.
123. Riederer BM. Microtubule-associated protein 1B, a growth-associated and phosphorylated scaffold protein. Brain Res Bull. 2007 Mar 30;71(6):541-58.
124. Dehmelt L, Halpain S. The MAP2/Tau family of microtubule-associated proteins. Genome Biol. 2005;6(1):204.
125. Sánchez C, Díaz-Nido J, Avila J. Phosphorylation of microtubule-associated protein 2 (MAP2) and its relevance for the regulation of the neuronal cytoskeleton function. Prog Neurobiol. 2000 Jun;61(2):133-68.
126. Goedert M, Crowther RA, Garner CC. Molecular characterization of microtubule-associated proteins tau and MAP2.Trends Neurosci. 1991 May;14(5):193-9.
127. Lian G, Sheen V.Cerebral developmental disorders. Curr Opin Pediatr. 2006 Dec;18(6):614-20.
128. Hoogenraad CC, Akhmanova A, Galjart N, De Zeeuw CI. LIMK1 and CLIP-115: linking cytoskeletal defects to Williams syndrome. Bioessays. 2004 Feb;26(2):141-50.
129. Hirokawa N, Takemura R. Molecular motors in neuronal development, intracellular transport and diseases. Curr Opin Neurobiol. 2004 Oct;14(5):564-73.
130. Goldstein LS, Yang Z. Microtubule-based transport systems in neurons: the roles of kinesins and dyneins. Annu Rev Neurosci. 2000;23:39-71.
131. Riederer B, Matus A Differential expression of distinct microtubule-associated proteins during brain development. Proc Natl Acad Sci U S A. 1985 Sep;82(17):6006-9.
132. Kosik KS, Finch EA. MAP2 and tau segregate into dendritic and axonal domains after the elaboration of morphologically distinct neurites: an immunocytochemical study of cultured rat cerebrum. J Neurosci. 1987 Oct;7(10):3142-53.
133. Alonso AD, Grundke-Iqbal I, Barra HS, Iqbal K. Abnormal phosphorylation of tau and the mechanism of Alzheimer neurofibrillary degeneration: sequestration of microtubule-associated proteins 1 and 2 and the disassembly of microtubules by the abnormal tau. Proc Natl Acad Sci U S A. 1997 Jan 7;94(1):298-303.
134. Dunn HG. Neurons and neuronal systems involved in the pathophysiologies of Rett syndrome. Brain Dev. 2001 Dec;23 Suppl 1:S99-S100.
135. Thyberg J, Moskalewski S. Microtubules and the organization of the Golgi complex. Exp Cell Res. 1985 Jul;159(1):1-16.
136. Letourneau PC, Wire JP. Three-dimensional organization of stable microtubules and the Golgi apparatus in the somata of developing chick sensory neurons. J Neurocytol. 1995 Mar;24(3):207-23.
137. Carson JH, Barbarese E. Systems analysis of RNA trafficking in neural cells. Biol Cell. 2005 Jan;97(1):51-62.

138. Pampaloni F, Florin EL. Microtubule architecture: inspiration for novel carbon nanotube-based biomimetic materials. Trends Biotechnol. 2008 Jun;26(6):302-10.
139. Gardner MK, Hunt AJ, Goodson HV, Odde DJ. Microtubule assembly dynamics: new insights at the nanoscale. Curr Opin Cell Biol. 2008 Feb;20(1):64-70.
140. Trickler WJ, Nagvekar AA, Dash AK. A novel nanoparticle formulation for sustained paclitaxel delivery. AAPS PharmSciTech. 2008;9(2):486-93.
141. Shulman RG, Rothman DL, Behar KL, Hyder F. Energetic basis of brain activity: implications for neuroimaging. Trends Neurosci. 2004 Aug;27(8):489-95.
142. Cheung EC, McBride HM, Slack RS. Mitochondrial dynamics in the regulation of neuronal cell death. Apoptosis. 2007 May;12(5):979-92.
143. Detmer SA, Chan DC. Functions and dysfunctions of mitochondrial dynamics. Nat Rev Mol Cell Biol. 2007 Nov;8(11):870-9.
144. Frederick RL, Shaw JM. Moving mitochondria: establishing distribution of an essential organelle. Traffic. 2007 Dec;8(12):1668-75.
145. Boldogh IR, Pon LA. Mitochondria on the move. Trends Cell Biol. 2007 Oct;17(10):502-10.
146. Crouch PJ, Cimdins K, Duce JA, Bush AI, Trounce IA. Mitochondria in aging and Alzheimer's disease. Rejuvenation Res. 2007 Sep;10(3):349-57.
147. Salnikov V, Lukyánenko YO, Frederick CA, Lederer WJ, Lukyánenko V. Probing the outer mitochondrial membrane in cardiac mitochondria with nanoparticles. Biophys J. 2007 Feb 1;92(3):1058-71.
148. Wasylycia JR, Sapelnikova S, Jeong H, Dragoljic J, Marcus SL, Harrison DJ. Nanobiopower supplies for biomolecular motors: the use of metabolic pathway-based fuel generating systems in microfluidic devices. Lab Chip. 2008 Jun;8(6):979-82.
149. Elder JB, Hoh DJ, Oh BC, Heller AC, Liu CY, Apuzzo ML. The future of cerebral surgery: a kaleidoscope of opportunities. Neurosurgery. 2008 Jun;62(6 Suppl 3):1555-79
150. Wang X, Su B, Perry G, Smith MA, Zhu X. Insights into amyloid-beta-induced mitochondrial dysfunction in Alzheimer disease. Free Radic Biol Med. 2007 Dec 15;43(12):1569-73.
151. Borges HL, Linden R, Wang JY. DNA damage-induced cell death: lessons from the central nervous system. Cell Res. 2008 Jan;18(1):17-26.
152. Soane L, Kahraman S, Kristian T, Fiskum G. Mechanisms of impaired mitochondrial energy metabolism in acute and chronic neurodegenerative disorders. J Neurosci Res. 2007 Nov 15;85(15):3407-15.
153. Trushina E, McMurray CT. Oxidative stress and mitochondrial dysfunction in neurodegenerative diseases. Neuroscience. 2007 Apr 14;145(4):1233-48.
154. Greenemeier L. Monkey see, robot do. Scientific American. January 15, 2008: http://www.sciam.com/article.cfm?id=monkey-think-robot-do
155. Hsiao MC, Chan CH, Srinivasan V, Ahuja A, Erinjippurath G, Zanos TP, Gholmieh G, Song D, Wills JD, LaCoss J, Courellis S, Tanguay AR Jr, Granacki JJ,Marmarelis VZ, Berger TW. VLSI implementation of a nonlinear neuronal model: a "neural prosthesis" to restore hippocampal trisynaptic dynamics. Conf Proc IEEE Eng Med Biol Soc. 2006;1:4396-9.
156. Zhou DD, Greenberg RJ. Microsensors and microbiosensors for retinal implants. Front Biosci. 2005 Jan 1;10:166-79.

157. Yumei Xing; Chun Hui; Yibo Wu; Ailan Xu; Gang Li; Jianlong Zhao; Qiushi Ren, Development of flexible neural microelectode arrays based on parylene for retinal prosthesis. Networking, Sensing and Control, 2008. ICNSC 2008. IEEE International Conference, vol., no., pp.548-551.
158. Häusler R, Stieger C, Bernhard H, Kompis M. A novel implantable hearing system with direct acoustic cochlear stimulation. Audiol Neurootol. 2008;13(4):247-56.
159. Sarpeshkar R, Salthouse C, Sit JJ, Baker MW, Zhak SM, Lu TK, Turicchia L, Balster S. An ultra-low-power programmable analog bionic ear processor. IEEE Trans Biomed Eng. 2005 Apr;52(4):711-27.
160. Liu Q, Xu Y, Cai H, Qin L, Ye X, Li R, Wang P. The study on bionic olfactory neurochip based on light-addressable potentiometric sensor. Conf Proc IEEE Eng Med Biol Soc. 2005;5(1):4862-4865.
161. He W, McConnell GC, Bellamkonda RV. Nanoscale laminin coating modulates cortical scarring response around implanted silicon microelectrode arrays. J Neural Eng. 2006 Dec;3(4):316-26.
162. Yu Z, McKnight TE, Ericson MN, Melechko AV, Simpson ML, Ill BM. Vertically aligned carbon nanofiber arrays record electrophysiological signals from hippocampal slices. Nano Lett. 2007 Aug;7(8):2188-95.
163. Swamy BE, Venton BJ. Carbon nanotube-modified microelectrodes for simultaneous detection of dopamine and serotonin in vivo. Analyst. 2007 Sep;132(9):876-84.
164. Keefer EW, Botterman BR, Romero MI, Rossi AF, Gross GW. Carbon nanotube coating improves neuronal recordings. Nat Nanotechnol. 2008 Jul;3(7):434-9.
165. Chow EY, Kahn A, Irazoqui PP. High data-rate 6.7 Ghz wireless asic transmitter for neural prostheses. Conf Proc IEEE Eng Med Biol Soc. 2007;1:6580-3.
166. Lien V, Lin H, Chuang J, Sailor M, Lo Y. A fiber-optic powered wireless sensor module made on elastomeric substrate for wearable sensors. Conf Proc IEEE Eng Med Biol Soc. 2004;3:2145-8.
167. Chang TM. 50th anniversary of artificial cells: their role in biotechnology, nanomedicine, regenerative medicine, blood substitutes, bioencapsulation, cell/stem cell therapy and nanorobotics. Artif Cells Blood Substit Immobil Biotechnol. 2007;35(6):545-54.
168. Björklund A. Cell therapy for Parkinson's disease: problems and prospects. Novartis Found Symp. 2005;265:174-86; discussion 187, 204-211.
169. Heese K, Low JW, Inoue N. Nerve growth factor, neural stem cells and Alzheimer's disease. Neurosignals. 2006-2007;15(1):1-12.
170. Green JJ, Zhou BY, Mitalipova MM, Beard C, Langer R, Jaenisch R, Anderson DG. Nanoparticles for gene transfer to human embryonic stem cell colonies. Nano Lett. 2008 Oct;8(10):3126-30.
171. Corot C, Robert P, Idée JM, Port M. Recent advances in iron oxide nanocrystal technology for medical imaging. Adv Drug Deliv Rev. 2006 Dec 1;58(14):1471-504.
172. Jan E, Kotov NA. Successful differentiation of mouse neural stem cells on layer-by-layer assembled single-walled carbon nanotube composite. Nano Lett. 2007 May;7(5):1123-8.
173. A, Kim JD, Lee KB. Nanotechnology for regenerative medicine: nanomaterials for stem cell imaging. Nanomed. 2008 Aug;3(4):567-78.

174. Cheon J, Lee JH. Synergistically integrated nanoparticles as multimodal probes for nanobiotechnology. Acc Chem Res. 2008 Aug 13.
175. Ferreira L, Karp JM, Nobre L, Langer R. New opportunities: the use of nanotechnologies to manipulate and track stem cells. Cell Stem Cell. 2008 Aug 7;3(2):136-46.
176. Sladek JR Jr, Bjugstad KB, Collier TJ, Bundock EA, Blanchard BC, Elsworth JD, Roth RH, Redmond DE Jr. Embryonic substantia nigra grafts show directional outgrowth to cografted striatal grafts and potential for pathway reconstruction in nonhuman primate. Cell Transplant. 2008;17(4):427-44.
177. Woolf NJ. Bionic microtubules: potential applications to multiple neurological and neuropsychiatric diseases. Journal of Nanoneuroscience 2008, 1:XXX-XXX.
178. Nechyporuk-Zloy V, Stock C, Schillers H, Oberleithner H, Schwab A. Single plasma membrane K+ channel detection by using dual-color quantum dot labeling. Am J Physiol Cell Physiol. 2006 Aug;291(2):C266-9.
179. Park KH, Chhowalla M, Iqbal Z, Sesti F. Single-walled carbon nanotubes are a new class of ion channel blockers. J Biol Chem. 2003 Dec 12;278(50):50212-6.
180. Mannix RJ, Kumar S, Cassiola F, Montoya-Zavala M, Feinstein E, Prentiss M, Ingber DE. Nanomagnetic actuation of receptor-mediated signal transduction. Nat Nanotechnol. 2008 Jan;3(1):36-40.
181. Yersin A, Hirling H, Kasas S, Roduit C, Kulangara K, Dietler G, Lafont F, Catsicas S, Steiner P. Elastic properties of the cell surface and trafficking of single AMPA receptors in living hippocampal neurons. Biophys J. 2007 Jun 15;92(12):4482-9.
182. Hillebrenner H, Buyukserin F, Stewart JD, Martin CR. Template synthesized nanotubes for biomedical delivery applications. Nanomed. 2006 Jun;1(1):39-50.
183. Dobson J. Magnetic micro- and nano-particle-based targeting for drug and gene delivery. Nanomed. 2006 Jun;1(1):31-7.
184. Goldberg M, Langer R, Jia X. Nanostructured materials for applications in drug delivery and tissue engineering. J Biomater Sci Polym Ed. 2007;18(3):241-68.
185. Gross PG, Kartalov EP, Scherer A, Weiner LP. Applications of microfluidics for neuronal studies. J Neurol Sci. 2007 Jan 31;252(2):135-43.
186. Zagnoni M, Sandison ME, Marius P, Lee AG, Morgan H. Controlled delivery of proteins into bilayer lipid membranes on chip. Lab Chip. 2007 Sep;7(9):1176-83.
187. Simmel FC. Towards biomedical applications for nucleic acid nanodevices. Nanomed. 2007 Dec;2(6):817-30.
188. Yurke B, Turberfield AJ, Mills AP Jr, Simmel FC, Neumann JL. A DNA-fuelled molecular machine made of DNA. Nature. 2000 Aug 10;406(6796):605-8.
189. Müller BK, Reuter A, Simmel FC, Lamb DC. Single-pair FRET characterization of DNA tweezers. Nano Lett. 2006 Dec;6(12):2814-20.
190. Dittmer WU, Kempter S, Rädler JO, Simmel FC. Using gene regulation to program DNA-based molecular devices. Small. 2005 Jul;1(7):709-12.
191. Bellizzi G, Bucci OM, Capozzoli A. On the energy transfer between the electromagnetic field and nanomachines for biological applications. Bioelectromagnetics. 2008 Jan 25
192. Ragusa A, García I, Penadés S. Nanoparticles as nonviral gene delivery vectors. IEEE Trans Nanobioscience. 2007 Dec;6(4):319-30. Review.
193. Luten J, van Nostrum CF, De Smedt SC, Hennink WE. Biodegradable polymers as non-viral carriers for plasmid DNA delivery. J Control Release. 2008 Mar 3;126(2):97-110.

194. Eliyahu H, Barenholz Y, Domb AJ. Polymers for DNA delivery. Molecules. 2005 Jan 31;10(1):34-64.
195. Konstan MW, Davis PB, Wagener JS, Hilliard KA, Stern RC, Milgram LJ, Kowalczyk TH, Hyatt SL, Fink TL, Gedeon CR, Oette SM, Payne JM, Muhammad O, Ziady AG, Moen RC, Cooper MJ. Compacted DNA nanoparticles administered to the nasal mucosa of cystic fibrosis subjects are safe and demonstrate partial to complete cystic fibrosis transmembrane regulator reconstitution. Hum Gene Ther. 2004 Dec;15(12):1255-69.
196. Taira S, Du YZ, Hiratsuka Y, Uyeda TQ, Yumoto N, Kodaka M. Loading and unloading of molecular cargo by DNA-conjugated microtubule. Biotechnol Bioeng. 2008 Feb 15;99(3):734-9.
197. Hess H, Bachand GD, Vogel V. Powering nanodevices with biomolecular motors. Chemistry. 2004 May 3;10(9):2110-6.
198. Dinu CZ, Opitz J, Pompe W, Howard J, Mertig M, Diez S. Parallel manipulation of bifunctional DNA molecules on structured surfaces using kinesin-driven microtubules.Small. 2006 Aug;2(8-9):1090-8.
199. Suh J, Wirtz D, Hanes J. Real-time intracellular transport of gene nanocarriers studied by multiple particle tracking. Biotechnol Prog. 2004 Mar-Apr;20(2):598-602
200. Kim T, Kao T, Hasselbrink EF, Meyhofer E. Nanomechanical Model of Microtubule Translocation in the Presence of Electric Fields. Biophys J. 2008 Jan 30;
201. van den Heuvel MG, de Graaff MP, Dekker C. Molecular sorting by electrical steering of microtubules in kinesin-coated channels. Science. 2006 May 12;312(5775):910-4.
202. Hughes S, El Haj AJ, Dobson J. Magnetic micro- and nanoparticle mediated activation of mechanosensitive ion channels. Med Eng Phys. 2005 Nov;27(9):754-62.
203. Dobson J. Remote control of cellular behaviour with magnetic nanoparticles. Nat Nanotechnol. 2008 Mar;3(3):139-43.
204. van den Heuvel MG, de Graaff MP, Lemay SG, Dekker C. Electrophoresis of individual microtubules in microchannels. Proc Natl Acad Sci U S A. 2007 May 8;104(19):7770-5.
205. Hutchins BM, Platt M, Hancock WO, Williams ME. Directing transport of CoFe2O4-functionalized microtubules with magnetic fields. Small. 2007 Jan;3(1):126-31.
206. Platt M, Muthukrishnan G, Hancock WO, Williams ME. Millimeter scale alignment of magnetic nanoparticle functionalized microtubules in magnetic fields. J Am Chem Soc. 2005 Nov 16;127(45):15686-7.
207. Ramachandran S, Ernst KH, Bachand GD, Vogel V, Hess H. Selective loading of kinesin-powered molecular shuttles with protein cargo and its application to biosensing. Small. 2006 Mar;2(3):330-4.
208. Sundberg M, Bunk R, Albet-Torres N, Kvennefors A, Persson F, Montelius L, Nicholls IA, Ghatnekar-Nilsson S, Omling P, Tĺgerud S, Mĺnsson A. Actin filament guidance on a chip: toward high-throughput assays and lab-on-a-chip applications. Langmuir. 2006 Aug 15;22(17):7286-95.
209. Martinez-Neira R, Kekic M, Nicolau D, dos Remedios CG. A novel biosensor for mercuric ions based on motor proteins. Biosens Bioelectron. 2005 Jan 15;20(7):1428-32.

210. Hess H, Clemmens J, Brunner C, Doot R, Luna S, Ernst KH, Vogel V. Molecular self-assembly of "nanowires" and "nanospools" using active transport. Nano Lett. 2005 Apr;5(4):629-33.
211. Prokop A. Bioartificial organs in the twenty-first century: Nanobiological devices. Ann N Y Acad Sci. 2001 Nov;944:472-90.

3

The Cytoskeleton as a Nanoscale Information Processor: Electrical Properties and an Actin-Microtubule Network Model

Summary

One of the major goals of nanotechnology is to advance the field of information processing. The central processing units of the future are likely to be quite different from those currently used. While biomolecular processors are unlikely to displace semiconductor processors for speed and accuracy, certain proteins may offer solutions to problems confronting logical processor design, including self-assembly and emergent computation. Cytoskeletal proteins may prove useful as biomolecular processors or may inspire hybrid designs. Actin filaments and microtubules, for example, have highly charged surfaces that enable them to conduct electric currents and process information. The biophysical properties of these filaments relevant to the conduction of ionic current include a condensation of counterions on the filament surface, the non-linear complex physical structure, and in the case of microtubules, nanopores that allow ions to pass between the outer environment to the microtubule lumen. Possible roles for cable-like, conductive filaments in neurons include intracellular information processing, regulation of developmental plasticity, and mediation of transport. Operating as an interconnected matrix, cytoskeletal proteins form a complex network capable of emergent information processing; moreover, they stand to intervene between inputs to and outputs from neurons. The cytoskeletal matrix receives information from the neuronal membrane and its intrinsic components (e.g., ion channels, scaffolding proteins, and adaptor proteins), especially at sites of synaptic contacts and spines, and in turn affects the output of the neuron. An information-processing model based on cytoskeletal networks is described, which may underlie certain types of learning and memory.

3.1 Electrical Properties of Actin and Actin Filaments

Actin filaments are highly concentrated just below the surface of the neuronal membrane in the cell cortex; nonetheless, these filaments also intermix with microtubules and neurofilaments found throughout dendrites, axons, and cell bodies [1]. Spines protruding from dendrites are particularly rich in actin filaments, which frequently span from the neuronal membrane deep into the subspine region [2].

These dynamic actin filaments are well known to play major roles in neural growth and development, and also participate in experience-related plasticity, such as the reorganization of spines following increased or decreased neural stimulation [3, 4].

Actin filaments also serve as anchors or tethers, physically stabilizing or constraining the location of signal transduction molecules, receptors, and ion channels [5]. Additional roles played by actin filaments may derive from being polyelectrolytes having charged groups that interact with counterions in the surrounding media.

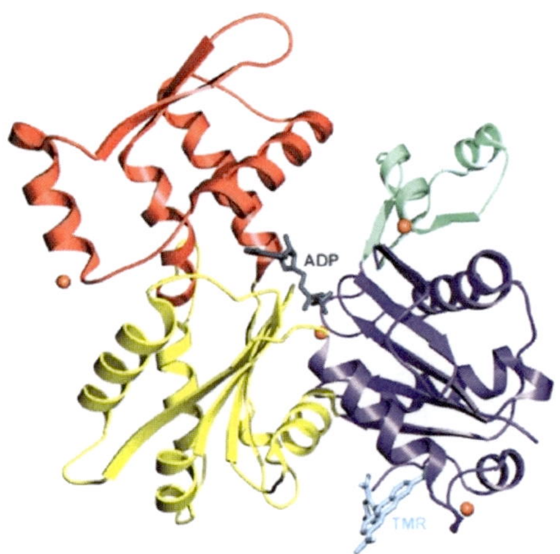

Fig. 3.1. Ribbon diagram of the actin monomer. Reprinted with permission; Otterbein et al., Science. 2001, 293:708-11 [8].

Actin filaments, as a result of being polyelectrolytes surrounded by counterions, possess the capacity of transmitting signals or sustaining ionic conductances [6, 7]. Although there is a growing number of biophysical studies demonstrating how actin filaments are able to act as "electric cables", this is a novel role proposed for these filaments and the underlying biophysical principles that enable this function are only beginning to be understood. At a minimum, properties such as the electric dipoles of actin monomers, counterion condensation of actin filaments, and the linear charge density along the longitudinal axis of the actin filament are critical to actin filaments sustaining ionic conductances.

3.1.1 The Actin Monomer: Structure, Surface Charge, and Electric Dipole

Roberto Dominguez and colleagues solved the crystal structure of actin bound to ADP to a resolution of 1.54 Å [8]. As shown in Figure 3.1, the actin monomer is a globular protein containing four subdomains. As discussed in Chapter 2, ADP versus ATP binding of the actin monomer affects its polymerization into filaments and its association to actin binding proteins. There are conformational changes to the actin monomer depending on whether it is bound to ADP versus ATP; however, these conformational shifts mainly occur at the active binding site and the sensor loop [9].

As shown in Figure 3.2, the surface of the G-actin monomer shows intermixed pockets of positive and negative electrostatic charge [10]. These surface charges influence the association of G-actin with actin-binding proteins, and with ADP or ATP, as well as affecting its polymerization into filaments. Magnetic dipoles can be induced in actin filaments sliding along myosin, as long as ATP is present. The energy of the magnetic dipole-dipole interaction between ATP-actin monomers was found to be very small but measureable at $\sim 1.1 \times 10^{-22} J$ (less than $1meV$), and the magnetization per actin monomer was measured at $\sim 1.7 \times 10^{-21} Am^2$ [11].

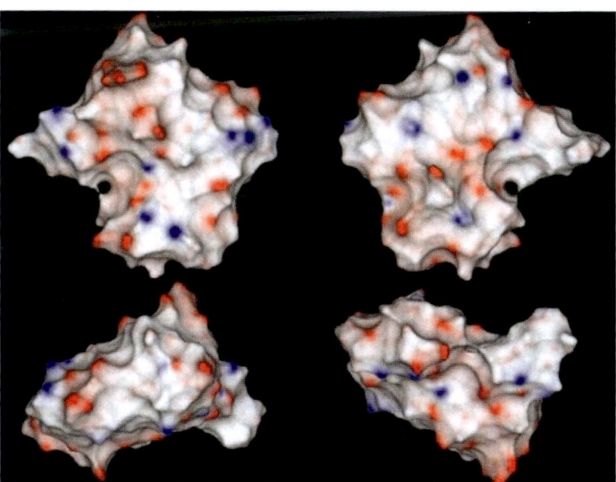

Fig. 3.2. Electrostatic surface charge on actin monomer, where red indicates negative charge, and blue indicates positive. From [10].

3.1.2 Actin Filaments: Counterions and Charge Density Waves

Rod-like polymers, such as actin filaments, are particularly likely to have counterions adsorbed to their surface. According to the Onsager-Manning-Oosawa condensation principle, any polyelectrolyte having a charge density over 1.0 electron per Bjerrum length lB will have counterions adsorbed to its surface [12]. These counterions greatly increase the potential for biophysical interactions.

In nature, actin filaments often bundle together, which seems to violate the Coulomb law of electrostatics that like charges repel. However, these protein filaments are not in vacuum but in solution surrounded by counterions that condense upon the charged surface of the actin filaments which contributes to actin filament bundling by overcoming the repulsive forces due to protein charges. This effect also leads to ionic conductance along actin filaments [13]. As illustrated in Figure 3.3, at low ionic concentrations ions disperse in the solution surrounding actin filaments, but at high ionic concentrations, such as are typical of the intracellular environment, ions densely adsorb to the surface of actin filaments due to charge complementarity. Clouds of counterions form charge density waves having a periodicity of 59.8 $\AA$, which follow the contours of the ridges that twist slightly along the longitudinal axes of actin filaments [13]. It appears that these "zipper-like" arrangements of counterions enable the close bundling of actin filaments together. Salt concentrations affect the distribution of counterions and the typical hexagonal packing of actin filaments demonstrated *in vitro* [14]. Thus, to the extent that ionic concentrations change inside neurons (e.g., during neural activity), actin filament rearrangements might be expected to occur in response to these changes. In dividing cells, an even more dramatic effect may be expected due to cell cycle changes in ionic concentrations as a function of time and a concomitant actin filament reorganization.

3.1.3 Actin Filaments: Electric Cable Properties

Horacio Cantiello and colleagues were first to discover that actin filaments generated electrical signals in an experiment evaluating actin filament responses to osmotic pressure [6]. This measurable conductive capability appeared to depend on the counterions that adsorb to the surface of the actin filament because of the following results. The electrical conductivity was pH dependent, no longer being observed at pH 5.5. Actin filaments in this experiment also responded to electric fields ranging between 500-2,000 V/cm. Inspired by this novel result, the Cantiello laboratory devised the experimental setup shown in Figure 3.4 to more fully elucidate this electrical phenomenon in actin filaments [15]. Further experiments using this experimental setup revealed that actin filaments act as electric charge conductors allowing ions to flow along the longitudinal axis of the filament [7]. This is likely due to a large linear charge density of $1.65 \times 10^2 e/nm$, of the actin filament [7]. This ionic conductance in

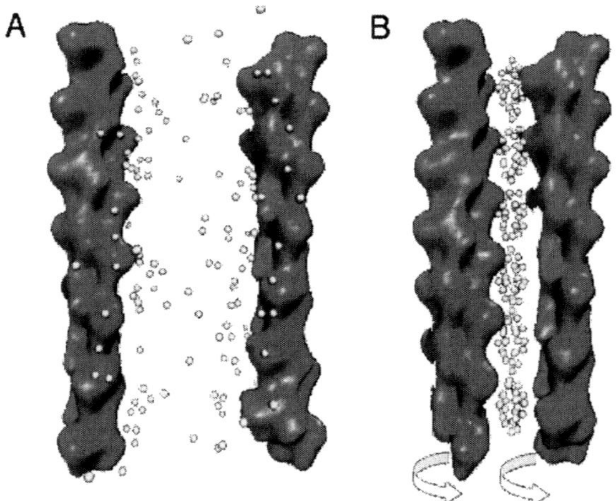

Fig. 3.3. Counterion charge density waves on actin filaments. In the presence of high concentrations, multivalent linear waves of counterions condense on the surface of actin filaments, as occurs on other polyelectrolytes. Reprinted with permission; Angelini et al., Proc Natl Acad Sci U S A. 2003, 100:8634-7 [13].

general depends on (1) a sufficiently high linear charge density, (2) a critical concentration of multivalent ions, and (3) a small dielectric constant of the surrounding medium. These criteria are met for actin filaments in neurons.

As illustrated in Figure 3.5, ions can in principle move along actin filaments without significantly affecting the surrounding environment [7]. This is because the ionic movement along the actin filaments is shielded. Also, actin filaments can store excess electrostatic charge forming a double layer with positive counterions surrounding the protein's negative charge; the capacitance per monomer being estimated as $\sim 96 \times 10^{-6} pF$. The calculated velocity of electrical signals along actin filaments is, depending on specific conditions between approximately 1 - 100 m/sec, which is in the range of the propagation of neural impulses, suggesting that concurrent propagation of electrical signals along actin filaments and electrochemical currents along the axonal membrane is possible in principle.

In addition to conducting electrical signals, actin filaments respond to electric and magnetic fields. When researchers applied electric current to actin filaments suspended in a solution-filled well located between two gold electrodes, the actin filaments aligned in parallel to the electric lines, resulting in a bridge extending over the gap between the two electrodes [16]. There are other examples and counterexamples of this effect, with some studies showing that actin filaments align parallel to the electric field, while other studies showing actin filaments align perpendicular to it [17, 18]. Williams Meggs suggested that electric fields could be responsible for the structural organization of actin filaments (as well as microtubules) [19]. His hypothesis was that dipole moments intrinsic to actin (and tubulin) aligned these polyelectrolytes parallel

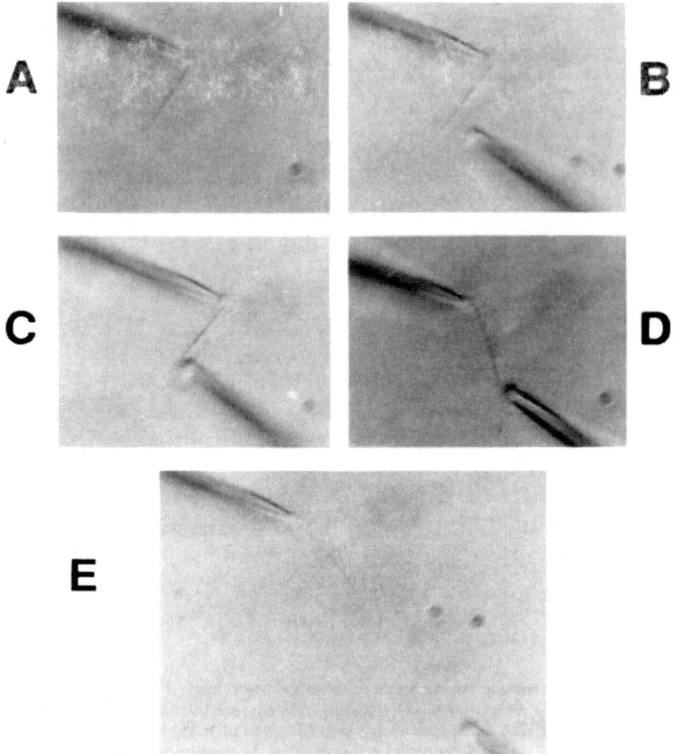

Fig. 3.4. A method for demonstrating electic cable properties of actin filaments. Reprinted with permission; Lin and Cantiello, Biophys J. 1993, 65:1371-8 [15].

to the main direction of the electric field. Experimental evidence supports the notion that actin-filament growth in cells depends on a balance of repulsive and attractive electrostatic interactions between these filaments [20]. The electric dipole moment of actin also affects its binding to various ligands [21]. As shown in Figure 3.5, the dipole in the actin filament is roughly perpendicular to the longitudinal axis of the filament [7].

3.2 Electrical Properties of Tubulin and Microtubules

Tubulin monomers, tubulin dimers, and microtubules possess fundamental electrical properties that govern their abilities to assemble, disassemble, bind associated proteins, orient in magnetic or electric fields, and propagate electrical signals. These properties include electric charge (particularly surface charge), dipole moment, and dipole vector orientations (see Table 3.1).

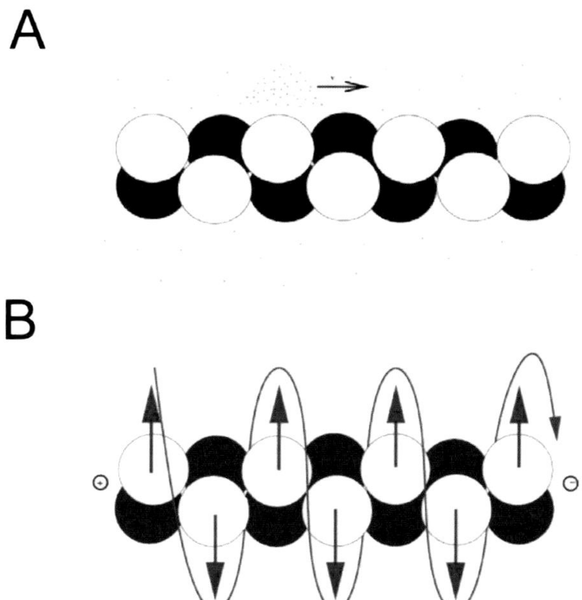

Fig. 3.5. Actin filaments support traveling ionic cloud (A), affected by the dipole of monomers (B). Adapted from [7].

Table 3.1. Electrostatic properties of α-tubulin and the tubulin dimer

Property	α-tubulin	tubulin dimer
Charge (electrons)	-5	-10
Dipole moment (Debye)		
Total dipole	566	1714
Dipole component p_x	115	337
Dipole component p_y	-554	-1669
Dipole component p_z	-6	198

Calculated from tubulin sequence [22]. For more details, see [23]-[25].

3.2.1 Structure, Surface Charge, and Electric Dipole of Tubulin

Nogales, Wolf, and Downing solved the structure of tubulin from an electron crystallographic image initially at a 3.7 Å resolution, as illustrated in Figure 3.6 [22]. Tubulin has a central β-sheet core, surrounded by multiple α-helices. The nucleotide-binding domain is located at the N-terminus near to the interface between α- and β-tubulin. Depending on whether GTP or GDP is bound to the tubulin nucleotide-binding site, a particular value of the dipole moment results for the tubulin dimer [26]. The intermediate domain of tubulin contains mixed four-strand β-sheets and three α-helices. A third domain contains two antiparallel α-helices that cross the first two domains. With the exception of poorly defined regions, such as that of the C-termini of tubulin monomers, the structure determined in [22] is a good fit to the Ramanchandran plot and is widely accepted as being accurate. Thus, there is a good understanding of the crystallized structure of tubulin except for the C-terminal region.

The main reason that the C-terminal regions of tubulin resist full structural characterization is their structural flexibility and a small size. These regions are nonetheless critical as they coincide with the binding sites between microtubules and MAPs [27, 28]. Depending on whether microtubules are bound or unbound to certain MAPs affects not only their stability (as discussed in Chapter 2), but also their electrostatic properties and their ability to propagate ionic signals.

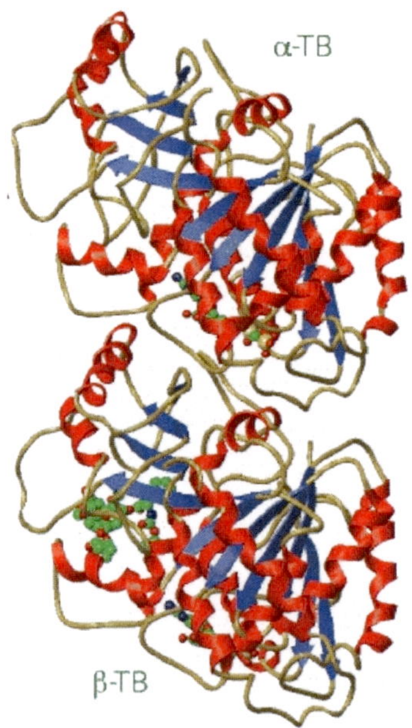

Fig. 3.6. The crystallographic structure of the tubulin dimer. Both α- and β-tubulins have β-sheet cores (blue) surrounded by α-helices (red). Reprinted with permission; Nogales et al., Nature, 1998, 391:199-203 [22].

As illustrated in Figure 3.7, the surface charge for the tubulin monomer, excluding the C-termini, is mostly electronegative; however, multiple pockets of electropositive charge exist [23]. The C-termini of tubulin, on the other hand, are almost exclusively electronegative [27]. Minoura and Muto present data they interpret as demonstrating that the C-termini of tubulin contain $\sim 50\%$ of the surface electronegative charge [29]. While this concentration of negative charge is critical to the distribution of counterions on the surface of the microtubule, Minoura and Muto used an electro-orientation method to measure the conductivity and dielectric constant of microtubules [29], and this raises some concerns that have been previously noted [30]. For example, it is not clear how the electric field used in the electroorientation method is calculated, and if this electric field can be considered constant within the viewing region. An effect of the field gradients on the electroorientation of microtubules requires future investigation.

Minoura and Muto found that when τ was calculated for different L values, it appeared to be proportional to the square of L, as is illustrated by the line of best fit given in the graph as $\tau = 4.1 \times 10^9 L^{2.0}$ [29]. When Craddock and colleagues did a power regression analysis on data from [29], they obtained a line of best fit for the relation between τ and L given by $\tau = 9.01 \times 10^8 L^{1.88}$ [30]. While the power exponent for L was within 10% with that obtained by Minoura and Muto, the prefactor was found to be only within 22% of their result [30] cf. [29]. The coefficient obtained in [30] was further inconsistent with [29] differing by 76% and yielding an intercept value comparable with the lower end values. When this intercept was constrained to be zero, then the regression analysis revealed a relationship of $\tau = 3.62 \times 10^9 L^2$ [30], which was still inconsistent with [29]. The coefficient value is of particular importance since the relation between it and the electric field is used in determining the effective polarization, conductivity and, crucially, the dielectric constant value for microtubules.

Other laboratories have obtained varying results when measuring the resistivity of microtubules (which partly depends on their state). Upon making electrical contacts to single microtubules following dry-etching with a substrate containing gold microelectrodes, 12 μm microtubules demonstrated an intrinsic resistance in the range of 500 MΩ , thereby giving a value of intrinsic resistivity of approximately 40 MΩ / μm or a conductivity of $\sigma = 38 S/m$ [31, 32]. Such studies also applied gold sputtering techniques to metalize microtubules (to a 30-nm gold layer), after which measurements on the gold-coated microtubules revealed resistance estimated to be below 50 Ω . In yet other studies, samples of tubulin, microtubules, and MAPs in buffer were assessed by RF reflectance spectroscopy and found to demonstrate DC resistances of 0.999 k Ω/m for the buffer solution, 0.424 kΩ/m for tubulin in buffer, and 0.883 kΩ/m microtubules in buffer, and finally 0.836 kΩ for microtubules and MAPs in buffer [33]. Making some assumptions about their geometrical arrangement and connectivity, such as all tubulin being polymerized and a uniform distribution of microtubules in parallel and series networks, one can estimate the resistance of a 10 μm microtubule, which forms the basic electrical element in such a circuit, to have approximately a 8 MΩ value. This estimate closely approximates an early theoretical estimate of microtubule conductivity based on the Hubbard model with electron hopping between tubulin monomers [34], which predicted the resistance of a 1-μm microtubule to be in the range of 200 kΩ (hence a 10-μm microtubule would be expected to have an intrinsic resistance of 2 MΩ). These estimates are well within the same order of magnitude as the result reported in [32] but are two orders of magnitude smaller than microtubules in dry state, which might be expected due to a large difference in the activation energy for electron hopping in a dry versus hydrated protein. Direct measurement of microtubule conductivity using micro-channels were only able to reveal an upper bound on the intrinsic conductivity of microtubules as $90S/m$ [35], which would correspond to 240 MΩ for a 10-μm microtubule, close to the result obtained for microtubules in the dry state [32].

The dipole moment of tubulin is another key determinant of its electrodynamic behavior. As shown in Table 3.1, the dipole moment of the α-tubulin monomer differs from that of the tubulin dimer due to modifications following bond interactions. The dipole moment, along with its specific x, y, and z components, affects tubulin binding not only to other tubulins, but also to MAPs, to other proteins, and to ions in the surrounding environment. Early experiments showed that high intensity electric fields (1,000 - 4,000 V/cm) perturbed tubulin indicating the presence of both a permanent and induced dipole; moreover, tubulin demonstrated a positive birefringence, which was altered upon the addition of MAPs [36].

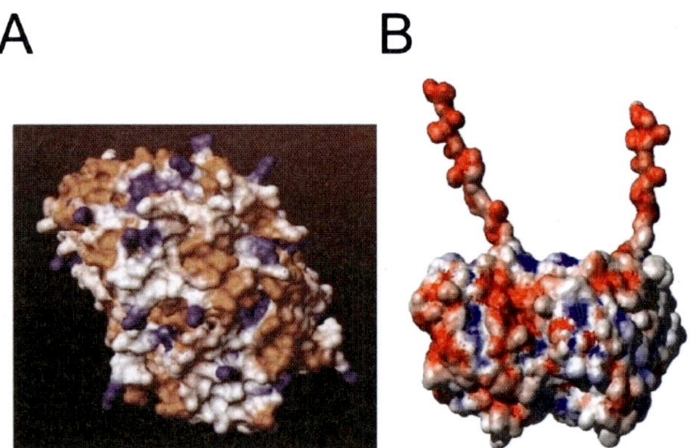

Fig. 3.7. The surface charge on tubulin is mostly electronegative (red) with a few pockets of electropositive charge. (A) The charge distribution of the tubulin monomer excluding the C-termini, as adapted from [23]. (B) The charge distribution of the tubulin monomer including the C-termini, as adapted from [27].

3.2.2 Distinct Tubulin Isoforms Differ in Their Biophysical Characteristics

The neuron, like most mammalian cells, contains $\alpha\beta$-tubulin dimers composed of different isotypes of α- and β-tubulins [37]. There is a high diversity of β-tubulin isotypes, and many of these isotypes are preferentially localized to the brain [38]. To date, there are at least ten major identified β-tubulin isotypes I, IIa, IIb, III, IVa, IVb, V, VI, VII, and VIII. These isotypes have high degrees of sequence homology, with most of the sequence differences being located in the C-termini, the region where MAPs bind [39]. These C-termini range from 9 to 25 residues in length, and because of multiple glutamate residues, contain most of the tubulin molecule's electrostatic surface charge. Table 3.2 illustrates selected C-termini sequences, some of which contain the full MAP2 binding site (EEAEEE), or the MAP2 binding motif (EEEV). These different sequences confer varying degrees of microtubule flexibility, which would be expected to affect their binding and function.

Depending on the amino acid sequence, various isotypes of tubulin exhibit different net electric charge, dipole moment, and dipole vector orientation [40]. Different tubulin isotypes will, therefore, exhibit differences in the dipole-dipole interaction energies, which affect the overall stability of any bonds formed within a microtubule lattice [41]. It is important to recall that the strength of the dipole-dipole interaction is proportional to the square of the dipole moment. Thus, tubulin isotypes having weaker dipole moments produce microtubules that are more stable and vice versa since the dipole-dipole interaction contributes a positive amount of energy to the tubulin-tubulin bond. A precision balance between stability and instability in microtubules is crucial to higher neuronal functions, especially those involving structural reorganization as will be discussed later in this chapter.

Table 3.2. β-Tubulin C-termini MAP2 binding sites, amino acid residues, and charge.

Isotype	MAP2 binding sequence or motif	Amino acids	Charge (e-)
I	EEAEEE	18	-12
II		19	-12
III		24	-12
IVa	EEAEEEV	18	-11
IVb	EEAEEEV	19	-12
V		20	-12
VI		25	-11
VII		9	-3
VIII	EEEV	18	-12

Adapted from [38].

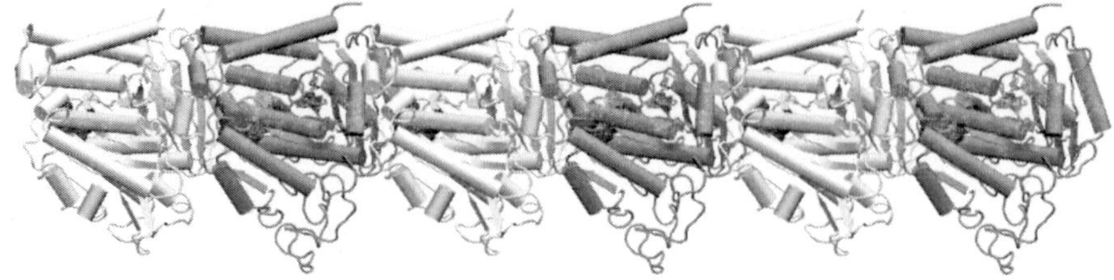

Fig. 3.8. A microtubule protofilament containing $\alpha\beta$-tubulin dimers [42].

Table 3.3. Elastic properties of microtubules.

Young's modulus	~2 GPa ($10^{-9}Nm^2$)
Shear modulus	low
Resilience	high
Flexural rigidity	$10^{-24}Nm^2$
Possion's ratio	0.3
Persistence length	$10^{-3}m$

Adapted from [42, 46].

3.2.3 Microtubules: Lattice Structure, Elastic Properties, Surface Charge, and Electric Dipole

A typical neuronal microtubule is composed of 13 protofilaments that have alternating α- and β-tubulin monomers as shown in Figure 3.8. In the intact microtubule, each tubulin monomer is bound to six other tubulins, and whether or not there is a seam determines either the A or B lattice structure [42, 43]. The bonds between tubulins are responsible for determining certain properties of intact microtubules, and the assembly of a microtubule follows a slightly different course depending on the availability of cations and other conditions [44]. According to cryo-electron microscopic analyses, the interdimer interface is configured to accommodate the tendency of α-tubulin to adopt a straight conformation and β-tubulin to adopt a curved conformation [45]. This is likely due to GTP hydrolysis (to GDP) only occurring for β-tubulin at the exchangeable site. The GTP attached to α-tubulin remains intact. The resultant effect is that, depending on the strength of lateral bonds, microtubule protofilaments will remain relatively straight or will curve, the latter of which favors disassembly. Longitudinal bonds between tubulins are stronger than lateral bonds between protofilaments by ~ 7 kcal/mole [42, 43]. Nonetheless, since tubulins are staggered in neighboring protofilaments, lateral contacts outnumber longitudinal contacts and can, under appropriate conditions, contribute greatly to lattice stability.

As given in Table 3.3, microtubules possess anisotropic elastic properties [42, 47]. These elastic properties contribute to the molecular forces governing longitudinal compression, lateral force, and shearing force [42]. Longitudinal bonds along a single protofilament are much stronger than lateral bonds linking protofilaments together; nonetheless, assembled microtubules exhibit exceptional strength. The stability and resilience of microtubules facilitates many of their biological roles, such as establishing and maintaining cell shape and mediating transport.

Microtubules are mostly electronegative on their outer surfaces, and this negative charge is particularly concentrated on the ridges of the protofilaments [48]. Positive surface is preferentially buried in the underlying regions between protofilaments. This longitudinal arrangement of charge is well suited to establish a sizable linear charge density. Minoura and Muto measured the linear charge density along microtubule protofilaments at 2.5 e/nm [29], as discussed earlier.

From a conductive point of view, microtubules can be modeled as two-dimensional lattices of tubulin dimers, each of which being represented as a resistor [46]. When the boundary conditions along the length of the microtubule are set as a very large torus, each tubulin dimer has six nearest neighbors, and can be configured as part of a large microtubule resistor network. Making certain assumptions about the network (such as all the tubulins being polymerized and a uniform distribution of microtubules), the resistance of a 10 micron-long microtubule (the basic electrical element in this network) has an estimated value of approximately 8MΩ as mentioned in the previous section.

3.2.4 Microtubules: Ferroelectric and Pyroelectric Properties

Assembled microtubules are ferroelectric, meaning they exhibit spontaneous dipole moments, the directions of which can be switched by the application of an external electric field [49]-[52]. This unique property further contributes to their conductive properties and has led to a number of theories regarding the biophysical basis for signaling in microtubules. The first consideration is how dipoles can be arranged in a microtubule lattice. There are three possible arrangements: (1) random, (2) strongly ferroelectric or parallel, and (3) weakly ferroelectric [49]. The weakly ferroelectric arrangement is the most interesting, since dipole-dipole interactions at the couplings of tubulin to its six nearest neighbors will invariably lead to conflicts or “frustrations”. Resolution of these conflicting dipole-dipole interactions leads to “kink-like” excitations that propagate down the microtubule [49]. This can also be termed a “solitary wave”, defined as traveling solitons or defects. The kink-like excitation is the defect, insofar as it represents a transient switch in the dipole moment of the tubulin monomer as it passes along the microtubule.

Another component of the kink-like excitation or traveling solitary wave relies on elastic coupling of the traveling wave and the energy of GTP hydrolysis calculated at $6.25 \times 10^{-20} J$ [50]. Accordingly, the solitary waves carry the free portion of this energy of GTP hydrolysis. Somewhat similarly, a pseudo-spin model posits GTP hydrolysis as a critical factor in determining the dipole state of the tubulin dimer [51]. There exists a double-well electrostatic potential within the tubulin dimer as experienced by a mobile electron that can be localized either to the α- or β-tubulin dimer [51, 52]. In the pseudo-spin model, the state change of this mobile electron is coupled to GTP hydrolysis. Yet another somewhat related model proposes dipole "flip waves" travel along microtubules as tubulins alternate between GTP and GDP states [26].

It has also been proposed that microtubules possess pyroelectric properties (i.e., the ability to generate an intrinsic electric field due to temperature changes in the medium [53]). The intrinsic electric field produced by microtubules may further regulate microtubule functions, such as transport and neural plasticity.

3.2.5 Conductance of Electrical Signals Along Microtubules

Both experimental and theoretical approaches have been used to study electrical signaling along microtubules. For experimental investigations, the dual patch-clamp set up was used, as described earlier in this chapter for use in measuring electrical conductance along actin filaments (see Figure 3.4). In the course of such experiments, electrical data collected from taxol-stabilized microtubules demonstrated that microtubules behave as biomolecular transistors when responding to brief pulses of electric potential ranging $\pm 200mV$ [54]. As illustrated in Figure 3.9, isolated microtubules amplified the applied electric current. The conclusion being microtubules exhibit a capability for ionic-based propagation that is likely to involve the condensed positive counterion cloud distributed along the length of the microtubule ($20e-$ per tubulin monomer). This cloud lays atop the negative surface charge of the microtubule. The basic mechanism of this effect is likely to be similar to the one discussed above for actin filaments with some specific differences such that microtubules are cylindrical, have a different surface charge density than actin and possess protruding C-termini structures which are lacking in actin.

Further measurements and calculations intended to model this phenomenon showed that microtubules support nonlinear wave propagation, much as would be expected for a biological transistor [54]. Following an electrical signal delivered to one end of a microtubule by a micropipette, the signal that arrived at the other end of the microtubule was more than twice as high as the signal recorded in a control experiment where identical pipettes were immersed in a solution without connecting microtubules. Conductivity along the longitudinal axis of microtubules was calculated to be on the order of $10nS$, a relatively high level of ionic conductivity. Certain cations clearly participate in the positive counterion cloud that serves as a basis for this conductivity. Bathing microtubules in solutions containing Ca^{2+} with a concentration ranging between 10^{-7} - 10^{-2} M increased the ability of microtubules to amplify electrical signals [55]. These data support the hypothesis that Ca^{2+} ions adsorbed to the surface of microtubules modulate the electrical properties of microtubules and contribute to a novel mode of cell signaling.

The conductive nature of microtubules has also been demonstrated using an electroorientation approach [29], as mentioned earlier. Conductance measured for intact microtubules was $157 \pm 7mS/m$, and for microtubules treated with subtilisin (i.e. lacking C-termini), $96 \pm 6mS/m$. These authors argue that counterions on the surface (many of which interact with the negatively charged C-termini) are responsible for the observed conductance. Their experimental results support this argument, as subtilism removes C-termini from microtubules.

To address electrical signaling from a theoretical perspective, Priel and Tuszyński modeled ionic conduction along microtubules as a nonlinear electrical circuit (see Figure 3.9a) [56]. According to this mathematical model, the microtubule cylinder core is separated from the rest of the ions in the extracellular fluid by the counterion condensation cloud. This would be expected to act as a dielectric medium, providing both resistive and capacitive components. Ion flow at a given radial distance from the center of the cylinder is approximately equal to the Bjerrum length. There is an inductive component to the electrical properties of ionic waves, because the helical structure of the microtubule induces a helical ionic flow resembling a solenoid. Each electrical component in this model was based, in part, on an earlier model for actin filaments [57].

Following a series of equations describing the potential difference, capacitance, and resistance of microtubules [27, 56], a nonlinear 3rd order differential equation can be integrated to give:

$$\begin{aligned} R_2 C_0 v a^2 V'' &+ (L C_0 v^2 - a^2) V' \\ &- v R_1 C_0 V + R_1 C_0 b v V^2 \\ &= V_0 (ka)^2 \cos(k\xi)/k + \alpha\xi + \text{const.} \end{aligned}$$

Here, the symbols used denote respectively: R_1 and R_2 resistive elements along and perpendicular to the tubulin dimer, C_0 a corresponding nonlinear capacitance of a dimer, L its inductance and $V(x)$ is the local electrostatic potential along the MT axis x.

This mathematically derived model of ionic conduction along microtubules supports the observations of amplified ionic signal propagation along microtubules and the possible existence of unstable solitary wave solutions [56] of the above equation.

Certain surface features of microtubules are likely to influence ionic wave conduction. It is probable that the strongly charged C-termini of tubulins interact with the ionic waves along microtubules [58]. Given their elastic properties and elastic degrees of freedom, multiple conformational states of C-termini are likely to occur *in situ*. The negatively charged C-termini are capable of interacting with (a) the surface of tubulin, (b) adjacent C-termini, or

(c) adjacent MAPs. The main two states for the C-termini are up and down. In the downward state, C-termini are attracted to pockets of positive charge located on the mostly negative surfaces of microtubules. Computer simulations based on molecular dynamics revealed the possibility of ionic waves being coupled to C-termini states changes. The probability of C-termini being in the downward state has been estimated at 15%, suggesting that up (away from the MT surface) is the preferred direction. As was previously suggested, multiple electronic and dipolar states of tubulin could be coupled to the conformations of their C-termini and to the counterion clouds that surround them, enabling a kind of nanoscale computational device [58].

Another consideration relevant to the propagation of ionic conductances along microtubules is the ability of ions and water to penetrate into the microtubule [59]. Although the open ends of microtubules are partly responsible for the penetration of ions and water into the microtubule inner core, nanopores also exist between tubulins assembled into microtubules that sustain ionic influxes and effluxes. There is no clear experimental or theoretical picture elucidating the role in ionic conductivity of these nanopores in the surface of microtubules but we expect them to be of major importance just like ion channels are in cellular membranes.

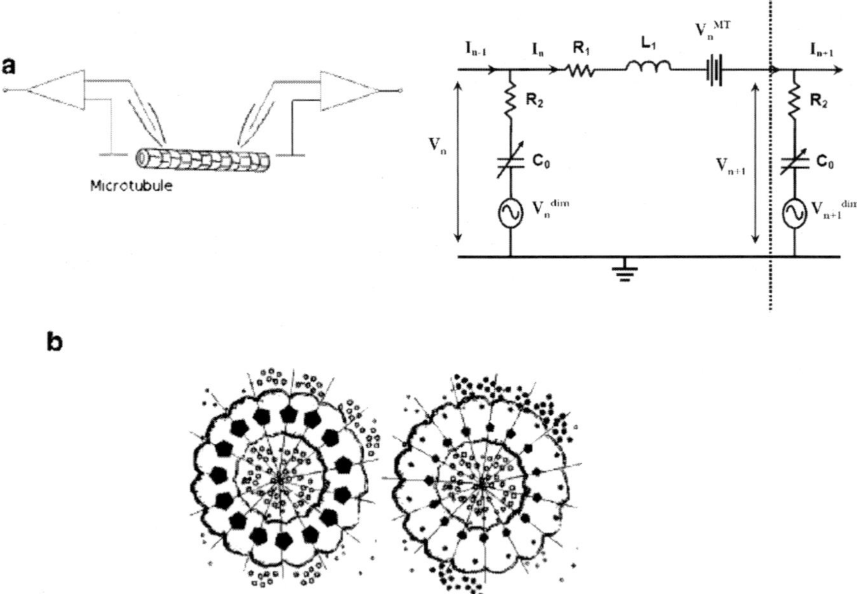

Fig. 3.9. The demonstration of current amplification by a microtubule as shown in [54]. a) Current is applied to one end of the microtubule and collected at the other end (electric model depicted at right). b) Electrostatic distribution of counterions inside and outside the microtubule responsible for propagation of the electrical signal as adapted from [48].

The notion of microtubules conducting electrical signals is a starkly novel concept. As discussed in Chapters 1 and 2, theoretical models of higher neural functions formulated thus far have focused almost exclusively on activity in neural circuits, a type of activity known to depend on ionic conductances mediated at the neuronal membrane. It is not yet clear how to reconcile these models with recent data supporting the conductance of electrical signal along actin filaments and microtubules. Models of higher cognitive function that involve propagation and subsequent processing of information within networks of microtubules and actin filaments have significant potential heuristic value, and these models are expected to have relevance for other macromolecule systems in neurons. Moreover, it is highly likely that these novel modes of intraneuronal conductance (and intraneuronal communication) are fundamentally enmeshed with neural membrane activities and neural activity correlates of higher cognitive functions.

3.3 Linking the Excitable Neuronal Membrane with the Cytoskeleton: Functional Implications

In light of these recent findings that microtubules are capable of transmitting electrical signals from one part of the neuron to another, the input sources of information need to be identified. When it comes to processing sensory-based information into higher cognitive states, the most significant and relevant source of input is from the neuronal membrane, in particular the neuronal membrane containing postsynaptic densities or sites receiving contacts from other neurons.

The linkage between the neuronal membrane and the underlying cytoskeleton is only beginning to become better understood. Using micropipettes to tug at the outer membrane of endothelial cells, Maniotis and colleagues generated a completely unanticipated result when they observed physical distortions in the cells' nuclei [60]. Their result indicated a complex cytoskeletal link spanning membrane-embedded receptors all the way to the nuclear membrane. These kinds of results imply cells may rely on the cytoskeleton to transfer mechanical, electrical and biochemical information throughout the cell.

3.3.1 Actin Filaments Connect the Neuronal Membrane with the Microtubule Matrix

As described earlier in this chapter and in Chapter 2, actin filaments are highly concentrated just below the membrane, in dendritic spines, and in the axon terminal. Referring back to Figure 2.3, actin filaments often bind to membrane scaffolding proteins and interact with signal transduction molecules. There are also recently documented interactions between ion channels and the cytoskeleton, in particular actin filaments [57]. Adding actin to patch-clamped membranes resulted in activation of voltage-gated K^+ channels.

When actin filaments were disrupted with the destabilizer, cytochalasin D, this also resulted in the activation of voltage-gated K^+ currents, an effect that was blocked by phalloidin, an actin stabilizer. These results, taken together, suggest that the cytoskeleton, particularly the actin filaments, regulate ion channels fundamental to neuronal activity.

In order for actin filaments to conduct an electrical signal to the microtubule matrix inside the neuron (or otherwise physically perturb by way of transmitting vibration or mechanical force), there must be a connection between these two cytoskeletal systems. There are at least three potential ways actin filaments and microtubules could interface: (1) through direct physical contact, (2) via cross-linking proteins, and (3) indirectly though signal transduction cascades.

Actin filaments are often observed in direct association with microtubules. Dual-wavelength fluorescent speckle microscopy analyses demonstrated that microtubules frequently migrate in conjunction with actin bundles [61]. Actin filaments interact with microtubules as part of neuronal migration, growth cone development, and neuronal receptor and ion channel transport [62]. While the underlying factors governing microtubule/ actin filament interaction still need to be completely elucidated, a recent surge of discoveries reveals a large number of cross-linker proteins.

Table 3.4. Linkers of actin filaments with microtubules.

Cross-linker protein	References
CLIP-115	[62]
CLIP-170	[62, 63]
CLASP1	[62, 64]
CLASP2	[62, 64]
Lis1	[62, 65]
EB family	[62, 66]
MAP2c	[67, 68]
Tau	[68]

As listed in Table 3.4, there are a number of cross-linker proteins that bind microtubules to actin filaments. Many of these are microtubule plus-end tracking proteins (+TIPs) [62]. Cytoplasmic linker proteins, CLIP-115 and CLIP-170, bind plus-ends of microtubules and are responsible for linking microtubules to cargo or to actin filaments though scaffolding protein intermediaries [63]. CLIP-associated proteins, CLASP1 and CLASP2, represent two additional +TIPs. CLASP2α exhibits an actual binding site for actin [64]. Lis1 appears to tether actin filament to microtubules via interactions with scaffolding proteins [65]. Similarly, the ERK-binding (EB) family of +TIPs, which include EB1 and EB2, rely on their interactions with other proteins (including CLASP1/2, CLIP-170) to bind microtubules to actin filaments [66]. Lastly, there are the microtubule-associated proteins MAP2 (especially MAP2c) and tau, which have long been to known to bind actin filaments, albeit at low affinity [67, 68]. signal transduction molecules, such as calmodulin, as well as Ca^{2+} ions and phosphorylation events further modulate the ability of MAP2 and tau to bind actin filaments and microtubules.

3.3.2 Does the Intracellular Cytoskeletal Matrix Compute and Determine Cell Structure and Function?

The neuronal cytoskeleton is largely responsible for determining structure in the context of (1) neural development, (2) neural regeneration, and (3) neural plasticity. Invariably, these processes involve the interplay between actin filaments and microtubules orchestrated by membrane interactions, binding proteins, and signal transduction molecules. It remains to be determined whether the propagation of signals, such as the transmittance of electrical pulses described in previous sections of this chapter, is fundamental to these processes.

Neural Development

Soon after its last mitotic division, the neuron begins to differentiate into its mature phenotype. The size and shape of the neuronal cell body, along with the organization of dendrites and of the axon, is what determines its structure and as a direct consequence, its function – a function usually characterized by a signature physiological response pattern (e.g., a bursting versus non-bursting type neuron, etc.). Initially, a neuron generates multiple neurites, thin protrusions that will elongate before committing to become either a dendrite or an axon. Differentiated neurons typically have one axon and multiple dendrites; however, there are exceptions. Current opinion in neurobiology emphasizes the effect of the environment, suggesting that the inputs a neuron receives or the externally derived neurotropic factors largely, if not exclusively, determine the neuron's fine structure during development [69]. Intrinsic mechanisms deriving from within the neuron, in particular from internal matrices of cytoskeletal proteins are largely overlooked. Recent evidence, however, suggests that the internal organization of microtubules may dictate future neuronal structure. It has been discovered that the position of the centrosome (which acts as a microtubule organizing center up until the time the neuron stops dividing) determines the position of the axon, or more precisely, the neurite closest to the centrosome becomes the axon of the mature neuron [70]. One of the reasons this result is important is that it hints at intrinsic cytoskeletal mechanisms governing neurite differentiation [71].

Neurite formation occurs at the earliest stages of development. Microtubules first bundle together at the cellular membrane creating an outward force that is opposed by an inward counter-force provided by actin filaments [72]. Thus, there is in principle an opportunity for electrical signals between microtubules and actin filaments to participate in early cell differentiation.

Electrical signals could be transmitted via cross-linking proteins like MAP2. In fact, MAP2c acting in concert with the microtubule motor, dynein, does facilitate the bundling and movement of microtubules outward [72]. In the typical neuron, dendrites possess not only multiple MAP2 isotypes, but also receptors like the GluR1 receptor that binds glutamate. Some signals, be they molecular, electric, or both, regulate the trafficking of these proteins to selectively reach either dendrites or axons.

Previous studies have further shown that when the axon of a neuron is cut, one of the dendrites of that neuron is capable of converting into a "substitute" axon; this is demonstrated experimentally by the converted axon losing its MAP2 and GluR1 [73]. Depletion of actin leads to dendrites converting into axons [73], further suggesting that cytoskeletal interactions are involved. Moreover, when dendrites are converted into axons the polarity of microtubules changes, more precisely, there is a loss of distally directed minus-end microtubules [74]. While other explanations are possible, a valid proposal is that actin filaments propagate some signal, conceivably an electrical signal, which is required to maintain dendrite integrity. By such still incompletely understood mechanisms, interactions between actin filaments and microtubules play fundamental roles in determining basic neuronal structures, such as whether a neurite develops into a dendrite or an axon.

Microtubules and actin filaments also interact during the migration of developing growth cones, a process that involves +TIPs, which tend to aggregate at the end of the axon shaft at the exact point where actin filaments abruptly reach high concentrations. In addition to +TIPs, the neurotrophin, NGF, and the signal transduction molecule, GSK-3β, mediate the fine balance between stable and unstable microtubules [75]. The dynamic relationships between actin filaments and microtubules during growth cone migration may well be mediated, at least in part, by electrical signals transmitted between these matrices. This suggestion has received some experimental support. Applied electric fields typical of what would be present in developing or regenerating axons were found to guide growth cones of Xenopus spinal neurons towards the positive current source, a result that depended on intact actin filaments and microtubules [76].

Neural Regeneration

Another situation in which actin filaments and microtubules interact to dramatically alter structure is during neural regeneration. Fully developed axons show marked structural changes when damaged; however, reorganization is usually possible following injury to peripheral axons, but not often following central axons [77]. Regeneration is mediated by molecular events triggered by the injury. Within 10-30s following peripheral nerve damage, researchers measured a massive increase in Ca^{2+} in the injured part of the axons, which subsided within 1 min [78]. This short surge of Ca^{2+} was then followed by an increase in proteolytic activity that continued for at least an hour. Proteases were responsible for degrading cytoskeletal proteins; and following degradation, there was an expected increase in reorganization of cytoskeletal protein [79]. In addition to molecular triggers, some results can be interpreted as being consistent with a role for electrical signaling in regeneration. Reports have been mixed regarding whether electric currents or the electric fields they produce can enhance nerve regeneration in mammalian CNS neurons, whereas such an effect is undisputed for Xenopus spinal neurons. Enhanced regeneration was not observed in rat or human spinal cord according to some recent reports [80, 81]; however, whether the electric field is applied in parallel versus perpendicular to the nerve may be critical [82]. A relationship between nerve growth in an electric field and electrical signal conductance by cytoskeletal proteins, to the extent that such a relationship exists, has yet to be measured experimentally.

Neural Plasticity

Dendrites in adult animals also show characteristic structural changes, and these changes involve both microtubules and actin filaments [83]. Proposed molecular mediators of these structural changes include the neurotrophin BDNF, the NMDA receptor, and the receptors and signal transduction cascades subsequently activated. Several situations produce dendritic changes, including: (1) enriched as opposed to impoverished environments (e.g., group versus isolation effects), (2) motor learning, and (3) sensory reorganization [83]. Some of the earliest literature on dendritic plasticity tackled this problem using enriched environment paradigms. In a series of classic studies by William Greenough and colleagues, basilar dendrites of pyramidal cells found in occipital, frontal, and temporal cortex were increased in animals exposed to complex environments as compared to those in animals reared in isolation [84, 85]. Recent studies have replicated these early results and have additionally observed increases in spine for animals reared in complex environments [86]. Thus, an elaboration of both microtubules and actin filaments would appear to underlie a neuron's response to the increased level of information processing likely to accompany a complex environment. Dendrite alterations that occur in response to enriched environments (and correlate with learning and memory) could be mediated, at least in part, by electrical signaling within the actin filament and microtubule matrices.

Nonlinear Dynamic Behavior Exhibited During Actin Filament Polymerization

Key biophysical properties of actin filaments and microtubules contribute to their ability to regulate neural growth, nerve regeneration, and neural plasticity. Actin filaments exhibit nonlinear dynamic behaviors during polymerization – a process necessary for neural growth and plasticity. Polymerization, initiated by an unfavorable nucleation process, is followed by rapid elongation.

Actin polymerization fits a previously proposed five-step model described as follows [87]:

$$\mathrm{A} + \mathrm{A} \underset{k_{-1}}{\overset{k_{+1}}{\rightleftharpoons}} \mathrm{A}_2 \qquad k_{+1} = 10\mu\mathrm{M}^{-1}s^{-1}, \quad k_{-1} = 10^6 s^{-1}$$

$$\mathrm{A} + \mathrm{A}_2 \underset{k_{-2}}{\overset{k_{+2}}{\rightleftharpoons}} \mathrm{A}_3 \qquad k_{+2} = 10\mu\mathrm{M}^{-1}s^{-1}, \quad k_{-2} = 10^3 s^{-1}$$

$$\mathrm{A} + \mathrm{A}_3 \underset{k_{-3}}{\overset{k_{+3}}{\rightleftharpoons}} \mathrm{A}_4 \qquad k_{+3} = 10\mu\mathrm{M}^{-1}s^{-1}, \quad k_{-3} = 10 s^{-1}$$

$$\mathrm{A} + \mathrm{A}_4 \underset{k_{-4}}{\overset{k_{+4}}{\rightleftharpoons}} \mathrm{N} \qquad k_{+4} = 10\mu\mathrm{M}^{-1}s^{-1}, \quad k_{-4} = 0 s^{-1}$$

$$\mathrm{A} + \mathrm{N} \underset{k_{-}}{\overset{k_{+}}{\rightleftharpoons}} \mathrm{N} \qquad k_{+} = 10\mu\mathrm{M}^{-1}s^{-1}, \quad k_{-} = 1 s^{-1}$$

where A, A_i, and N represent the concentrations of actin monomers, polymers composed of i actin monomers, and filaments, respectively, and the rate constants for the corresponding reactions derive from [88].

Even though nucleation of new F-actin filaments is required for determining and maintaining neuronal shape during development, regeneration, or plastic rearrangement, actin filament formation can progress following a *de novo* nucleation or as a catalytic nucleation from pre-existing filaments. Filaments resulting from catalytic nucleation are initiated by the fragmentation of pre-existing filaments or by filament branching in the presence of the Arp2/3 protein complex. During actin polymerization, there is capping of the branched ends by gelsolin. All these processes induce strong nonlinearities in their kinetic description.

Assuming that the nucleation core of F-actin is a trimer, and nucleation is G-actin dependent, *de novo* nucleation has been mathematically modeled in [87] as follows:

$$3\mathrm{G} \underset{\nu_-}{\overset{\nu_+}{\rightleftharpoons}} \mathrm{F} \tag{3.1}$$

where G is a G-actin monomer and F is a free F-actin filament.

The reverse of actin filament polymerization, debranching, is modeled as:

$$\mathrm{F_N} \overset{\delta}{\rightleftharpoons} \mathrm{F} + \mathrm{A} \tag{3.2}$$

where the turnover between ATP G-actin and ADP G-actin is assumed to be instantaneous. It is possible to use the proposed model to describe the self-aggregation of F-actin filaments into dendritic networks.

Assuming constant concentrations of inactive Arp2/3 complex, A, and WASP protein concentration, W, and that there are 4 state variables (G, Y, F, and F_N) and 7 parameters (ν_+, ν_-, α, β, π_+, π_-, and δ), the corresponding kinetic equations governing the process can be written as

$$\dot{G} = \underbrace{3\nu_- F}_{Elimination} - \underbrace{3\nu_+ G^3}_{Nucleation} \underbrace{- \pi_+ G(F+F_N) +}_{Polymerization} \underbrace{\pi_-(F+F_N)}_{Depolymerization} - \underbrace{\alpha G}_{Activation}$$

$$\dot{Y} = \underbrace{\alpha G}_{Activation} \underbrace{- \beta Y(F+F_N)}_{Branching}$$

$$\dot{F} = \underbrace{\nu_+ G^3}_{Nucleation} - \underbrace{\nu_- F}_{Elimination} + \underbrace{\delta F_N}_{Debranching} - \underbrace{\beta FY}_{Branching}$$

$$\dot{F_N} = \underbrace{2\beta FY}_{Branching} + \underbrace{\beta F_N Y}_{Branching} - \underbrace{\delta F_N}_{Debranching}$$

where ν_+ is the rate constant of filament nucleation and ν_- is the rate constant of nucleation core elimination. Rate constants π_+ and π_- represent the assembly and disassembly of a G-actin monomer with an F-actin filament, respectively. Rate constant α represents the Arp2/3 complex activation by WASP proteins, rate constant β, filament branching, and rate constant δ the dissociation of a filament from the network (i.e., the rate of debranching).

These kinds of mathematical models enable quantitative predictions to be made regarding the precise behavior of actin filaments during neural growth, nerve regeneration, and neural plasticity. Mathematical models can also be applied to microtubule polymerization to better describe that subcellular phenomenon in similar contexts.

3.3.3 Information Storage in the Intracellular Cytoskeletal Matrix: A Role in Memory

There is increasing evidence that many forms of learning and memory involve alterations in the neuronal cytoskeleton. Learning is the process by which an organism (in particular its nervous system) better adapts to its environment. Memory, on the other hand, is an entity – often referred to as the "memory trace". It is the something that is remembered or stored in neurons or in the connections between neurons. Biological memory may operate somewhat similarly to content-addressable computer memory [89, 90]; however, biological memory is notably more robust as is capable of retrieval based on incomplete input.

Since learning is the process that leads to memory storage, researchers typically employ a learning task and subsequently test for improved performance on that task. That improvement is then "operationally defined" as memory, usually a specific type of memory, which is related (1) to the task at hand (e.g., spatial memory for navigation tasks), (2) to the time intervening between training and testing (e.g., short-term versus long-term memory), or (3) to the demand characteristics of the task (e.g., cued versus non-cued recall memory).

Given that the memory trace is hypothesized to be a physical substrate, a feasible neural correlate for memory is the reorganization of microtubules and actin filaments in neurons, since these protein structures fundamentally underlie the spatial distribution of synapses and synaptic strength. Altered cytoskeletal arrangements can in principle lead to modifications both of the inputs to a given neuron and the postsynaptic receiving apparatus. Due to neural plasticity mediated by cytoskeletal rearrangements, the overall number of inputs might increase or decrease, or the distribution of inputs might become skewed to favor activation of certain dendrites over others. The postsynaptic membrane, its receptors, scaffolding and adaptor proteins, in conjunction with cytoskeletal protein rearrangements, can be altered as a result of new learning. An even more profound notion is that some biophysical parameter of the neuronal cytoskeleton permanently encodes information in a durable form. A number of experimental studies support these structural-functional relationships or are consistent with the more hypothetical suggestions noted here.

Cytoskeletal Protein Correlates of Learning and Memory

Empirical evidence supporting the idea that microtubules are reorganized with learning and memory is found in the 3-fold increase in microtubules following passive avoidance training [91]. There are also numerous studies showing that the microtubule toxin, colchicine, impairs performance on several learning paradigms, including the Morris water maze, radial-arm maze, aversive conditioning, and operant conditioning [92, 93, 94, 95]. Since colchicine binds to tubulin dimmers and blocks the polymerization of microtubules, these studies suggest that learning relies on dynamic microtubules – those capable of shortening and lengthening and of interacting with other microtubules and actin filaments.

Gene expression and protein binding analyses also support the proposal that cytoskeletal proteins figure prominently in the consolidation (i.e., the conversion) of short-term memory into long-term memory [96, 97]. Actin, tubulin, and F-actin capping protein were among 16 brain proteins shown to increase their binding to other proteins during memory consolidation [97]. Although intrinsically dynamic in nature, cytoskeletal proteins appear to rely on modulation by MAPs and growth-associated proteins, such as GAP-43, stathmin, and SCG10 [98]. Not unexpectedly, MAP2 expression is perturbed and stathmin binding increased with memory consolidation [96, 97].

Correlates of Learning and Memory Are Localized to Dendrites and Dendritic Spines

Multiple studies evaluating different types of learning and memory have demonstrated the dendrite-specific binding protein MAP2 to be particularly important in memory consolidation. Social isolation, which results in recognition memory deficits, decreased levels of α-tubulin and MAP2 in the hippocampus [99]. Likewise, cerebral hypoperfusion that decreased MAP2, GAP-43, and synaptophysin correlated with impaired performance on the Morris water maze [100]. Also, the senescence-accelerated (SAMP10) mouse strain, which is known to exhibit learning and memory deficits, was shown to express less cortical MAP2 and fewer apical dendrites [101]. Finally, fear conditioning to tone or to the training context (i.e., the training chamber) was shown to induce marked changes in MAP2 immunohistochemical staining in precise regions of the cerebral cortex or hippocampus, with the region being related to the type of training [79, 102, 103]. For example, auditory cortex, which is where information about different tones is stored, showed evidence of MAP2 breakdown (i.e., proteolysis) when animals were trained to a tone signal preceding floor grid shock stimuli, or as illustrated in Figure 3.10, enhanced MAP2 breakdown was noted in pyramidal cells of the hippocampus with contextual learning. Western blots revealed an increase in breakdown products with an approximate molecular weight of $\sim$90 kD in similarly trained animals, indicating that enhanced MAP2-immunostaining was caused by heightened immunoreactivity of MAP2 breakdown products, due to an increased accessibility of antigenic binding sites. As is apparent in Figure 3.10, this breakdown of MAP2 was noted throughout the cell body and apical dendrites of selectively stained neurons. Approximately 15% of cortical and hippocampal neurons contained elevated levels of MAP2 and many of these were the large pyramidal cell type [104]. Similar learning-related changes in MAP2 followed avoidance training, and were accompanied by altered staining patterns for muscarinic receptor and protein kinase C (PKC) [105].

There are multiple interpretations as to why MAP2 is broken down with memory consolidation. The simplest explanation is that the existing microtubule matrix must be broken down before a new structural matrix can be built. Another possibility is that the cytoskeleton may be essential for certain signal transduction molecules to function properly. Transgenic mice, in which the N-terminus of MAP2 was truncated, showed marked impairments on contextual memory, as well as deficits in binding cAMP-dependent kinase (PKA) and reduced capacity for phosphorylation of MAP2 [106]. An intact cytoskeletal matrix, sustaining proper interactions between MAPs, microtubules, actin filaments, and signal transduction molecules, appears to be critical to the learning process.

Tau also participates in learning and memory, but in a different way from that of MAP2. Studies to date suggest that adding excess tau by way of a transgene results in a measurable impairment in olfactory learning and spatial reference memory [107, 108]. It is hypothesized that excess tau overly increases the stability of microtubules and that this is a deterrent to normal microtubule dynamics ordinarily at play in learning [107]. This interpretation is consistent with recent reports of increased tau in P301 transgenic mice having improved memory before tau has a chance to aggregate, the latter state being invariably

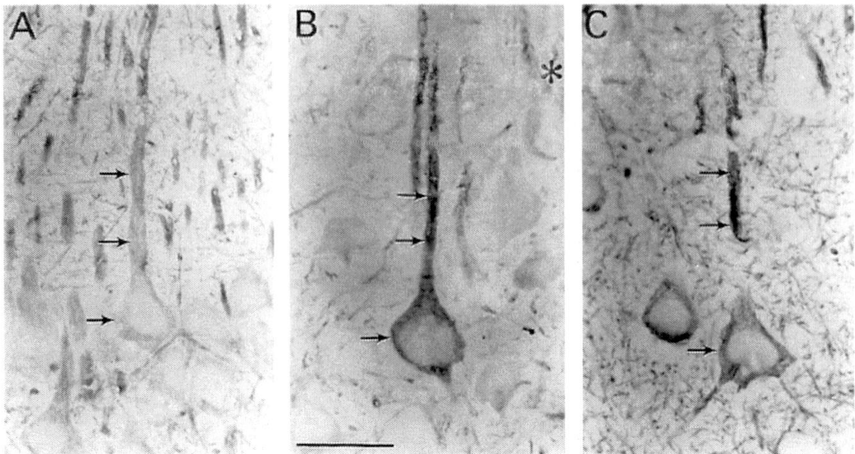

Fig. 3.10. MAP2 immunohistochemistry in large pyramidal cells in the hippocampus of control (A), tone-contextual trained (B), and contextual trained (C) rats. MAP2 immunostaining is increased due to increase in breakdown products, which have more antigenic binding sites. From [104].

associated with impaired memory [109]. A delicate balance between stability and instability for microtubules (as well as for actin filaments) is undoubtedly essential to normal learning.

Actin filament plasticity is another correlate of learning, occurring preferentially in spines. Recently, it has been shown that extinction to contextual fear (i.e., learning that reduces an existing fear response) depends on actin filament rearrangements in spines [110]. Actin-rich spines also show distinct types of morphological change following long-term potentiation (LTP) – an experimental paradigm that is often recorded from hippocampal tissue slices but can be observed in behaving animals [111]. LTP, which can be divided into an initial, intermediate, and late phase, is a widely accepted neural correlate of learning. During the intermediate phase of LTP there is an increase in the local concentration of mRNA for MAP2 and for the Ca^{2+}-calmodulin-dependent kinase II (CaMK II) [112]. CaMK II subsequently is responsible for phosphorylating MAP2 and the AMPA receptor, leading to enhanced synaptic response [113]. There is also evidence that increased MAP1B phosphorylation occurs within minutes of the induction of LTP [114].

There is a critical window lasting approximately 10 minutes after learning during which CaMK II dissociates from actin filaments and translocates to the postsynaptic density whereupon it tethers AMPA receptors to NR2B subunits of the NMDA receptor [115]. CaMK II is a holoenzyme that is well known for its association with the cytoskeleton; and it phosphorylates serine residues on MAPs leading to biophysical alterations of these sidearms [116, 117, 118]. Given that the nanoscale dimensions of the CaMK II holoenzymatic complex and its hexagonal shape fit nicely onto sets of six neighboring microtubules, it is conceivable that this holoenzyme could attach to the microtubule surface during learning, and that this mechanism could be utilized in long-term memory storage.

Cytoskeletal Electrical Signaling and Memory

While learning-related changes in microtubules, actin filaments, and their related MAPs and signal transduction molecules are well documented, little is known about how electrical signaling in the microtubule and actin filament matrix might participate in memory. Current opinion in neurobiology is that memory is represented by altered connections or modified synaptic strengths among large assemblies of interconnected neurons [119]. This is an extension of the "Hebbian synapse" idea put forth by Donald Hebb, in which synaptic knobs continuously extend and retract [120], much the way spines are now known to do.

Changes in synaptic strength represented in the form of LTP can be transient or durable – lasting anywhere from minutes to weeks or longer [121]. There is evidence that LTP expression is near maximal following each isolated learning experience; however, the range shifts upwards with subsequent learning experiences [122]. Since this shift cannot continue to rise indefinitely, an alternative permanent representation of the memory trace that does not rely on potentiation may be necessary. Memory storage deeper within the neuron in the microtubule and actin filament matrix could serve this purpose [123]. This storage arrangement would enable synaptic physiology to return to baseline nullifying a continuing need to shift the range and maximal activity levels upwards.

Coincident activation of the pre- and postsynaptic cell leading to enhanced synaptic strength or potentiation is a popular model for associative memory. In this model, the postsynaptic cell must be depolarized before the NMDA receptor will permit flow of Ca^{2+} across the membrane [124]. LTP also obeys the spatiotemporal learning rule indicating that neurons show heightened sensitivity to repeated stimulus intervals and to inputs that are synchronous but spatially distinct [125, 126]. Electrical signaling by actin filaments and microtubules may figure prominently in coincidence detection and storage of spatiotemporal patterns of inputs, and signaling within the cytoskeletal matrix may be particularly critical to information storage that lasts longer than LTP is able to persist (if indeed there is eventual LTP decay, as seems likely). The initial route to the microtubule matrix would be through the actin filaments concentrated in the spines. Inputs to sites virtually anywhere on the neuron can be transmitted from the neuronal membrane to actin filaments in spines by way of scaffolding proteins and signal transduction molecules. Next putative electrical signals can be transmitted by way of actin filament crosslinker proteins to microtubules, and by way of MAPs and signal transduction molecules to other microtubules.

3.4 A Dendritic Cytoskeleton Information Processing Model

Mathematical models of microtubules and actin filaments in information processing may shed light on the participation of these cytoskeletal proteins in higher cognitive functions. Traditional models of computation are based on the concept of Turing machines. The practical consequence is that every task the machine is about to process has to be pre-programmed, namely an *algorithm* has to be designed and written to instruct the *processor* how to manipulate the data (symbols) in each and every step. The computational process runs until a halt-state is introduced. In this type of computation the input is available at the beginning of the operation; however, no trace of this computation is left when the same machine carries out a subsequent computation for another input. One of the paradigms proposed to overcome the apparent difficulties in solving complex problems (e.g. pattern recognition, temporal sequences processing, etc.) were artificial neural networks. These computational models rely both on massively interconnected parallel networks of simple units (i.e., neurons) and on the availability of some learning algorithm to train, or adapt, the parameters of the model given enough data regarding the problem at hand. Depending on the type of artificial neural network, the parameters of the model may be the strength of connections between the model's interacting units, the number of units required, their connectivity, and the details of the artificial neuron [127]-[129].

Artificial neural network models have been widely used to solve various problems that are intrinsically "static" (i.e., the task is in effect time-independent). In order to deal with time-dependent problems, some of the models were extended to incorporate the time variable, or new models were proposed [130]. Nevertheless, one of the main difficulties of these models remains in the realm of continuous streams of data. Models that are based on attractor dynamics are not suitable in most cases, due to the huge number of attractors required to represent the information, as well as the time required for the dynamic system to converge to a solution attractor. Additionally, these models lack a true memory of recent inputs, and are therefore unable to process the current information in the context of the recently observed data. Lastly, these models are unfit as a tool to study real neurons, in particular with respect to the highly dynamic behavior observed in synapses, both in plasticity and time scales of integration [131, 132]. While synapses in artificial neural networks vary slowly during learning, or are taken to be static altogether after the learning phase is over, synapses in real neurons are highly dynamic and activity-dependent. This means that the neuron itself is much more than a simple processing unit, unlike the model used in artificial neural networks.

A new concept for real-time neural computation of temporal processing has been recently proposed to explain the existence and function of microcircuits in the brain, in particular in the cortex. These brain-wide neural modules, or microcircuits, were found to be highly generic (i.e. they are not task dependent), and their dynamics appears to change continuously – that is they do not seem to converge to a particular attractor. This means that the computation is ongoing and the results do not converge to a particular dynamic state (i.e., the input information does not arrive in one batch). This concept is based on a non-specific, high-dimensional dynamical system, serving as a source of trajectories, called a "liquid state machine" [133, 134]. Another group of scientists independently proposed essentially the same idea under the name "echo state networks" [135]. The basic structure of a liquid state machine is composed of an excitable medium (hence "liquid") and an output function that maps the current liquid state (illustrated in Figure 3.11). The liquid module must be sufficiently complex and dynamic to guarantee a universal computational power. This is sufficient to ensure that different input excitations will lead to separate trajectories in the internal states of the machine. This statement has been rigorously proven [136]. The output function, f^M, is trained on a specific task. Some of the proposed examples of a "liquid" include a network of spiking neurons and a recurrent neural network. The output function, or readout, in these examples has been implemented by simple perceptrons, threshold functions or even linear regression functions. Clearly, simpler readout functions restrict the ability of the whole system to capture complex nonlinear dependencies.

Inspired by the liquid state machine idea and its applications on the one hand, and by the experimental and theoretical results regarding nonlinear wave propagation along microtubules and actin on the other hand, it has been hypothesized that the cytoskeletal biopolymers comprising the backbone for ionic wave propagation may also behave as a *"sub-neural liquid state machine"*. The core concept is that the cytoskeleton matrix interacts with, and regulates neural membrane components (e.g., ion channels or scaffolding proteins). Figure 3.12 diagrams the cytoskeleton at the neural cell level, whereas Figure 3.13 depicts a portion of the dendritic shaft where microtubules are interconnected by MAP2. Connections between microtubules and actin filaments are shown as well. The analogy to the liquid state machine is based on the following observations:

- A typical dendritic cross-section may contain more than 100 microtubules [137].
- The microtubules are highly interconnected by MAP2, hence creating a network.
- Input/output connections to the network are conveyed by actin filaments.
- Each of the network's elements, and in particular actin filaments and microtubules, behave as a nonlinear electrical component [7, 27, 56].

According to this hypothesis, there exists a mechanism by which actin filaments and the microtubule matrix directly regulate ion channels and thus the synaptic strength, thereby controlling the electrical response of the neuron at

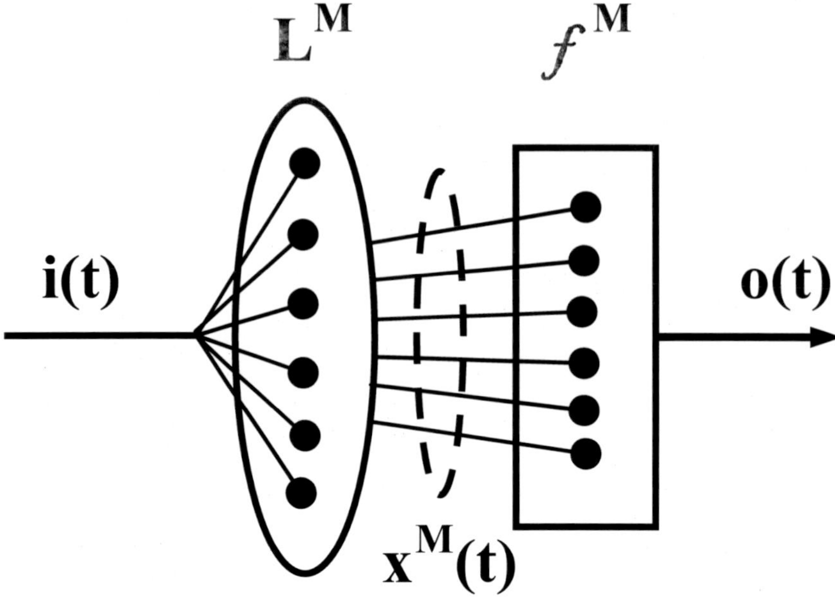

Fig. 3.11. The structure of a liquid state machine. Continuous stream of input data $i(t)$ is injected to the liquid module L^M which evolves its internal state $x^M(t)$; the internal state is transformed by the readout module $f^M(t)$ to generate the output stream $o(t)$.

large. In this scenario, microtubules receive electrical signals from synaptic elements, such as ion channels linked to actin filaments [138], which are in turn connected to microtubules by MAP2 [139], or via direct microtubule connections to postsynaptic density proteins by molecules such as CRIPT [140, 141]. Consequently, the microtubule matrix may act as a high-dimensional dynamic system, or the liquid module, where the main degrees of freedom are related to the electrical flow along each microtubule. The input signals perturb the current state of the system that continues to evolve. As has been previously suggested [142], an integration of the above ideas is outlined in Figure 3.13. Neural inputs arriving at the postsynaptic density produce electrical signals, which in turn transmit ionic waves along the associated actin filaments at the synapse (see Figure 3.13, frame A). These input signals are further propagated in the form of ionic waves through actin filaments to the microtubule matrix (see Figure 3.13, frame B). At this point, the microtubule network operates as a high-dimensional state machine, evolving these input states by dynamically changing the flow associated with individual microtubules (see Figure 3.13, frame C), or by supporting non-linear wave collisions. The output from the microtubule matrix can be measured as the state of the system at a time T. This state could, under certain circumstances, be transmitted by actin filaments to remote ion channels. One output function could be to regulate the temporal gating state of voltage-sensitive channels (see Figure 3.13, frame D). This would subsequently regulate the conductivity of the neuronal membrane and could potentially modulate the axon hillock membrane potential by changing the distribution and topology of opened versus closed channels (see Figure 3.12).

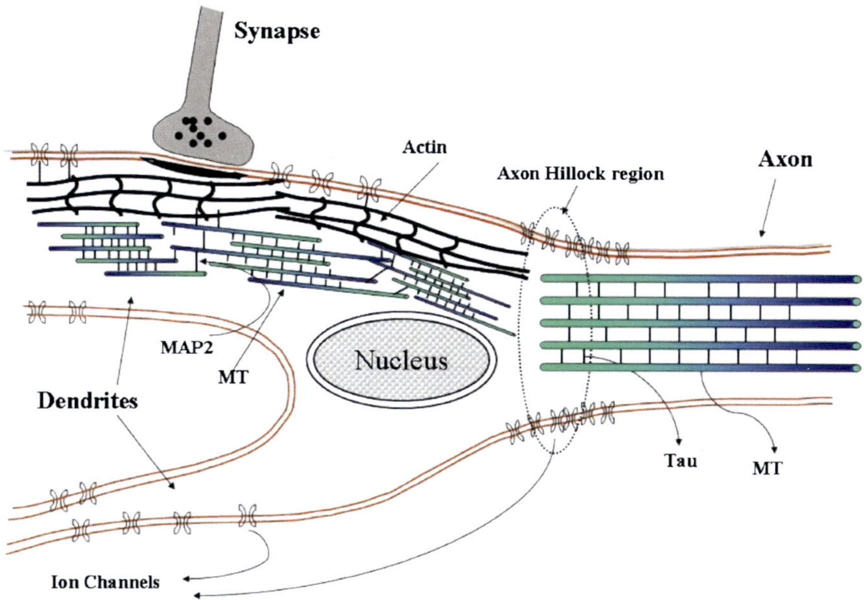

Fig. 3.12. Adapted from [143]

The concept that cytoskeletal structures may behave as a liquid state machine is intriguing as it provides a means for real-time computation without the need for stable attractors. Moreover, the output is relatively insensitive to small variations in either the microtubule matrix or the input stream. Recent perturbations have a long-term effect on the dynamic trajectories, and as a result there is a memory effect inherent to this system. The output from the matrix may be a linear function that converges at or near ion channels to regulate their temporal behavior.

In the context of neuronal function, with focus on synaptic strengthening, LTP, and memory enhancement, the output function may simply reflect an effect of the microtubule network on synaptic channel function, such that the desired state of the channel appears in a higher open probability. One possibility is a Hebbian-based response where more frequent activity of certain subdomains of the microtubule network output states gives rise to a higher (or lower) density of actin filaments connecting to corresponding channels.

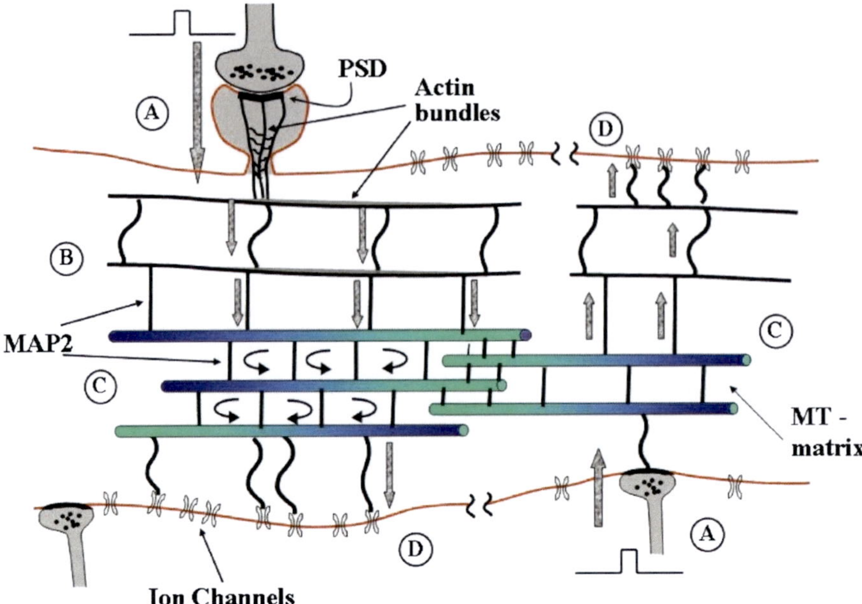

Fig. 3.13. A portion of the dendritic shaft where microtubules are interconnected by MAP2. Connections between microtubules and actin filaments are shown as well. In this example, actin bundles bind to the postsynaptic density (PSD). On the upper left hand side a spiny synapse is shown where actin bundles enter the spine neck and bind to the PSD. Adapted from [142].

Although an attractive hypothesis, a number of questions remain unanswered with respect to real time computation using microtubule matrices.

1. All experimental and most theoretical results are available for single components. It is imperative to extend the knowledge base to assemblies of such elements (e.g., microtubules interconnected by MAP2).
2. The transition to the network level will require a completely new set of experimental tools. It is expected that nanotechnology-based ideas and techniques will play a crucial roll in aspects such as high-resolution spatio-temporal measurements taken simultaneously on a microscopic grid. Understanding the electrical properties of the matrix is essential to the realization of the computational power of such a network, and in particular its ability to function as a liquid state machine.
3. The microtubule matrix is connected to membrane constituents, in particular ion channels and synapses. Actin, MAP2, and possibly other bridging molecules mediate these connections. It has been observed that actin regulates certain ion channels [11]. Such a connection has to be established at a level of greater detail and specificity. It is necessary to further characterize the specific links between actin filaments and individual ion channels and

to assess how these filaments remotely regulate the electrical properties of those channels.

These and other questions need to be more fully addressed both theoretically and experimentally.

References

1. Goldman JE. Immunocytochemical studies of actin localization in the central nervous system. J Neurosci. 1983 Oct;3(10):1952-62.
2. Honkura N, Matsuzaki M, Noguchi J, Ellis-Davies GC, Kasai H. The subspine organization of actin fibers regulates the structure and plasticity of dendritic spines. Neuron. 2008 Mar 13;57(5):719-29.
3. Star EN, Kwiatkowski DJ, Murthy VN. Rapid turnover of actin in dendritic spines and its regulation by activity. Nat Neurosci. 2002 Mar;5(3):239-46.
4. Matus A. Actin-based plasticity in dendritic spines. Science. 2000 Oct 27;290(5492):754-8.
5. Gomez LL, Alam S, Smith KE, Horne E, Dell'Acqua ML. Regulation of A-kinase anchoring protein 79/150-cAMP-dependent protein kinase postsynaptic targeting by NMDA receptor activation of calcineurin and remodeling of dendritic actin. J Neurosci. 2002 Aug 15;22(16):7027-44.
6. Cantiello HF, Patenaude C, Zaner K. Osmotically induced electrical signals from actin filaments. Biophys J. 1991 Jun;59(6):1284-9.
7. Tuszyński JA, Portet S, Dixon JM, Luxford C, Cantiello HF. Ionic wave propagation along actin filaments. Biophys J. 2004 Apr;86(4):1890-903.
8. Otterbein LR, Graceffa P, Dominguez R. The crystal structure of uncomplexed actin in the ADP state. Science. 2001 Jul 27;293(5530):708-11.
9. Rould MA, Wan Q, Joel PB, Lowey S, Trybus KM. Crystal structures of expressed non-polymerizable monomeric actin in the ADP and ATP states. J Biol Chem. 2006 Oct 20;281(42):31909-19.
10. Watson GS, Cahill C, Blach J, Myhra S, Alexeeva Y, Ivanova EP, Nicolau DV (2004). Actin Nanotracks for Hybrid Nanodevices Based on Linear Protein Molecular Motors. In Advanced Microsystems-Integration with Nanotechnology and Biomaterials. In: J.T. Borenstein, P. Grodzinski, L.P. Lee, J. Liu, Z. Wang ed(s) Nanoengineered Assemblies and Advanced Micro/Nanosystems. MRS, San Francisco pp O2.4
11. Matsuno K. The internalist enterprise on constructing cell motility in a bottom-up manner. Biosystems. 2001 Jul-Aug;61(2-3):115-24.
12. Trizac E, Téllez G. Onsager-Manning-Oosawa condensation phenomenon and the effect of salt. Phys Rev Lett. 2006 Jan 27;96(3):038302.

13. Angelini TE, Liang H, Wriggers W, Wong GC. Like-charge attraction between polyelectrolytes Induced by counterion charge density waves. Proc Natl Acad Sci U S A. 2003 Jul 22;100(15):8634-7.
14. Shikinaka K, Kwon H, Kakugo A, Furukawa H, Osada Y, Gong JP, Aoyama Y, Nishioka H, Jinnai H, Okajima T. Observation of the three-dimensional structure of actin bundles formed with polycations. Biomacromolecules. 2008 Feb;9(2):537-42.
15. Lin EC, Cantiello HF. A novel method to study the electrodynamic behavior of actin filaments. Evidence for cable-like properties of actin. Biophys J. 1993 Oct;65(4):1371-8.
16. Arsenault ME, Zhao H, Purohit PK, Goldman YE, Bau HH. Confinement and manipulation of actin filaments by electric fields. Biophys J. 2007 Oct 15;93(8):L42-4.
17. Borejdo J, Ortega H. Electrophoresis and orientation of F-actin in agarose gels. Biophys J. 1989 Aug;56(2):285-93.
18. Kobayasi S, Asai H, Oosawa F. Electric birefringence of actin. Biochim Biophys Acta. 1964 Nov 29;88:528-40.
19. Meggs WJ. Electric fields determine the spatial organization of microtubules and actin filaments. Med Hypotheses. 1988 Jul;26(3):165-70.
20. Kwon HJ, Kakugo A, Shikinaka K, Osada Y, Gong JP. Morphology of actin assemblies in response to polycation and salts. Biomacromolecules. 2005 Nov-Dec;6(6):3005-9.
21. Sakata T, Yan Y, Marriott G. Optical switching of dipolar interactions on proteins. Proc Natl Acad Sci U S A. 2005 Mar 29;102(13):4759-64.
22. Nogales E, Wolf SG, Downing KH. Structure of the alpha beta tubulin dimer by electron crystallography. Nature. 1998 Jan 8;391(6663):199-203.
23. Stracke R, Böhm KJ, Wollweber L, Tuszyński JA, Unger E. Analysis of the migration behaviour of single microtubules in electric fields. Biochem Biophys Res Commun. 2002 Apr 26;293(1):602-9.
24. Hagan S, Hameroff SR, Tuszyński JA. Quantum computation in brain microtubules: decoherence and biological feasibility. Phys Rev E Stat Nonlin Soft Matter Phys. 2002 Jun;65(6 Pt 1):061901.
25. Brown JA. A study of the interactions between electromagnetic fields and microtubules: ferroelectric effects, signal transduction and electronic conduction. Ph.D. Thesis, University of Alberta, Edmonton, 1999.
26. Mershin A, Kolomenski AA, Schuessler HA, Nanopoulos DV. Tubulin dipole moment, dielectric constant and quantum behavior: computer simulations, experimental results and suggestions. Biosystems. 2004 Nov;77(1-3):73-85.
27. Priel A, Tuszyński JA, Woolf NJ. Transitions in microtubule C-termini conformations as a possible dendritic signaling phenomenon. Eur Biophys J. 2005 Dec;35(1):40-52.
28. Di Noto L, DeTure MA, Purich DL. Disulfide-cross-linked tau and MAP2 homodimers readily promote microtubule assembly. Mol Cell Biol Res Commun. 1999 Jul;2(1):71-6.
29. Minoura I, Muto E. Dielectric measurement of individual microtubules using the electroorientation method. Biophys J. 2006 May 15;90(10):3739-48.
30. J. A. Tuszyński, T. J. A. Craddock and E. J. Carpenter, Bioferroelectricity at the Nanoscale, Journal of Theoretical and Computational Nanoscience 5, no.10, 2022-2032 (2008)

31. Fritzsche W, Boehm K, Unger E, Koehler JM, Making electrical contact to single molecules. Nanotechnology 9, 177 (1998)
32. Fritzsche W., J.M. Koehler, K. Boehm, E. Unger, T. Wagner, R. Kirsch, M. Mertig and W. Pompe, Nanotechnology 10, 331-335 (1999).
33. Goddard G, Whittier JE, Biomolecules as nanomaterials: interface characterization for sensor development, Proceedings of SPIE – Volume 6172 Smart Structures and Materials 2006: Smart Electronics, MEMS, BioMEMS, and Nanotechnology, Vijay K. Varadan, Editor, 617206 (Mar. 31, 2006)
34. Tuszyński JA, Brown JA and Hawrylak P, Dielectric Polarization, Electrical Conduction, Information Processing and Quantum Computation in Microtubules, Are They Plausible?, Transactions of the Royal Society A 356, 1897-1926 (1998)
35. Umnov M, Palusinski OA, Deymier PA, Guzman R, Hoying J, Barnaby H, Yang Y, Raghavan S. Experimental evaluation of electrical conductivity of microtubules Journal of Materials Science, Volume 42, Number 1, January 2007 , pp. 373-378(6)
36. Mithieux G, Chauvin F, Roux B, Rousset B. Association states of tubulin in the presence and absence of microtubule-associated proteins. Analysis by electric birefringence. Biophys Chem. 1985 Oct;22(4):307-16.
37. Ludueña RF. Multiple forms of tubulin: different gene products and covalent modifications. Int Rev Cytol. 1998;178:207-75.
38. Huzil JT, Ludueña RF, Tuszyński JA. Comparative modelling for human β tubulin isotypes and implications for drug binding, Nanotechnology 17, S90-S100.
39. Luchko T, Huzil JT, Stepanova M, Tuszyński J. Conformational analysis of the carboxy-terminal tails of human beta-tubulin isotypes. Biophys J. 2008 Mar 15;94(6):1971-82.
40. Tuszyński JA, Carpenter EJ, Huzil JT, Malinski W, Luchko T, Luduena RF. The evolution of the structure of tubulin and its potential consequences for the role and function of microtubules in cells and embryos. Int J Dev Biol. 2006;50(2-3):341-58.
41. Carpenter EJ, Huzil JT, Ludueña RF, Tuszyński JA. Homology modeling of tubulin: influence predictions for microtubule's biophysical properties. Eur Biophys J. 2006 Dec;36(1):35-43.
42. Tuszyński JA, Luchko T, Portet S, Dixon JM. Anisotropic elastic properties of microtubules. Eur Phys J E Soft Matter. 2005 May;17(1):29-35.
43. Sept D, Baker NA, McCammon JA. The physical basis of microtubule structure and stability. Protein Sci. 2003 Oct;12(10):2257-61.
44. Diaz JF, Andreu JM, Diakun G, Towns-Andrews E, Bordas J. Structural intermediates in the assembly of taxoid-induced microtubules and GDP-tubulin double rings: time-resolved X-ray scattering. Biophys J. 1996 May;70(5):2408-20.
45. Krebs A, Goldie KN, Hoenger A. Structural rearrangements in tubulin following microtubule formation. EMBO Rep. 2005 Mar;6(3):227-32.
46. Dixon JM, Chelminiak P, Tuszyński JA. Surface conductance in seamless microtubules. Physica A: Statistical Mechanics and its Applications, 2008, 387: 4183-4194.
47. Pampaloni F, Florin EL. Microtubule architecture: inspiration for novel carbon nanotube-based biomimetic materials. Trends Biotechnol. 2008 Jun;26(6):302-10.

48. Baker NA, Sept D, Joseph S, Holst MJ, McCammon JA. Electrostatics of nanosystems: application to microtubules and the ribosome. Proc Natl Acad Sci U S A. 2001 Aug 28;98(18):10037-41.
49. Tuszyński, J., S. Hameroff, M.V. Sataric, B. Trpisova & M.L.A. Nip. 1995. Ferroelectric behavior in microtubule dipole lattices: implications for information processing, signaling and assembly/disassembly. J. Theor. Biol. 174, 371 (1995).
50. Sataric MV, Tuszyński JA. Relationship between the nonlinear ferroelectric and liquid crystal models for microtubules. Phys Rev E Stat Nonlin Soft Matter Phys. 2003 Jan;67(1 Pt 1):011901.
51. Chen Y, Qiu XJ, Dong XL. Pseudo-spin model for the microtubule wall in external field. Biosystems. 2005 Nov;82(2):127-36.
52. Sataric MV, Tuszyński JA, Zakula RB. Kinklike excitations as an energy-transfer mechanism in microtubules. Phys Rev E Stat Phys Plasmas Fluids Relat Interdiscip Topics. 1993 Jul;48(1):589-597.
53. Sataric MV, Budinski-Petkovic L, Loncarevic I, Tuszyński JA. Modelling the Role of Intrinsic Electric Fields in Microtubules as an Additional Control Mechanism of Bidirectional Intracellular Transport. Cell Biochem Biophys. 2008;52(2):113-24.
54. Priel A, Ramos AJ, Tuszyński JA, Cantiello HF. A biopolymer transistor: electrical amplification by microtubules. Biophys J. 2006 Jun 15;90(12):4639-43.
55. Priel A, Ramos AJ, Tuszyński JA, Cantiello HF. Effect of calcium on electrical energy transfer by microtubules. J Biol Phys, 2008, DOI 10.1007/s10867-008-9106-z
56. Priel A, Tuszyński JA. A nonlinear cable-like model of amplified ionic wave propagation along microtubules. European Physics Letters. 2008;83/68004:1-5.
57. Tuszyński JA, Portet S, Dixon JM, Luxford C, Cantiello HF. Ionic wave propagation along actin filaments. Biophys J. 2004 Apr;86(4):1890-903.
58. Tuszyński JA, Priel A, Brown JA, Cantiello HF, Dixon JM. Electronic and ionic cconductivities of microtubules and actin filaments, their consequences for cell signaling and applications to bioelectronics. In: Lyshevski SE, ed. Nano and Molecular Electronics, CRC Press, 2007, pp. 18:1-45.
59. Pokorny J, Excitation of vibrations in microtubules in living cells, 2003, The Bioelectrochemical Society XVIIth International Symposium on Bioelectrochemistry and Bioenergetics Abstracts, p. 108.
60. Maniotis AJ, Chen CS, Ingber DE. Demonstration of mechanical connections between integrins, cytoskeletal filaments, and nucleoplasm that stabilize nuclear structure. Proc Natl Acad Sci U S A. 1997 Feb 4;94(3):849-54.
61. Salmon WC, Adams MC, Waterman-Storer CM. Dual-wavelength fluorescent speckle microscopy reveals coupling of microtubule and actin movements in migrating cells. J Cell Biol. 2002 Jul 8;158(1):31-7.
62. Jaworski J, Hoogenraad CC, Akhmanova A. Microtubule plus-end tracking proteins in differentiated mammalian cells. Int J Biochem Cell Biol. 2008;40(4):619-37.
63. Watanabe T, Wang S, Noritake J, Sato K, Fukata M, Takefuji M, Nakagawa M, Izumi N, Akiyama T, Kaibuchi K. Interaction with IQGAP1 links APC to Rac1, Cdc42, and actin filaments during cell polarization and migration. Dev Cell. 2004 Dec;7(6):871-83.
64. Tsvetkov AS, Samsonov A, Akhmanova A, Galjart N, Popov SV. Microtubule-binding proteins CLASP1 and CLASP2 interact with actin filaments. Cell Motil Cytoskeleton. 2007 Jul;64(7):519-30.

65. Kholmanskikh SS, Koeller HB, Wynshaw-Boris A, Gomez T, Letourneau PC, Ross ME. Calcium-dependent interaction of Lis1 with IQGAP1 and Cdc42 promotes neuronal motility. Nat Neurosci. 2006 Jan;9(1):50-7.
66. Lansbergen G, Akhmanova A. Microtubule plus end: a hub of cellular activities. Traffic. 2006 May;7(5):499-507.
67. Kotani S, Nishida E, Kumagai H, Sakai H. Calmodulin inhibits interaction of actin with MAP2 and Tau, two major microtubule-associated proteins. J Biol Chem. 1985 Sep 5;260(19):10779-83.
68. Selden SC, Pollard TD. Interaction of actin filaments with microtubules is mediated by microtubule-associated proteins and regulated by phosphorylation. Ann N Y Acad Sci. 1986;466:803-12.
69. Bouquet C, Nothias F. Molecular mechanisms of axonal growth. Adv Exp Med Biol. 2007;621:1-16.
70. de Anda FC, Pollarolo G, Da Silva JS, Camoletto PG, Feiguin F, Dotti CG. Centrosome localization determines neuronal polarity. Nature. 2005 Aug 4;436(7051):704-8.
71. Knoblich JA. Neurobiology: getting axons going. Nature. 2005 Aug 4;436(7051):632.
72. Dehmelt L, Nalbant P, Steffen W, Halpain S. A microtubule-based, dynein-dependent force induces local cell protrusions: Implications for neurite initiation. Brain Cell Biol. 2006 Feb;35(1):39-56.
73. Bradke F, Dotti CG. Differentiated neurons retain the capacity to generate axons from dendrites. Curr Biol. 2000 Nov 16;10(22):1467-70.
74. Takahashi D, Yu W, Baas PW, Kawai-Hirai R, Hayashi K. Rearrangement of microtubule polarity orientation during conversion of dendrites to axons in cultured pyramidal neurons. Cell Motil Cytoskeleton. 2007 May;64(5):347-59.
75. Gordon-Weeks PR. Microtubules and growth cone function. J Neurobiol. 2004 Jan;58(1):70-83.
76. Rajnicek AM, Foubister LE, McCaig CD. Growth cone steering by a physiological electric field requires dynamic microtubules, microfilaments and Rac-mediated filopodial asymmetry. J Cell Sci. 2006 May 1;119(Pt 9):1736-45.
77. Vargas ME, Barres BA. Why is Wallerian degeneration in the CNS so slow? Annu Rev Neurosci. 2007;30:153-79.
78. Spira ME, Oren R, Dormann A, Ilouz N, Lev S. Calcium, protease activation, and cytoskeleton remodeling underlie growth cone formation and neuronal regeneration. Cell Mol Neurobiol. 2001 Dec;21(6):591-604.
79. Woolf NJ. A structural basis for memory storage in mammals. Prog Neurobiol. 1998 May;55(1):59-77.
80. Robinson KR, Cormie P. Electric field effects on human spinal injury: Is there a basis in the in vitro studies? Dev Neurobiol. 2008 Feb 1;68(2):274-80.
81. Walker JL, Kryscio R, Smith J, Pilla A, Sisken BF. Electromagnetic field treatment of nerve crush injury in a rat model: effect of signal configuration on functional recovery. Bioelectromagnetics. 2007 May;28(4):256-63.
82. Greenebaum B, Sisken BF. Does direction of induced electric field or current provide a test of mechanism involved in nerve regeneration? Bioelectromagnetics. 2007 Sep;28(6):488-92.
83. Hickmott PW, Ethell IM. Dendritic plasticity in the adult neocortex. Neuroscientist. 2006 Feb;12(1):16-28.
84. Greenough WT, Volkmar FR. Pattern of dendritic branching in occipital cortex of rats reared in complex environments. Exp Neurol. 1973 Aug;40(2):491-504.

85. Greenough WT, Volkmar FR, Juraska JM. Effects of rearing complexity on dendritic branching in frontolateral and temporal cortex of the rat. Exp Neurol. 1973 Nov;41(2):371-8.
86. Leggio MG, Mandolesi L, Federico F, Spirito F, Ricci B, Gelfo F, Petrosini L. Environmental enrichment promotes improved spatial abilities and enhanced dendritic growth in the rat. Behav Brain Res. 2005 Aug 30;163(1):78-90.
87. Tuszyński JA, Portet S, Dixon JM. Nonlinear assembly kinetics and mechanical properties of biopolymers. Nonlinear Analysis, 2005, 63:915-925.
88. Pollard TD, J. Cell Biol. 103 (1986), p. 2747.
89. Ma J. A hybrid neural network of addressable and content-addressable memory. Int J Neural Syst. 2003 Jun;13(3):205-13.
90. Swindale NV. Feedback decoding of spatially structured population activity in cortical maps. Neural Comput. 2008 Jan;20(1):176-204.
91. O'Connell C, O'Malley A, Regan CM. Transient, learning-induced ultrastructural change in spatially-clustered dentate granule cells of the adult rat hippocampus. Neuroscience. 1997 Jan;76(1):55-62.
92. Nakayama T, Sawada T. Involvement of microtubule integrity in memory impairment caused by colchicine. Pharmacol Biochem Behav. 2002 Jan-Feb;71(1-2):119-38.
93. Bensimon G, Chermat R. Microtubule disruption and cognitive defects: effect of colchicine on learning behavior in rats. Pharmacol Biochem Behav. 1991 Jan;38(1):141-5.
94. Di Patre PL, Oh JD, Simmons JM, Butcher LL. Intrafimbrial colchicine produces transient impairment of radial-arm maze performance correlated with morphologic abnormalities of septohippocampal neurons expressing cholinergic markers and nerve growth factor receptor. Brain Res. 1990 Jul 23;523(2):316-20.
95. Mileusnic R, Lancashire CL, Rose SP. Recalling an aversive experience by day-old chicks is not dependent on somatic protein synthesis. Learn Mem. 2005 Nov-Dec;12(6):615-9.
96. Cavallaro S, D'Agata V, Manickam P, Dufour F, Alkon DL. Memory-specific temporal profiles of gene expression in the hippocampus. Proc Natl Acad Sci U S A. 2002 Dec 10;99(25):16279-84.
97. Nelson TJ, Backlund PS Jr, Alkon DL. Hippocampal protein-protein interactions in spatial memory. Hippocampus. 2004;14(1):46-57.
98. Mori N, Morii H. SCG10-related neuronal growth-associated proteins in neural development, plasticity, degeneration, and aging. J Neurosci Res. 2002 Nov 1;70(3):264-73.
99. Bianchi M, Fone KF, Azmi N, Heidbreder CA, Hagan JJ, Marsden CA. Isolation rearing induces recognition memory deficits accompanied by cytoskeletal alterations in rat hippocampus. Eur J Neurosci. 2006 Nov;24(10):2894-902.
100. Liu HX, Zhang JJ, Zheng P, Zhang Y. Altered expression of MAP-2, GAP-43, and synaptophysin in the hippocampus of rats with chronic cerebral hypoperfusion correlates with cognitive impairment. Brain Res Mol Brain Res. 2005 Sep 13;139(1):169-77.
101. Shimada A, Tsuzuki M, Keino H, Satoh M, Chiba Y, Saitoh Y, Hosokawa M. Apical vulnerability to dendritic retraction in prefrontal neurones of ageing SAMP10 mouse: a model of cerebral degeneration. Neuropathol Appl Neurobiol. 2006 Feb;32(1):1-14.

102. Woolf NJ, Zinnerman MD, Johnson GV. Hippocampal microtubule-associated protein-2 alterations with contextual memory. Brain Res. 1999 Mar 6;821(1):241-9.
103. Woolf NJ, Young SL, Johnson GV, Fanselow MS. Pavlovian conditioning alters cortical microtubule-associated protein-2. Neuroreport. 1994 May 9;5(9):1045-8.
104. Woolf NJ. Cholinoceptive cells in rat cerebral cortex: somatodendritic immunoreactivity for muscarinic receptor and cytoskeletal proteins. J Chem Neuroanat. 1993 Nov-Dec;6(6):375-90.
105. Van der Zee EA, Douma BR, Bohus B, Luiten PG. Passive avoidance training induces enhanced levels of immunoreactivity for muscarinic acetylcholine receptor and coexpressed PKC gamma and MAP-2 in rat cortical neurons. Cereb Cortex. 1994 Jul-Aug;4(4):376-90.
106. Khuchua Z, Wozniak DF, Bardgett ME, Yue Z, McDonald M, Boero J, Hartman RE, Sims H, Strauss AW. Deletion of the N-terminus of murine map2 by gene targeting disrupts hippocampal ca1 neuron architecture and alters contextual memory. Neuroscience. 2003;119(1):101-11.
107. Mershin A, Pavlopoulos E, Fitch O, Braden BC, Nanopoulos DV, Skoulakis EM. Learning and memory deficits upon TAU accumulation in Drosophila mushroom body neurons.Learn Mem. 2004 May-Jun;11(3):277-87.
108. Pennanen L, Wolfer DP, Nitsch RM, Götz J. Impaired spatial reference memory and increased exploratory behavior in P301L tau transgenic mice. Genes Brain Behav. 2006 Jul;5(5):369-79.
109. Boekhoorn K, Terwel D, Biemans B, Borghgraef P, Wiegert O, Ramakers GJ, de Vos K, Krugers H, Tomiyama T, Mori H, Joels M, van Leuven F, Lucassen PJ. Improved long-term potentiation and memory in young tau-P301L transgenic mice before onset of hyperphosphorylation and tauopathy. J Neurosci. 2006 Mar 29;26(13):3514-23.
110. Fischer A, Sananbenesi F, Schrick C, Spiess J, Radulovic J. Distinct roles of hippocampal de novo protein synthesis and actin rearrangement in extinction of contextual fear. J Neurosci. 2004 Feb 25;24(8):1962-6.
111. Yuste R, Bonhoeffer T. Morphological changes in dendritic spines associated with long-term synaptic plasticity. Annu Rev Neurosci. 2001;24:1071-89.
112. Roberts LA, Large CH, Higgins MJ, Stone TW, O'Shaughnessy CT, Morris BJ. Increased expression of dendritic mRNA following the induction of long-term potentiation. Brain Res Mol Brain Res. 1998 May;56(1-2):38-44.
113. Fukunaga K, Muller D, Miyamoto E. CaM kinase II in long-term potentiation. Neurochem Int. 1996 Apr;28(4):343-58.
114. Zervas M, Opitz T, Edelmann W, Wainer B, Kucherlapati R, Stanton PK. Impaired hippocampal long-term potentiation in microtubule-associated protein 1B-deficient mice. J Neurosci Res. 2005 Oct 1;82(1):83-92.
115. Wang H, Feng R, Wang LP, Li F, Cao X, Tsien JZ. CaMKII Activation State Underlies Synaptic Labile Phase of LTP and Short-Term Memory Formation. Curr Biol. 2008 Oct 28;18(20):1546-54.
116. Hudmon A, Schulman H. Neuronal CA2+/calmodulin-dependent protein kinase II: the role of structure and autoregulation in cellular function. Annu Rev Biochem. 2002;71:473-510.
117. Easley CA, Faison MO, Kirsch TL, Lee JA, Seward ME, Tombes RM. Laminin activates CaMK-II to stabilize nascent embryonic axons. Brain Res. 2006 May 30;1092(1):59-68.

118. Steiner B, Mandelkow EM, Biernat J, Gustke N, Meyer HE, Schmidt B, Mieskes G, Söling HD, Drechsel D, Kirschner MW, et al. Phosphorylation of microtubule-associated protein tau: identification of the site for Ca2(+)-calmodulin dependent kinase and relationship with tau phosphorylation in Alzheimer tangles. EMBO J. 1990 Nov;9(11):3539-44.
119. Golomb D, Hansel D. The number of synaptic inputs and the synchrony of large, sparse neuronal networks. Neural Comput. 2000 May;12(5):1095-139.
120. Hebb, DO. The Organization of Behaviour: A neuropsychological approach. John Wiley and Sons, New York (1949).
121. Korol DL, Gold PE. Epinephrine converts long-term potentiation from transient to durable form in awake rats. Hippocampus. 2008;18(1):81-91.
122. Rioult-Pedotti MS, Donoghue JP, Dunaevsky A. Plasticity of the synaptic modification range. J Neurophysiol. 2007 Dec;98(6):3688-95.
123. Woolf, NJ In: Consciousness, Neurobiology and Quantum Mechanics In: The Emerging Physics of Consciousness, (Ed.) Tuszyński,J., 2006.
124. Bender VA, Bender KJ, Brasier DJ, Feldman DE. Two coincidence detectors for spike timing-dependent plasticity in somatosensory cortex. J Neurosci. 2006 Apr 19;26(16):4166-77.
125. Tsukada M, Pan X. The spatiotemporal learning rule and its efficiency in separating spatiotemporal patterns. Biol Cybern. 2005 Feb;92(2):139-46.
126. Tsukada M, Aihara T, Saito HA, Kato H. Hippocampal LTP Depends on Spatial and Temporal Correlation of Inputs. Neural Netw. 1996 Nov;9(8):1357-1365.
127. Bishop CM, Pattern recognition and machine learning. New York: Springer, 2006.
128. Duda, Hart and Stork, Pattern Classification, Wiley & Sons, 2000
129. Haykin S, Neural Networks and Learning Machines. Third Edition. Prentice Hall, 2008.
130. Mandic D, Chambers J, Recurrent Neural Networks for Prediction: Learning Algorithms, Architectures and Stability, Wiley; 2001
131. Abbott LF and Regehr WG, Synaptic computation. Nature 431, 796-803, 2004.
132. Natschlager, T., Maass, W. and Zador, A. Efficient Temporal Processing with Dynamic Synapses. Network, 12: 75-87, 2001
133. Maass W, Natschläger T, and Markram H. Real-time computing without stable states: A new framework for neural computation based on perturbations. Neural Computation, 14(11):2531-2560, 2002.
134. Maass W, Natschläger T, and Markram H. A model for real-time computation in generic neural microcircuits. In S. Becker, S. Thrun, and K. Obermayer, editors, Proc. of NIPS 2002, Advances in Neural Information Processing Systems, volume 15, pages 229-236. MIT Press, 2003
135. Jaeger H., The "echo state" approach to analysing and training recurrent neural networks. GMD Report 148, GMD - German National Research Institute for Computer Science, 2001
136. Maass W, and Markram H. On the computational power of recurrent circuits of spiking neurons. Journal of Computer and System Sciences, 69(4):593-616, 2004.
137. Leclerc N, Baas PW, Garner CC, and Kosik KS, Juvenile and mature MAP2 isoforms induce distinct patterns of process outgrowth. Mol Biol Cell 7(3): 443-455 , 1996

138. Maguire G, Connaughton V, Prat AG, Jackson GR Jr, Cantiello HF. Actin cytoskeleton regulates ion channel activity in retinal neurons. Neuroreport.9(4):665-70. 1998
139. Rodriguez, O. C., A. W. Schaefer, C. A. Mandato, P. Forscher, W. M. Bement, and C. M. Waterman-Storer. 2003. Conserved microtubule-actin interactions in cell movement and morphogenesis. Nature Cell Biol. 5:599-609.
140. Passafaro, M., C. Sala, M. Niethammer, and M. Sheng. 1999. Microtubule binding by CRIPT and its potential role in the synaptic clustering of PSD-95. Nature Neurosci. 2:1063-1069.
141. Niethammer M, Valtschanoff JG, Kapoor TM, Allison DW, Weinberg RJ, Craig AM, Sheng M. CRIPT, a novel postsynaptic protein that binds to the third PDZ domain of PSD-95/SAP90. Neuron 20 (1998), pp. 693-707.
142. Priel, A, Tuszyński, JA. and Cantiello, HF.(2005) Electrodynamic Signaling by the Dendritic Cytoskeleton: Toward an Intracellular Information Processing Model',Electromagnetic Biology and Medicine,24:3,221 - 231.
143. Priel A, Tuszyński JA and Cantiello H, Ionic Waves Propagation Along the Dendritic Cytoskeleton as a Signaling Mechanism, "Molecular Biology of the Cell", Vol. 37, E. Bittar and S. Khurana (eds), Elsevier, 2006

4

Nanocarriers and Intracellular Transport: Moving Along the Cytoskeletal Matrix

Summary

The cytoskeleton of neurons is the nanoscale matrix along which organelles, proteins, mRNAs, or signaling complexes are guided to their final destinations inside the cell. Nanotechnology and molecular biology have enabled precision study of these biomolecular machines, in some cases down to the level of single molecules. Motor, linker, and adaptor proteins are essential to the transport process – the three main motors being kinesin, dynein, and myosin, each of which give rise to families of related motor proteins. Neurons are unique in that they possess two distinct transport systems: one in the axon and the other in the somatodendritic compartment. Microtubules are the main tracks for transport in the axon shaft, with neurofilaments (also concentrated within axons) stabilizing the microtubule network. Synaptic vesicles, containing biosynthetic enzymes that are responsible for manufacturing and releasing neurotransmitters, are routinely transported down along axonal microtubules towards actin-rich axon terminals. Endosomes incorporating neurotrophins typically travel in the reverse direction, from axon terminal to the cell body. These transport processes have been tracked with quantum dot nanoparticles attached to single motor proteins or individual cargo molecules. Microtubules also fill the somatodendritic compartments of neurons where they are pivotal to the transport of neurotransmitter receptor subunits and mRNAs from the cell body to postsynaptic sites, in particular to spines – the postsynaptic specializations enriched with actin filaments. Levels of synaptic activity affect the transport of neurotransmitter receptors and mRNA, and permanent changes in synaptic strength partly depend on transport to postsynaptic sites. Alterations in axonal and dendritic transport underlie neuronal responses to injury, regeneration and morphogenesis, as well as learning and memory. Modifications of transport tracks may constitute a subcellular memory mechanism by which the altered intraneuronal connectivity contributes to the memory trace. Elucidation of this mechanism of memory will come with a greater understanding of the biophysics of transport and motor protein mechanics. Biophysical models detailing the nanoscale mechanisms of cellular transport have already increased our understanding of how biological motors operate mechanistically, providing fundamental guiding principles for nanotechnological advancements. Potential nanotechnologies expected to result from biophysical studies of biological transport include bioengineered motors and biomimetic nanocarrier devices, both promising to be useful in biomedicine or as analytical devices. Cytoskeletal and motor proteins, or hybrid designs including these proteins, stand to contribute to a wide variety of potential nanostructured products.

Table 4.1. Types of transport in neurons.

Type of transport	Function
Axonal	
Anterograde	Supplies axon terminals with synaptic vesicles and other needed materials
Retrograde	Delivers retrograde messengers to cell (e.g., neurotrophins); removes debris
Dendritic	
Anterograde	Provides dendrites with receptor proteins, mRNA, and other needed materials; important for plasticity and learning
Retrograde	Transports transcription factors from synapse to nucleus

4.1 Types of Transport in Neurons

Transport occurs in both the axons and dendrites of neurons; however, different materials are carried in accordance with the specialized functions of these neuronal compartments (see Table 4.1). Axon transport is bi-directional: it can be anterograde (from the cell body to the axon terminal) or retrograde (from the axon terminal to the cell body). Transported materials or "cargos" are carried in one or the other direction, depending on the particular motor, adaptor, and scaffolding proteins [1]. During the lifetime of the organism, the structure of the cytoskeletal matrix in neurons changes with neural development and as a consequence of activity, experience, or learning. The structure of the cytoskeleton is in essence a roadmap of the past – a record of the biomechanical operations of the neuron.

Because of their scale, nanotechnological methods are optimally suited to the study of axonal transport. Nanosecond lasers have been used to track both anterograde and retrograde axonal transport *in vivo* [2]. Semiconductor nanocrystals, called quantum dots, have the advantage of being able to track transport of various cargoes without significantly interfering with the transport process [3]. Quantum dot nanoparticles have been used to track neurotrophins and neurotransmitter receptors being transported along axons as cargo or being distributed along dendrites. Single molecules of neurotrophins conjugated individually to quantum dots were found to travel along axons predominately in the retrograde direction, but also registered some anterograde mobility [4, 5]. Quantum dot labeling has been used in combination with immunohistochemistry to track a limited number of neurotransmitter receptors, and can in principle track virtually any neurotransmitter receptor [6].

4.2 Motor, Adaptor, and Scaffolding Proteins

Motor proteins are classic nanoelectromechanical systems (NEMS) that convert chemical energy into mechanical force [7]. Recent studies of these molecules include probing them with nanotechnological tools such as optical "tweezers" and "wrenches" that enable high resolution tracking of their movements and measurement of forces produced [8, 9].

Similar to other eukaryotic cells, neurons exhibit three families of motor proteins (see Table 4.2). Members of the kinesin, dynein, and myosin families (see Figure 4.1) are responsible for transporting particular cargoes, using specific adaptor proteins, in certain neuronal and sensory cell populations [10]-[15]. With some notable exceptions, kinesins are responsible for fast anterograde transport, dyneins for fast retrograde transport (both over long distances along microtubule tracks), whereas myosin typically directs cargo over short distances along actin filaments. While various classification schemes focus on the differences between the motor proteins, in the living neuron these motors work together, and a given cargo is often passed from one cytoskeletal track to another, and in some cases this involves a switch in the motor protein carrier [16].

4.2.1 Kinesins

The majority of kinesins are plus-end-directed microtubule motors. Each member of the kinesin family has head, stalk, and tail domains that vary in size [17]. The two globular heads of conventional kinesin (kinesin-1 or KIF5A) are 10 nm in diameter (see Figure 4.1). These two heads represent the microtubule-binding portion, and the amino acid sequence of the head region is conserved across kinesin superfamilies. Kinesin-1 and kinesin-2 have the longest stalk domains. The stalk of kinesin-1 measures 10 nm in diameter and 80 nm in length. Other kinesin family members have considerably shorter stalks, indicative of the varied functions served by these motors. The tail regions of kinesin motors are highly diversified in accordance with their binding to a variety of cargoes. The amino acid sequences of the tail region determine what type of cargo each kinesin will bind [18].

Some 35 identified proteins bind non-covalently to kinesin-1, including cargo proteins, signaling proteins, and molecular chaperones [19]. Adaptor proteins intervene in one of four ways: they can (1) directly link the motor to the cargo, (2) chaperone the binding of the motor to another adaptor protein, (3) bind a protein kinase or phosphatase that in turn regulates binding, or (4) participate in the regulation of the binding between motor protein and cargo [20].

The two heads of kinesin make coordinated, alternating, 8-nm steps on the microtubule dimer (corresponding to the lattice periodicity determined by the size of each tubulin dimer), and are thus responsible for transporting the cargo down the microtubule [21]. The rate of consecutive steps made by kinesin before detaching is defined as its processivity. Recent tracking studies of single kinesin molecules in living cells using high resolution fluorescence microscopy show that kinesin has a maximum processivity of $0.78 \pm 0.11\ \mu\text{m/s}$ (which depends on the concentration of ATP, among other factors) and travels an average of slightly over a one-micron distance on a microtubule track before detaching [22]. Similar results have been obtained using quantum dot and gold nanoparticles [23, 24, 25], and it has furthermore been demonstrated that a crowded environment (as is typical of a living cell) does not necessarily impede kinesin processivity or transport rate. Nonetheless, a high backwards load can induce backwards processivity [26]. Other nanoscale studies of microtubule transport have revealed that molecular “road blocks”, rather than friction, can decrease transport rates of cargo along microtubules [27]. Nanoparticle-coated microtubules have also been used to track the collective movements of kinesin-1, with this method demonstrating a lack of synchrony among individual motors transporting a common cargo [28].

Transport along microtubule tracks by kinesin is driven by the energy released in the hydrolysis of ATP [29]. Microtubules are needed to stimulate the ATPase activity intrinsic to kinesin, which converts energy from ATP hydrolysis into mechanical work (in this case the hinge motion that governs stepping) [30, 31]. As illustrated in Figure 4.2, ATP hydrolysis results in a hypothetical “unlatching” of the hinge mechanism between the kinesin head and the neck of the stalk, thereby enabling kinesin to advance along the microtubule track. Other mechanisms that account for the forward movements of kinesin heads include the power stroke model and Brownian ratchet mechanisms, both of which adequately describe different aspects of kinesin processivity [32] but are still deficient in some aspects. More detailed biophysical models of kinesin walking will be presented later in this chapter.

The dynamical interactions between kinesin and microtubules are governed both by binding and processivity. Regarding processivity, after kinesin binds to tubulin, it tends to stay on one microtubule protofilament until it dissociates from the microtubule [33]. The structural conformation of kinesin changes when it binds to the microtubule, and this is independent of its being bound to ATP versus ADP [34]. The microtubule track is essential for the conformational changes that enable kinesin to move cargo. Moreover, it

Table 4.2. Motor proteins found in the CNS and in primary sensory cells.

Motor protein	Cargo transported	Adaptor or scaffolding protein	Possible function or location
Kinesins			
Kinesin-1 (KIF5A)	Vesicles, mitochondria, lysosomes, tubulin, neurofilament proteins	APP, JIP, GRIP1	Axons throughout the nervous system
Kinesin-2 (KIF3)	Vesicles, choline acetyltransferase	α-Fodrin	Ribbon synapses in retina
Kinesin-3 (KIF1)	Synaptic vesicle precursors, mitochondria	Membrane lipids	
Kinesin-4 (KIF4)	Vesicles	Adhesion molecule L1	Facilitates mitosis
Kinesin-6 (MKLP1)			Facilitates dendrite formation
Kinesin-13 (KIF2, KIF13A)	Vesicles	AP1-adaptor complex, PSD95	Induces microtubule depolymerization; transports vesicles to plasma membrane
Kinesin-14 (NCD, KIFC)	Retrograde vesicles, C-terminal motors		Localized at ribbon synapses in the retina (KIFC3)
KIF17		mLin2, mLin 7, mLin10	Dendritic transport of NMDA receptor
KIF21	Large protein complexes		Localized to dendrites
Dyneins			
Cytoplasmic dynein-1	Lysosomes, endosomes, protein and mRNA-complexes	Dynactin	
Cytoplasmic dynein-2			Cilia of photoreceptor cells
Myosins			
I, II	Actin		Neuronal migration and growth, hearing and balance
III			Vision
V, VI	Vesicles, small organelles, CaMKII, M4 receptor	Kinesin (V)	Neuronal morphogenesis, axonal transport, hearing and balance
VII, IX, XV	Actin		Hearing and balance, neuronal morphogenesis (IX)

Adapted from [10]-[15]. Nomenclature converted as described in [15].

appears that the C-termini of microtubules are particularly important for kinesin binding [35]. Removing the C-termini of microtubules with subtilisin increases the stability of the kinesin-microtubule bond, which as expected, interferes with normal dissociation and impedes transport rate. MAP2 and tau also bind to the C-termini of microtubules, and as a result, tend to decrease binding and processivity of kinesin. Experiments evaluating kinesin processivity in the presence of MAP2C and tau have shown that these MAPs decrease the rate of association between kinesin and microtubules, but once attached, these motors exhibit normal run lengths and rates [36]. Other studies were not able to replicate tau inhibiting the binding of kinesin in squid axon, suggesting earlier studies may not properly reflect *in vivo* interactions [37]. The stability of the microtubule is an issue. Kinesin binds with higher affinity to detyrosinated microtubules, and may act as a cross-bridge between these more stable microtubules and neurofilaments [38]. Acetylation of microtubules also increases their binding to kinesin and transport [39].

4.2.2 Dynein

Dynein is a minus-end-directed microtubule motor protein [11, 40]. Cytoplasmic dynein (also known as MAP1C) is found in neurons and plays particularly crucial roles in axonal transport. Of the two types, cytoplasmic dynein-1 has more diversified roles than cytoplasmic dynein-2, and is found in association with the Golgi apparatus, endosomes, as well as protein and mRNA complexes within neurons. Cilia of photoreceptor cells also contain cytoplasmic dynein-1.

Given that microtubules are homogeneously arranged being plus-ends directed towards the axon terminal, it is not surprising that cytoplasmic dynein, which is a minus-end-directed motor, gives rise to retrograde transport from the axon terminal to the cell body [41]. Retrograde transport rates of 100-400 mm/day have been measured for dynein. Nonetheless, it has been demonstrated that cytoplasmic dynein is responsible not only for fast retrograde transport, but also for slow anterograde transport [42]. Similar to kinesin, dynein takes 8-nm steps along microtubule tracks (equivalent to the size of the tubulin dimer). However, unlike kinesin, dynein is additionally capable of taking longer steps, moving sideways, and stepping backwards [43]. These varied types of movements were observed for single-molecule analyses of dynein processivity. Quantum dot nanoparticles were used to show that dynein-mediated transport of vesicles stays on individual microtubule protofilaments [44]. Gold nanoparticles have also been used to track dynein steps along microtubule tracks [25].

There is ATPase activity localized to the inner ring-shaped core of dynein [45]. Two levers extend from this core: one binds the cargo and the other binds to the microtubule. Nucleotide binding induces conformational changes in dynein enabling it to advance along the microtubule surface.

Dynein frequently associates with dynactin, a protein complex containing actin-related protein-1 (Arp1) among other subunits [46]. Dynactin is required for bi-directional movement of dynein along the microtubule track; however, dynactin is not necessary for dynein to attach to the microtubule [47]. The dynein/dynactin complex has further been demonstrated to interact with tau in the formation of linkages between microtubules and actin filaments [48]. The dynein/dynactin complex also works in conjunction with kinesin to mediate bi-directional transport of neurofilaments along microtubules [49]. In addition to dynactin, dynein binds to Lis1, a protein that is deficient in lissencephaly [50].

4.2.3 Myosin

Myosin is another prominent motor protein family, and it plays multiple roles in neurons and sensory cells (see Table 4.2; Figure 4.1). Unlike kinesin and dynein, which typically serve as motors traveling along microtubules, myosin primarily transports cargo along actin filaments. Thus, the functions of myosin are best understood in terms of what actin filaments do and to which cellular domains actin filaments are compartmentalized.

The myosin isotypes I, V, and VI have been shown to be essential to axonal transport of vesicles, small organelles, and endoplasmic reticulum along actin filament networks extending from the cell body to the axon terminal [51]. These fine networks of actin filaments fill the entire length of individual axons, and are particularly concentrated at the inner membrane surface. The axon terminal is also rich in actin filaments. Myosin transport appears to work in tandem with kinesin- and dynein-mediated transport enabling individual cargoes the opportunity to change from microtubule tracks to actin filament tracks [13]. The route to the final destination of the cargo is frequently provided by an actin filament track.

In addition to a role in axonal transport, myosin isotype Vb appears to play a specific role in mobilizing materials necessary for signal transduction in dendrites and dendrite spines [51]. Spines are actin-rich structures that have relatively few microtubules (see Figure 2.3 in Chapter 2). Functioning as highly specialized post-synpatic structures, spines hold receptor proteins coupled to scaffolding and adaptor proteins, which are in turn coupled to actin filaments. Their macromolecular complexity has caused them to be likened to "nanomachines" [52]. Spines also contain large endoplasmic reticular sacs filled with Ca^{2+} ions. This internal store of Ca^{2+} is released in response to IP_3 receptor activation, and subsequently acts as a second messenger activating both calmodulin and Ca^{2+}/calmodulin-dependent kinase II (CaMKII). Myosin contributes to this spine machinery by transporting the endoplasmic reticulum, scaffolding protein PSD95, CaMKII, and the GluR1 subunit of the α-amino-3-hydroxy-5-methyl-4-isoxazolepropionic acid (AMPA) glutamate receptor. Transport of receptors within the spine region is a dynamic process. Quantum dot nanoparticles have been used to track the ongoing mobilization of AMPA receptor subunits [53]. Additional receptor subtypes are mobilized or recycled by actin filaments. Myosin Vb, for example, regulates the recycling of the M4 muscarinic receptor [54].

In addition to being pivotal to transport in both axons and dendrites of neurons, myosin plays essential roles in specialized sensory cells of the auditory, vestibular, and visual systems. Hearing and the sense of equilibrium rely on shearing forces that deflect stereocilia located on hair cells in the cochlea, vestibular organ, and semicircular canals. These stereocilia are filled with actin filaments, which are attached to mechanotransducer channels via the myosin isotype Ic [55, 56]. Sound waves (vibration) and physical acceleration create hydrodynamic waves in the viscous fluids of the inner ear which results in tension on actin filaments of hair cells. This is in turn amplified by myosin Ic, thereby enabling the opening of multiple cation-sensitive channels. This myosin-mediated amplification is quite remarkable in that it enables hair cells to respond to movements of only a few angstroms. Photoreceptor cells, on the other hand, depend on myosin isotype III, which is responsible for transporting arrestin in and out of the microvilli [57]. This transport is necessary for transduction of visual inputs. Myosin is also critical to the process of adaptation, which is crucial for sensory systems being finely tuned to changes in stimuli and stimulus intensity, while decreasing response rates to constant stimuli.

Like kinesin and dynein, myosin is an ATPase. The power stroke of myosin stepping on actin filaments is driven by ATP hydrolysis [58]. The coupling of chemical energy obtained from ATP hydrolysis with the mechanical force exerted by myosin has been measured with nanoscale precision at the single molecule level [59]. Myosin typically advances by 4-nm steps on the actin filament track, but can take longer steps if the situation so requires. While all these motors are viewed largely in the context of transporting materials, it is important to note that like kinesin and dynein, myosin does under certain circumstances participate in stabilizing neuronal structure by linking actin with physically constrained adaptor and scaffolding proteins [60].

4.3 Mechanisms of Axonal Transport and Nanotechnology

Researchers interpreted results from early studies as axonal transport occurring at either fast or slow rates. It is now known that axonal transport rate is based to a large extent on the length of time that the molecular cargo pauses rather than how fast it is able to move [62]-[64]. As schematized in Figure 4.2, different axonal transport rates correspond to different motor proteins and cargoes. Synaptic vesicles containing neurotransmitters and their biosynthetic enzymes, for example, travel from the cell body to the axon terminal where they are stored. Neurotrophins encased in endosomes, conversely, travel from the axonal terminal to the cell body where they ultimately affect nuclear transcriptional factors.

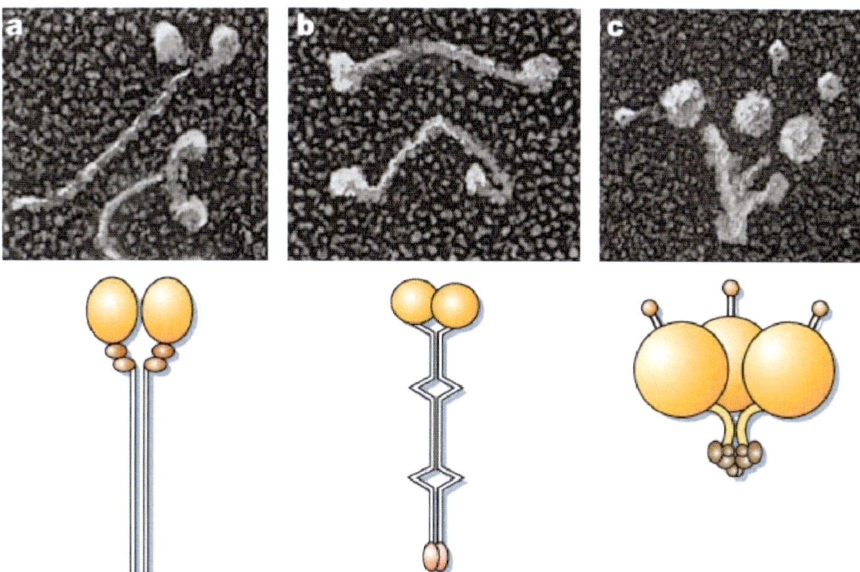

Fig. 4.1. Motor proteins of the kinesin, dynein, and myosin families. Myosin II (A), conventional kinesin (B), and dynein (C). Reprinted with permission; Schliwa and Woehlk, Nature, 2003, 422: 759-765 [61].

Nanotechnological methods are currently being applied to the study of cytoplasmic transport. Quantum dot nanoparticles have been used, for example, to track the movements of endosomes along microtubules, confirming that transport occurs in discrete 8-nm steps as expected for cargo attached to kinesin and dynein motors [65]. Magnetic nanoparticles possessing a cobalt ferrite oxide core (CoFe2O4) have been used to track different motor proteins that are responsible for mobilizing early versus late endosomes [66]. Nanotechnology has also provided tools capable of modifying cytoplasmic transport and functionally altering cellular function. Nanoparticles coated with polyethelene glycol, for example, are capable of increased cytoplasmic transport rates due to this coating decreasing steric interference with other structures in the cell [67]. Among the many potential biomedical applications, optimizing cytoplasmic transport would be expected to increase the efficiency of nanoscale drug-delivery devices.

As nanotechnological interventions increase the capacity to manipulate cytoplasmic transport, such manipulation can also be used to alter the structural morphology of neurons, i.e., to administer reconstructive "cell therapy". In addition to transporting membrane constituents, proteins, and mRNA, the cytoskeletal matrix contributes to the continuous building and maintaining of its own structure by transporting cytoskeletal protein subunits. As listed in Table 4.3, one component of slow axonal transport, called slow component a

(SCa), mobilizes short segments of microtubules, tubulin monomers, neurofilament subunits, and a host of MAPs, including tau and spectrin, whereas its counterpart, slow component b (SCb) mobilizes actin among other proteins. These two components of slow axonal transport, along with transport of structural constituents to dendrites, are inextricably linked to neural plasticity.

Table 4.3. Axonal transport modes for different cargoes.

Transport Type (motor)	Rate (mm/day)	Cargo
Fast Transport		
Anterograde (*kinesin*)	200-400	Synaptic vesicles, dense-core vesicles, secretory vesicles (*containing neurotransmitter-related biosynthetic enzymes, vesicular transporters, and neuropeptide precursors*)
Bidirectional (*kinesin, dynein*)	50-100	Mitochondria (*energy production, synthesis of glutamate*)
Retrograde (*dynein*)	200-400	Endosomes, multivesicular bodies, lysosomes (*containing neurotrophins, retrograde signal transduction molecules*)
Slow Transport		
Slow component a (SCa) (*kinesin, dynein*)	0.3-3	Microtubules, tubulin, neurofilament polymers and monomers, tau, spectrin
Slow component b (SCb) (*kinesin, dynein*)	2-8	Actin, clathrin, dynactin

Adapted from [62, 64].

4.3.1 Axonal Transport of Neurotransmitter-Related Proteins

One of the primary roles of fast anterograde axonal transport is to mobilize vesicles storing the machinery to manufacture neurotransmitters to the presynaptic terminal in preparation of release. Different neuronal populations rely on different processes in preparing for neurotransmitter release. Cholinergic cell bodies, for example, manufacture the biosynthetic enzyme choline acetyltransferase and the vesicular acetylcholine transporter, which are subsequently packaged into synaptic vesicles and transported to the axon terminal [68]-[70]. Similarly, noradrenergic and dopaminergic neurons produce and package their biosynthetic enzymes (tyrosine hydroxylase, dopa-decarboxylase, and for noradrenergic neurons, dopamine hydroxylase) and vesicular transporters into small synaptic vesicles and large dense core vesicles and then transport them to the axon terminal [68, 71]. Serotonergic neurons transport the biosynthetic enzyme tryptophan hydroxylase and the serotonin transporter to their axon terminals [72]. Neuropeptides, on the other hand, are often synthesized in the cell body as inactive precursors, packaged into secretory vesicles, transported to the axon terminal, and at some point during transport converted into active agents [73].

The synthetic enzyme phosphate-activated glutaminase is directly responsible for synthesizing glutamate from glutamine, and in some axon terminals, providing precursor glutamate from which to synthesize GABA [74, 75]. The mitochondria of astrocytes produce the majority of neurotransmitter glutamate via the tricarboxylic acid cycle, using approximately 20% of glucose intake for this purpose [76]. In a typical glutamatergic axon terminal, glutamine reuptake from the synaptic gap and from surrounding glial cells is available to be converted into glutamate; nonetheless, some of the original glutamate pool originates from the tricarboxylic acid cycle associated with mitochondria in the cytosol of the presynaptic neuron [77]. Glutamate, whether taken into the presynaptic axon terminal or synthesized in the axon terminal cytosol, is free to become incorporated into synaptic vesicles via specific vesicular glutamate transporters [78]. These vesicular glutamate transporters are carried from the cell body to the axon terminal in association with synaptic vesicles.

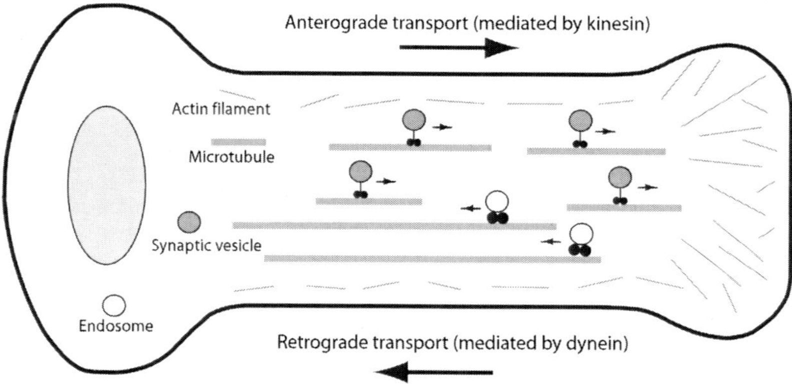

Fig. 4.2. Axonal transport is bi-directional partly because different motor proteins carry cargo in a preferred direction. Kinesin motors carry cargos such as synaptic vesicles anterogradely, whereas dynein carries cargos retrogradely. Myosin carries cargos along actin filaments (not shown).

Mitochondria, organelles critical to glutamatergic neurotransmission as well as providing ATP to serve metabolic needs, are transported bidirectionally along microtubule tracks and can also be transferred onto actin filaments [79]. Both mobile and static mitochondrial pools exist, and the cytoskeleton filaments along with various motor, linker, and docking proteins are responsible for moving or keeping mitochondria in place [79]. Axon terminals are particularly rich in mitochondria, as might be expected. Mitochondria are plastic, similar to other neuronal compartments. The cytoskeleton components and the various associated proteins also allow existing mitochondria to change shape, usually elongating to accommodate increased energy needs [80].

It is worth noting that biomolecular-nanoparticle hybrid systems have been devised that are capable of detecting enzymatic activity [81]. This approach and other nanotechnologies may be useful to nanoscale precision studies of the transport, release, and activity of neurotransmitter-related enzymes. Nanotechnological advances have also made it possible to selectively target precise cellular compartments, such as mitochondria in living cells. Nanocarriers made from dequalinium chloride encapsulating paclitaxel have been shown to selectively target mitochondria and induce apoptotic cell death [82]. Liposome-based nanocarriers have also been used to selectively target mitochondria [83]. Multifunctional envelope-type nanodevices are non-viral gene delivery systems that have additionally been used to target mitochondria, in this case delivering oligonucleotides or siRNA [84]. Applications of these nanotechnologies to neurobiology and neurological disorders have yet to be explored but already offer great promise of future advances in therapy and prevention.

4.3.2 Axonal Transport of Neurotrophins

Neurotrophins, such as nerve growth factor (NGF), brain-derived neurotrophic factor (BDNF), neurotrophin-3 (NT3), neurotrophin-4 (NT4), and glial-cell-line-derived neurotrophic factor (GDNF), are retrogradely transported from axon terminals to the cell body in endosomes (see Figure 4.2). The first step of this process is the binding to tyrosine kinase (Trk) receptors [85]. NGF is the classic example [86]. While there are additional modes of uptake, NGF typically binds to TrkA receptor, and then the ligand-receptor complex in taken up into the neuron by way of a clathrin-coated vesicle. These clathrin-coated vesicles contain Ras/mitogen-activated protein kinase (MAPK) and other components of signaling pathways that determine the fate of neurotrophin transport. In the presence of Rab5 (a small GTPase), clathrin-coated vesicles transform into early endosomes. Multiple endosomes may coalesce to form a multivesicular body, a specialized type of endosome containing smaller vesicles and an outer membrane that associates with Rab7, but not Rab5. In some cases, Rab7 endosomes evolve into late endosomes or lysosomes. Quantum dot nanoparticles have been used to show that endosomal uptake is a cytoskeletal-dependent process [87].

The internalized NGF-TrkA complex is subsequently transported to the cell body along microtubule tracks by the motor protein dynein [88]. Transport rates of neurotrophins have been calculated by a number of methods. Radioiodinated NGF, for example, travels at 10-20 mm/hr in cultured sympathetic neurons [89], which is consistent with the retrograde transport rates generally (see Table 4.3). Quantum dots tracking endocytosis and redistribution of TrkA receptors in developing PC12 cells reveal that both retrograde and anterograde transport of these receptors occurs at average rates of approximately 0.24 m/s [90]. This bi-directional transport of TrkA receptors depends on microtubules, as it is blocked by the microtubule-disrupting agent nocodazole. In other experiments using PC12 cells, TrkA receptors labeled with quantum dots were shown to accumulate mainly in the cell body and growing tips of neurites [90]. Within the cell body, Trk receptors activate the MAPK family member (i.e., extracellular signal-related kinase), ERK5, which then translocates to the cell nucleus and phosphorylates cAMP response element-binding protein (CREB) as needed to stimulate or inhibit transcription during the course of cell maintenance and survival [91].

In addition to retrograde transport, anterograde transport of neurotrophins BDNF, NT3, and GDNF also occurs, as does transcytosis, which enables a given neurotrophic factor to affect a series of interconnected neurons [92]. Transcytosis may enable particular neural networks to structurally adapt in response to heightened neural activity. Neurotrophins are transported throughout different compartments in the neuron, including dendrites as well as axons. Thus, neurotrophins mediate both presynaptic and postsynaptic structural changes that underlie growth and plasticity. Nanotechnologies, such as microfluidic chambers, enable the dispersion of small quantities of neurotrophins to specifically direct neurite outgrowth and study these phenomena [93].

Neurotrophin activation relies on neural activity. The internalization of BDNF and activation of TrkB receptors in hippocampal neurons, for example, depends on depolarization as well as on internal Ca^{2+} levels [94]. The Ras/MAPK pathway is responsible for keeping TrkB phosphorylated during the time it is transported to the cell body. Ligand-bound TrkB rapidly reaches the cell nucleus, whereupon it induces transcriptional changes by phosphorylating CREB [95]. CREB-mediated transcriptional changes support neuronal survival and structural adaptations. Application of BDNF to organotypic slices produces markedly increased branching in cortical pyramidal cells, a response that also requires ongoing neural activity and L-type Ca^{2+} channel activity [96]. In addition to mediating neural plasticity, BDNF results in increased neurotransmitter release from presynaptic hippocampal cells [97].

Microtubules, along with actin filaments, found throughout the neuron provide cables on which neurotrophin-filled endosomes travel. These tracks also provide potential pathways for signaling between neuronal compartments. As described in Chapter 3, electric signals are propagated along cytoskeletal tracks, which may in turn direct ionic fluxes. Of potential relevance to this mode of signaling, some neurotrophic actions involve the rapid propagation of Ca^{2+} or phosphorprotein waves traveling between the axon terminals to the cell body [86, 98]-[100]. Protein kinase/phosphatase cascades are calculated to span from the cell membrane to deep in the cytoplasm at rates of several hundred m/s [100]. Cytoskeletal proteins, by virtue of their placement in the cell and their high numbers of phosphorylation sites, may be the scaffolding upon which such phosphoprotein cascades spread.

4.3.3 Axonal Transport of Cytoskeletal Proteins

In addition to transporting neurotransmitter-related materials, organelles, and neurotrophins, components of the cytoskeletal tracks are transported as part of maintenance and growth. While components of microtubules, actin filaments, and neurofilament proteins have long been categorized as cargo of "slow" axonal transport (see Table 4.3), it is now apparent that these components move bi-directionally in the axon, giving rise to a net "slow" movement in the predominantly anterograde direction [101]. A variety of techniques have been used to track axonal movement of cytoskeletal proteins, including injection of radioisotope-labeled amino acids and fluorescent-labeled proteins. Researchers using fluorescent labeling demonstrated that both polymerized and soluble tubulins, as well as actin, travel along microtubule tracks in the axon [102]. Nonetheless, most microtubule proteins are not in motion. Compared to other cells in the human body, axons are particularly rich in βII- and βIII-tubulin, and these two isotypes are largely stationary in the axon [103].

Motor proteins participate in transporting cytoskeletal proteins. Short segments of microtubules (i.e., polymerized tubulin) are transported by cytoplasmic dynein on both microtubule and actin tracks, and by kinesins along microtubule tracks [104]. Axonal transport of tubulin depends additionally on collapsin response mediator protein-2 [105]. Neurofilaments, which are mainly found in axons, are anterogradely transported in association with plus-end directed kinesins, and are retrogradely transported by dynein [106]. MAPK is responsible for switching the direction of transport through its capacity to phosphorylate the C-terminus of neurofilament proteins. Highly phosphorylated neurofilaments tend to associate with dynein, whereas less phosphorylated neurofilaments are more likely to bind kinesin.

The lack of opportunity to bind to motor proteins may be responsible for stationary cytoskeletal components. Many older neurofilaments remain stationary, due in part to their binding to motors being blocked by newly transported neurofilaments [107]. As is the case with microtubular proteins, the majority of neurofilament proteins in the axon are stationary. Only about 5% of the total neurofilament protein moves at a time, and based on green fluorescent protein (GFP)-label studies, those neurofilament proteins move rapidly and asynchronously along stretches of microtubule tracks [64].

Some MAPs are transported along axonal microtubule tracks along with other slow components, and their phosphorylation state appears to regulate this process [108]. Unphosphorylated MAP1B tends to be transported with SCa, whereas phosphorylated MAP1B travels with SCb. Transport of MAPs would be expected to increase during axonal growth and collateral sprouting of new axon branches. A hypothetical framework for collateral sprouting called "cut and run" illustrates how microtubule segments might redistribute throughout neurons [109]. Severing enzymes, such as katanin and spastin, break microtubules into short segments, and these polymers are then able to move in any direction to nucleate longer microtubules. The assumption is that short segments produce less drag than longer microtubules and would thereby be more readily transported bi-directionally by plus-end and minus-end kinesins, and retrogradely by dynein. Assuming that individual microtubule "snippets" consist of particular sequences of tubulin isotypes, transport of these intact segments to different parts of the neuron might be a means of transmitting encoded information.

In addition to axonal transport of cytoskeletal components, there is evidence that small quantities of mRNA for β-actin are transported from cell bodies to axon terminals [110]. This has been noted for developing axons in the hippocampus, sympathetic ganglia, hypothalamus, and olfactory bulb. The most likely function of this mRNA is to manufacture actin in response to local signals.

4.4 Dendritic Transport

Dendrites differ from axons, both morphologically and functionally, due in large part to the transport of different proteins and organelles [111]. Axons are typically long and thin, especially at the initial segment, whereas dendrites taper and the transition from cell body to dendrite is gradual. Because of their shape dendrites can, to some degree, be viewed as extensions of the cell body, whereas axons are clearly distinct. Dendrite-specific proteins, including neurotransmitter receptors, scaffolding proteins, or select cytoskeletal proteins, further illustrate this distinction.

Certain cytoskeletal proteins are preferentially transported from the cell body to dendrites rather than to axons. MAP2 is a dendrite-specific protein that is blocked entry to the axon, in contrast to tau, which is preferentially transported into the axon [112]. After its synthesis in the cell body, MAP2 is sorted at the Golgi apparatus before being routed to dendrites. MAP2 regulates microtubule dynamics in dendrites dictating how microtubules will assemble or rearrange in accordance with activity patterns [113].

Different organelles are also transported to dendrites as opposed to axons. In contrast to axons, dendrites contain smaller versions of the Golgi apparatus, which nonetheless function similarly to those in the cell body. These scaled down organelles, called Golgi outposts, provide central sites for sorting proteins and governing dendritic growth locally within the confines of the dendrite [114]. Golgi outposts associate with the rough and smooth endoplasmic reticulum and with enzymes responsible for glycosylation. Golgi outposts and their enzymatic complexes are responsible for local dendritic processing and packaging of proteins into exocytotic vesicles. These are functions apparently reserved for the cell body and dendrites, since axons do not have Golgi outposts.

There is also evidence for retrograde transport of transcriptional factors from synapses to the nucleus. Activation of synapses located on distal dendrites has been shown to trigger the retrograde transport of CREB to the cell nucleus as part of memory-related plasticity [115].

As will be discussed in the next two sections, much of the transport that occurs within dendrites relates to their role in receiving inputs from other neurons. For one, there is abundant dendritic transport of vesicles containing neurotransmitter receptor proteins, scaffolding proteins, and certain cytoskeletal proteins specific to postsynpatic sites along the dendritic membrane and at spines. Secondly, dendrites also transport a select group of mRNA types, ribosomes, and organelles needed for *de novo* synthesis near to postsynaptic sites.

4.4.1 Transport of Neurotransmitter Receptors into Dendrites

Transport of glutamate receptors is pivotal to neuronal information processing, since the majority of synapses in the CNS are glutamatergic [116]. Glutamate receptors of the AMPA, kainate, N-methyl D-aspartate (NMDA), and metabotropic glutamate (mGlu) varieties are synthesized at the endoplasmic reticulum in the cell body, and then routed to the Golgi apparatus where they are packaged into vesicles and transported to postsynaptic sites in dendrites [117]. In addition to vesicular transport, glutamate receptors travel in the cytosol, with substantial lateral movement of glutamate receptors in the dendritic membrane. According to one proposed path, glutamate receptors are released in the cell body cytosol, then become incorporated into the membrane and subsequently move laterally into the dendritic membrane to their final position. Other scenarios include vesicles traveling some distance down the dendrite until receptors are released into the cytosol and then penetrate the membrane or move first into spines and then penetrate the postsynaptic membrane on the spine head. Individual receptors presumably take a particular path based on their final destination and other factors.

The vesicles transporting neurotransmitter receptors attach to kinesin motors. There are specific kinesins known to travel in dendrites, and some of these have been directly linked to specific glutamate receptors [118]. The NR2B subunit of the NMDA receptor is transported by KIF17 [119, 120]. The NR2B subunit binding to KIF17 occurs in the presence of the adaptor/scaffolding complex, consisting of mLin2, mLin7, and mLin10. Kinesin-1 (isoform KIF5A), as well as KIF5B and KIF5C, are responsible for transporting AMPA receptor subunits, in a process that involves scaffolding protein, GRIP1, and additional signaling molecules [121].

At the postsynaptic site, scaffolding proteins such as PSD95 and gephyrin (a $GABA_A$ receptor scaffolding protein) are responsible for linking neurotransmitter receptors to the underlying cytoskeletal structure [122]. Although these linkages anchor and partially stabilize receptors, many receptors nonetheless retain the ability to move laterally [123]. Scaffolding proteins and the underlying cytoskeleton, somewhat unexpectedly, both promote and restrict this lateral movement. Gephyrin scaffolding protein binds $GABA_A$ receptors to microtubules as well as to actin-interacting proteins [124]. This scaffolding protein is critical to concentrating $GABA_A$ receptors at active postsynaptic sites. actin filaments and microtubules exert antagonistic effects, insofar as actin filaments support lateral movements of gephyrin complexes and microtubules function to anchor them [125]. The lateral movements of gephyrin are furthermore activity-dependent. Structural reorganization has also been noted for other scaffolding proteins, including PSD95, which similarly depends on both synaptic activity and an intact actin cytoskeleton [126, 127]. Actin polymerization provides the propulsive force responsible for reorganizing the structure of the postsynaptic density of glutamatergic synapses, acting to stably position PSD95 [128]. As described in Chapter 3, electric or ionic signaling along actin filaments might play a role in orchestrating the overall patterns of receptor reorganization actuated by dynamic interplay between actin filaments and microtubules.

In order to study transport at the level of single molecules, nanotechnological approaches can be used to label receptor proteins and track their distribution and redistribution in response to synaptic input. Quantum dot nanoparticles are universal labels capable of detecting virtually any surface receptor [129]. Quantum dot labeling has revealed that individual $GABA_A$ receptors distribute along the postsynaptic membrane of dendrites in a pattern that correlates with the location of synaptic inputs [130]. Statistical modeling of simulated data matches experimental data; both of which affirm that the cytoskeleton plays a critical role in redistributing neurotransmitter receptors that mediate the neuronal response to external signals [131].

4.4.2 Transport of mRNA into Dendrites

Dendrites of neurons possess the ability to synthesize new proteins, although this capacity is more limited in extent than protein synthesis in the soma and it appears to occur selectively for certain mRNA types, many of which are involved in building, maintaining, and regenerating the postsynaptic apparatus. The capacity for *de novo* protein synthesis in dendrites was first suggested by electron microscope studies showing polyribosomes being transported to dendrites where they accumulate at the base of spines, particularly during synaptogenesis [132]. These initial observations were followed up by studies tracing the migration of $^3H - RNA$, which indicated a transport rate of 0.5 mm/day for mRNA entering and traveling along the extended length of individual dendrites [133]. We now know that the motor proteins responsible for transporting mRNAs along dendrites belong to the KIF5 family [134].

The mRNA for MAP2 was among the earliest to be localized to the dendritic compartment [135]. Since that initial observation, mRNAs for numerous proteins have been found in dendrites [136, 137]. Some of the most thoroughly studied dendritic mRNAs are listed in Table 4.4. In addition to MAP2 mRNA,

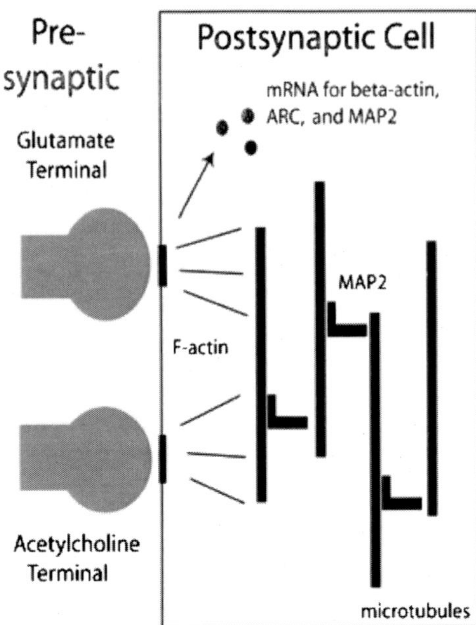

Fig. 4.3. The postsynaptic site contains mRNAs for cytoskeletal proteins. These mRNAs are in some cases stimulated by glutamate inputs and may also be modulated by neuromodulators such as acetylcholine. Altered levels of mRNA for cytoskeletal proteins can lead to restructuring of the postsynaptic site and the transport tracks to and from the site.

at least two other cytoskeletal protein mRNAs are transported into dendrites. These are the mRNAs for activity-related cytoskeletal protein (Arc) and β-actin [138, 139]. Arc mRNA travels bi-directionally in dendrites at rates consistent with movement along microtubule tracks and along actin filaments [140]. As illustrated in Figure 4.3, inputs to the neuron stimulate mRNAs for these cytoskeletal proteins at local postsynaptic sites, providing newly synthesized cytoskeletal proteins for constructing and restructuring the physical matrix surrounding postsynaptic sites. Given the cytoskeletal nature of these mRNAs present, transport routes to and from such endowed postsynaptic sites are likely under continuous reconstruction, with the internal cytoskeletal structure of the neuron serving as a reservoir of roadmaps for where to move what and under what circumstances.

In addition to mRNAs for cytoskeletal proteins, mRNAs for neurotransmitter receptors, neurotrophins, and signal transduction molecules are transported into dendrites. Multiple glutamate receptor subunits have mRNAs localized within dendrites. This phenomenon has been particularly well studied in hippocampal neurons, due in part to the large amount of literature on synaptic plasticity in the hippocampus. The mRNA for the NR1 subunit of NMDA receptor is found in dendrites of hippocampal neurons grown in culture during all stages of development [141]. The dendrites of hippocampal cells also contain mRNA for GluR1 and GluR2 subunits of AMPA receptors [142]. Levels of mRNA for GluR1 and GluR2 are decreased by NMDA receptor activation and increased by metabotropic glutamate (mGlu) receptor activation. The NMDA-mediated decrease in new transcription of mRNA for GluR1 and GluR2 subunits is Ca^{2+}-dependent, ERK/MAPK-dependent, and involves a synapse-to-nucleus signal that requires intact microtubules. Again, it is conceivable that the ionic or electric signals described in Chapter 3 might be involved in propagating a synapse-to-nucleus signal along microtubules or actin filaments directing the transport of the mRNAs for glutamate receptors subunits. Alternatively, it has been suggested that ERK is the probable molecular signal that influences transcription in the cell nucleus [142]. Rather than this kinase having to actually travel from the synapse to the cell nucleus, it is possible, as described earlier in this chapter, that microtubules and actin filaments carry phosphoprotein waves or protein kinase/phosphatase cascades, thereby enabling synaptic activity to rapidly affect transcription at the cell nucleus [98]-[100].

Hippocampal cell dendrites also contain appreciable levels of mRNA for BDNF and its receptor TrkB [143]. Both BDNF and TrkB mRNA is increased upon stimulation, demonstrated experimentally by exposure to 10 mM KCl. Upon release, this neurotrophic factor also upregulates its own levels. Stimulation of hippocampal cells causes the release of BDNF, which in turn activates the phosphatidylinositol-3 kinase to increase dendritic accumulation of both BDNF and TrkB mRNAs [144]. BDNF also increases translation of mRNAs for the NR1 subunit of NMDA receptor and the type α isoform of CaMK II [145]. All these actions contribute to the role BDNF plays in modulating synaptic responsiveness in accordance with synaptic activity. BDNF mRNA – its axonal transport, local trafficking, and transcription in the synaptic region – is critical to long-term potentiation (LTP), a classic model of synaptic plasticity of likely relevance to learning and memory [146].

The mRNAs for at least two protein kinases are transported to dendrites. Significant amounts of α-CaMK II mRNA have been localized to dendrites of hippocampal cells [147]. NMDA receptors regulate the translational capability of α-CaMKII mRNA by catalyzing its polyadenylation [148]. Thus, this mRNA is positioned to participate in synaptic plasticity. An isoform of protein kinase C, called protein kinase Mζ (PKMζ) is also transported into dendrites where it participates in synaptic modulation [149]. Local translation of PKMζ mRNA may also be crucial to memory, since PKMζ plays a role in maintenance of LTP, a possible neural activity correlate of memory [150].

Table 4.4. Dendritic transport of different mRNAs and their regulation by synaptic activity and neurotrophin stimulation.

mRNA type	Synaptic modulation
Cytoskeletal proteins	
MAP2	
β-actin	NMDA receptor and BDNF increase dendritic transport
Arc	NMDA receptor and BDNF increase dendritic transport
Neurotransmitter receptors	
NR1 subunit of NMDA receptor	BDNF increases dendritic translation
GluR1 subunit of AMPA receptor	mGlu receptor increases translation; NMDA receptor decreases levels
GluR2 subunit of AMPA receptor	mGlu receptor increases translation; NMDA receptor decreases levels
mGlu receptor	
Neurotrophins and receptors	
BDNF	
TrkB	
Signal transduction molecules	
α-CaMK II	NMDA receptor and BDNF increase dendritic translation; mGlu receptor increases transport
PKMζ	

Adapted from [136, 137]. See text for additional details.

4.5 Cytoskeleton Transport Dynamics with Neural Injury, Regeneration, and Morphogenesis

Cytoskeletal transport adapts to the ongoing conditions of the neuron. Situations that alter transport include neural damage, regenerative responses to damage, and neural growth. Neural responses to damage can be categorized as acute or permanent. Acute responses are transient, beginning immediately after the insult and subsiding days to weeks later. Permanent effects may include regenerative responses, which occur more gradually and vary greatly in the extent to which they occur, if at all. Dynamic cytoskeletal reorganization is particularly robust in dendrite spines. Spine morphogenesis occurs not only during early development, but also throughout the lifetime of the neuron in conjunction with learning and memory, as was reviewed in Chapter 3.

Nanoscale precision methods employing optical imaging and nanoparticles can be used to study single molecule events at nascent nerve contacts, as well as the underlying cytoskeletal protein dynamics [151]. In addition to furthering our understanding of axonal and dendritic transport, experiments on damaged neurons are often aimed at establishing new clinical approaches for treating brain and nerve damage following traumatic brain injury, stroke, tumors, and degenerative diseases. Nanotechnology offers new approaches to treating nervous system injury and guiding neural growth. Gold nanoparticles, for example, have been found to enhance the ability of certain materials to guide axonal growth [152]. Nanostructures can be used to deliver drugs or genes. Polyethylenimine/DNA nanocomplexes are as efficient as adenoviruses for delivering gene vectors, and they reliably transport gene vectors through neurites [153]; however, other nanoparticles demonstrate much more efficient expression of the gene vector [154]. The ideal nanoparticle for gene therapy – one that is efficiently transported along neurites and leads to gene expression – may yet to be discovered. These and other potential nanotechnological treatments for nervous system disorders will be discussed in Chapter 5.

4.5.1 Acute Responses to Neuronal Insult

When a neuron is damaged, there is major reorganization of the cell, which often involves increased transport of cytoskeletal proteins and signaling proteins. Such neuronal responses have been studied following axotomy, nerve crush, hypoxia, ischemia, and toxic insult. Traumatic brain injury causes both immediate and delayed acute effects, which include the generation of calcium-activated proteases, breakdown of cytoskeletal proteins, mitochondrial dysfunction, and free radical generation [155]. Cerebral blood flow is compromised within minutes of a serious head injury, and the resulting hypoxia and ischemia are largely responsible for the longer lasting effects of mechanical injury [156]. The neuronal cytoskeleton is disrupted not only by the mechanical injury, but also by the hypoxia and ischemia [157]. Traumatic brain injury causes a rapid and long-term accumulation of β-amyloid precursor protein, neurofilament proteins, and synuclein in damaged axons [158].

Depending on the severity of the trauma, mechanical injury can also disrupt the axonal membrane [159]. Nerve crush leads to axonal swelling within 4 - 24 hours, followed by axonal detachment at the site of injury and Wallerian (i.e., anterograde) degeneration. In the case of moderate to severe trauma, many of these changes may be initiated by damage to the bilipid membrane and the attached submembranous spectrin, actin, ankyrin, and β1-integrin.

There is additional evidence that changes at the site of injury within the axon are relayed back to the cell body and trigger chromatolysis (i.e., the breaking apart of chromatin) [160]. This retrograde signal has been identified as a complex consisting of vimentin, dynein, importin β, and ERK1 and ERK2-the latter being members of the MAPK family [161, 162]. Vimentin is an intermediate filament from the same family as neurofilament proteins that is ordinarily expressed early in development but nonetheless appears to be synthesized *de novo* in injured axons. Before being transported, vimentin is cleaved by the calcium-activated protease, calpain. Cleaved vimentin then binds to phosphorylated ERK1 and ERK2, and then to cytoplasmic dynein and importin β. The complex is retrogradely transported to the cell body, where phosphorylated ERK2 and ERK2 stimulate transcription factors. As previously noted, the electrical signaling mechanisms outlined in Chapter 3 should enable microtubules and actin filaments to carry phosphoprotein waves or protein kinase/phosphatase cascades back to the nucleus, which could effectively act much more rapidly than transporting signaling molecules.

Traumatic brain injury affects microtubules and MAPs in addition to the neurofilament proteins, indicating compromise to the entire transport matrix. Measures of tubulin demonstrate microtubules are affected by nerve injury. In one experiment employing sciatic nerve crush, β-tubulin isotypes II and III were increased at both 4 and 14 days post-injury [163]. Multiple laboratories employing various animal models of head injury have additionally demonstrated that trauma changes the pattern of the dendritic MAP2 [164]-[167]. In one study, a time course analysis showed moderate traumatic brain injury in mice produced profound decreases of MAP2 in the cortex and in parts of the hippocampus at 5 minutes post-injury, whereas MAP2 in neurons that survived the trauma began to recover at 90 minutes and continued to show signs of recovery at 24 hours after injury [168]. Conversely, other researchers examining MAP2-stained hippocampal neurons from animals with brain injury found long-lasting deficits, describing hippocampal cell dendrites as "fragmented, scattered, unaligned, and consisting of irregularly spaced and darkly stained swollen segments" [169]. Thus, both transient and conceivably permanent effects on MAP2 dendrites have been described.

To what extent changes in MAP2 represent altered transport of this cytoskeletal protein or its mRNA in dendrites is unclear. Mild traumatic brain injury effects on MAP2 were found by one group to endure longer than transient changes to β-amyloid precursor protein [170]. These researchers interpreted altered β-amyloid precursor protein as indicative of abnormal axonal transport, which was temporary and correlated with transient cognitive impairments in the affected animals. MAP2 changes, on the other hand, were more persistent than the cognitive impairments. Another research group, however, found that head injury caused long-lasting (21 day) alterations to β-amyloid precursor protein, whereas reduced MAP2 normalized over that time frame [171]. These controversies surrounding MAP2 responses to injury have not been fully resolved.

Tau is another microtubule stabilizing protein that has been observed to change with traumatic brain injury. Patients who suffered traumatic brain injury and showed higher increases in tau levels in the ventricular cerebrospinal fluid (CSF) had a poorer prognosis and outcome 1 year later [172]. These researchers interpreted increased CSF tau as being a marker for axonal damage. Tau also influences axonal transport and its improper processing leads to neurodegenerative changes, such as neurofibrillary tangles. Neurofibrillary tangles, composed of hyperphosphorylated tau, accumulate following repetitive head injury and in Alzheimer's disease, but not after single-episode traumatic brain injury [173]. Rather than being similar to the neuropathological cascade that occurs during tangle formation, traumatic brain injury seems to produce a large (1.5 - 8 fold) increase in a cleaved form of tau [174]. Thus, single-episode traumatic brain injury appears to more narrowly influence the breakdown or proteolysis of tau. The protease calpain is increased following traumatic brain injury, and the cleavage of some, but not all, cytoskeletal proteins seems to be mediated by calpain [175, 176].

There are behavioral consequences of an impaired cytoskeleton; in particular, there are deleterious effects on functions that rely on transport of materials to dendritic compartments of the neuron. Animal studies verify that traumatic brain injury impairs memory [177]. Impairment on the Morris water following traumatic brain injury, moreover, correlates with the extent of damage to MAP2-stained dendrites throughout the hippocampus [178]. Accordingly, memory impairment appeared to vary in severity in proportion to damage to dendritic transport systems.

4.5.2 Transport Regulation in Regeneration and Morphogenesis

Transport of cytoskeletal proteins is fundamental to both axonal regeneration and dendrite morphogenesis. Various regulatory factors, signal transduction molecules, and neurotrophic factors regulate these processes. The localization of some of these factors accounts, at least in part, for the difference in regenerative capacity within the CNS versus the peripheral nervous system.

During axonal regeneration, the rebuilding of the internal structure of the axon relies pivotally on the anterograde transport of tubulin and actin proteins, which produce a forward-moving protrusive force. Supporting glial cells found in the CNS (particularly the oligodendrocytes) contain a number of factors that limit regeneration. These factors include Nogo-A, myelin-associated glycoprotein, oligodendrotcyte-myelin glycoprotein, and tenascin-R [179]. Such factors are at least partly responsible for the limited capacity for nerve regeneration within the CNS. In the absence of these factors, axons regenerate reasonably well in the peripheral nervous system – that is unless the damage is severe.

There are also factors that enhance regeneration, both centrally and peripherally. The second messenger cAMP stimulates the upregulation of tubulin isotypes associated with axonal growth, and enables sensory axons to partially reinnervate CNS targets [180, 181]. Nonetheless, the increase in cAMP produced by a conditioning lesion is not sufficient to overcome the effects of CNS inhibitory factors or physical barriers, such as glial scarring. Various experimental approaches have been devised to overcome these barriers. In one approach, transplanting fibroblasts that were genetically modified to secrete BDNF into the zone of damage in the spinal cord of rats resulted in partial behavioral recovery [182]. Increasing neural activity may also facilitate recovery because BDNF release is increased when nerves are active. Electrical stimulation applied to axotomized femoral nerve has been shown to increase BDNF and TrkB mRNAs, along with tubulin and growth-associated protein-43 mRNAs, but to decrease genetic expression of neurofilament protein NF-M [183]

Axonal regeneration differs from axonal development in a number of key ways, as has been shown in cultured rat cortical neurons [184]. The growing tips of regenerating axons have less elaborate filopodia, the rate of growth is less, and there are more frequent pauses, as compared to that of developing axons. Regenerating axons also are less sensitive to neurotrophins such as BDNF and glial-derived growth factor. This last result may be specific for cortical neurons under these precise experimental conditions, since other axons clearly respond to this neurotrophin. Despite these and other experimentally observed differences, there are many similarities in growth patterns and the cytoskeletal transport responses observed during development and regeneration.

MAPs also appear to regulate axonal development and regeneration. MAP1B regulates axons attaining their appropriate morphology during development and regeneration [185]. Knockout mice null for the MAP1B gene have axons with excess collateral branches and deficits in axonal turning to find targets. These mice also have reduced acetylated tubulin, suggesting that MAP1B might exert its action by enabling posttranslational modifications of microtubules. A stabilized cytoskeletal matrix is required for normal transport function. MAP2 and MAP1B, which are responsible for dendrite stability, are substrates of c-Jun N-terminal kinase 1 (JNK1), and this member of the MAPK family regulates dendrite architecture [186, 187]. Knockout mice null for the JNK1 gene show signs of axonal degeneration in the spinal cord, as well as dendritic alterations in the hippocampus. Pyramidal cells in the hippocampus of JNK1-/-mice are reduced in overall extent and microtubules that fill these dendrites are shortened. Thus, the phosphorylation of MAP2 and MAP1B by JNK1 is a crucial step in the attainment of the cortical architecture upon which higher cognition depends.

During early development, both axons and dendrites are subject to major reorganization. As has been shown in developing hippocampal cells, neurites destined to become either dendrites or axons go through steps including: (1) an increase in the production and transport of the plasma membrane, (2) an increase in the synthesis and transport of signaling molecules, (3) enhanced actin dynamics, and (4) enhanced microtubule polymerization [111]. In the absence of injury, adult CNS axons are relatively stable in terms of their morphology. This is not the case for dendrites, and in particular, for dendritic spines. Morphological changes are noted in adult dendrites and spines, and these changes may contribute to memory storage. Actin filaments provide the main structure of spines, but other proteins, including drebrin A, are critical for recruiting actin and stimulating actin filament assembly [188]. Just as actin filaments form the growing tips of axons, filopodia are the precursors of nascent spines. Drebrin A initially appears at the submembranous contact site preceding the development of a spine.

Dendritic spines receive predominantly glutamatergic inputs, and are studded with AMPA, kainate, NMDA, and mGlu receptors. Glutamate receptors possess the capability to modulate spine morphology. Prolonged stimulation of NMDA receptors (5 min) has been shown to cause a widespread disappearance of spines [189]. AMPA and L-glutamate produces a similar response. This action appears to be mediated by actin depolymerization. NMDA effects on spine density have also been noted. Knockout mice having the NR1 subunit of the NMDA receptor deleted had decreased spine densities, while individual spine head size was increased for layers 2 and 3 pyramidal cells [190]. The working hypothesis accounting for these results is that spines grow in search of input, and then retract upon receiving excess stimulation, the final result being a finely tuned, precision-sculpted receiving channel.

The overall structure of the individual dendrite is also malleable in the adult animal, and this appears to be related to transport along the cytoskeleton and the cytoskeletal structure. In a Drosophila model, mutations of dynein and kinesin genes led to decreased dendrite arbors and a tendency for dendrites to form proximal to the cell body [191]. This effect was due in part to down-regulation of transport of early endosomes.

Pyramidal cells have been shown to possess intrinsic programs that determine their characteristic morphologies [192]. Apical (i.e., primary or principal) dendrites and basilar (i.e., secondary) dendrites have different developmental courses. The larger primary dendrite expresses a sudden dynamic surge in growth that is not noted for secondary dendrites. Key cytoskeletal proteins or regulators of the cytoskeleton are suggested to serve as the mechanism responsible for this intrinsic program, thereby implicating transport mechanisms. Apical dendrites demonstrate increased anterograde transport of Golgi outposts and AMPA receptor subunits as compared to basilar dendrites, and Golgi outposts tend to aggregate at branch points in apical dendrites [193, 194]. These latter results indicate that selective transport along cytoskeletal tracks plays a critical role in dendrite morphogenesis and differentiation. Given that information is computed in and stored by the cytoskeletal matrix as described in Chapter 3, neuronal structure is an arguably valid correlate of cognitive function and memory.

4.6 Cytoskeletal Transport in Learning and Memory

Transport of mRNAs, cytoskeletal proteins, scaffolding proteins, and neurotransmitter receptors play key roles in dendritic and spine reorganization, and this kind of reorganization is largely believed to underlie learning and memory [195]. LTP and long-term depression (LTD), respectively, lead to enlarged and shrunken spine heads [196]. This partly reflects alterations in the transport and translation of mRNAs, as described earlier in this chapter [136, 146]. Polyribosomes, the site of mRNA translation, move from the dendrite shaft into spines of hippocampal neurons upon induction of LTP [197]. There is morphological reorganization of spines with LTP consolidation, which is dependent on actin polymerization. Small stubby spines are converted into large mushroom-shaped spines as a result of increased transport of f-actin and scaffolding proteins, such as actin depolymerizing factor (ADF)/cofilin and PSD95 [198, 199].

LTP-inducing stimuli have been shown to stimulate the trafficking of "recycling endosomes" into the spine region [200]. These mobilized endosomes contribute to spine morphogenesis, providing a direct link between LTP and spine enhancement. Experiments using knockout transgenic mice have demonstrated that the actin motor, myosin Vb, is essential to LTP because of its role in local trafficking of recycled endosomes [201]. Myosin Vb was found to mobilize recycled endosomes (which are enriched with AMPA receptors) in response to LTP-inducing stimuli, thereby facilitating the insertion of these receptors into the postsynaptic membrane. Mechanistically, myosin Vb mediates endosomal recycling via a conformation shift from a folded state, in which myosin Vb is inactive, to the unfolded conformational state of the protein, which is active.

There is also direct evidence that motor proteins play a role in animal learning and memory. The overexpression of KIF17, which is known to transport the NR2 subunit of NMDA receptors, results in enhanced memory performance [202]. Other kinesins that are likely to play a role in memory are KIF5 (which is responsible for transporting the GluR2 subunit of the AMPA receptor) and KIF2 (which participates in microtubule depolymerization) [203]. It does not appear, however, that GluR1 receptor is required for LTP-dependent spine enlargement [204]. LTP-induced spine enlargement does depend on the neurotrophin BDNF [205]. Moreover, certain patterns of activity, such as the theta rhythm, stimulate actin assembly in enlarging spines following the induction of LTP [206].

The role of cytoskeletal transport in learning and memory can be summarized as follows:

> That transport of synaptic proteins and organelles along microtubules and actin filaments is critical to animal learning and memory suggests intraneuronal connectivity is as fundamental to memory as is interneuronal connectivity.

As shown in Figure 4.4, most current models for synaptic plasticity focus on morphological changes at synapses (or spines). A popular model for learning posits coincident input leads to altered synaptic activity and modified connectivity within a larger neural assembly [207]. Accordingly, the neural correlate of learning is a change to the synapse, its response, and the activity patterns in the overall network of interconnected neurons often spanning many cortical areas. Spatial and temporal relationships existing between the coincidently activated synapses are stored within the larger neural network. Figure 4.5 illustrates the current proposal of intraneuronal connectivity within the cytoskeleton contributing to the memory trace. Actin filaments in spines connect with microtubules in dendrite shafts, enabling communication between synaptic sites. These cytoskeletal tracks route necessary proteins and mRNAs to synapses in response to changes in neural activity. There exist numerous sites, or choice points, throughout the interior of the neuron where the relationships between coincidently activated postsynaptic sites could be stored. This type of intraneuronal storage is presently hypothesized to operate in tandem with alterations in connectivity within the overall neural assembly.

The currently proposed intraneuronal storage model has experimental support. The underlying microtubule matrix is extensively reorganized with learning and memory, as was discussed in Chapter 3. Proteolysis of MAP2 occurs with memory processing, implicating breakdown followed by rebuilding of the cytoskeletal matrix [208, 209]. Other MAPs participate in learning, tau protein, for example [210]. Thus, an extensive reorganization of the internal neuronal matrix occurs with memory formation, supporting the notion that relationships between synaptic inputs could be stored in the transport tracks. According to this hypothesis, the neuron coordinates simultaneous transport to coincidently activated synaptic sites. This translates into our memories and basic mental representations deriving from the very basic mechanical need for neurons to transport materials to the right place and at the right time. This proposal is consistent with a nanomechanical view of the entire neuron, instead of focusing solely on the synapse.

4.7 Biophysical Models of Transport

Biophysical mechanisms operating in nanoscale dimensions are responsible for the transport mediated by motor proteins along cytoskeletal tracks. A number of biophysical models have been proposed to account for the mechanism of kinesin walking along microtubule tracks and myosin walking along actin filaments. Additional models describe the overall dynamics of the cytoskeletal network in relation to transport function.

For mechanically based models, fluctuations in the surrounding environment can be dealt with by Langevin and Fokker-Plank equations. Langevin dynamics approaches biomechanics using stochastic differential equations. For a single mass point the Langevin equation in one dimension is [215]:

$$\xi \frac{dx}{dt} = -\partial_x V(x) + F(t) \tag{4.1}$$

where x is the spatial coordinate along the microtubule long axis, $V(x)$ is the effective potential, ξ is the friction coefficient, and $F(t)$ the uncorrelated random force [216].

The Fokker-Planck equation relates the probability density $P(x, t)$ to the particle at the position x at the time t, as follows [217]:

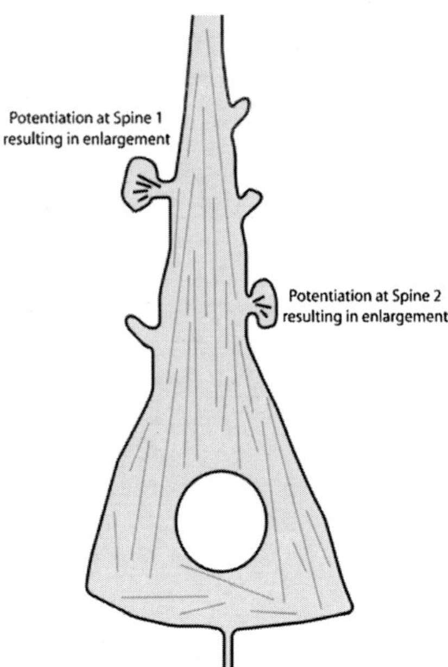

Fig. 4.4. Actin filament-dependent changes in spine size and shape follow certain patterns of input and provide experimental support for the long-held notion of modified interneuronal connectivity attendant with learning and memory (see text). Diagramed here are two enlarged spine heads (i.e., the conversion of nascent spines to a mushroom-shaped spines), similar to those detected following the application of an intense stimulation or tetanus. Synapse-oriented hypotheses assume the spatial and temporal relationships are stored as a memory trace in the form of changes in connectivity within the larger neural assembly.

$$\frac{\partial P(x,t)}{\partial t} = \left(\frac{\partial}{\partial dx} V(x) + \frac{\partial^2}{\partial x^2} D \right) P(x,t) \tag{4.2}$$

where $V(x)$ is the substrate potential and D is the diffusion coefficient. These types of stochastic ratchet-based models are attractive first order approximations due to their simplicity but have numerous drawbacks such as their unrealistic one-dimensional representation of the motor and a lack of orientational specificity for binding between the motor and the filament.

Bolterauer and colleagues, following an exhaustive critique of the existing alternative models, have proposed a novel physical mechanism for the processive motion of two-headed kinesins, in particular conventional kinesin [211]. This detailed physical model describes how the binding of kinesin heads to tubulin leads to a twisting of the neck linker region of kinesin which stores elastic energy and how unbinding relaxes the twist leading to processive motion. The model predicts two torsional springs per kinesin head that could potentially store elastic energy originating from the chemical energy of ATP hydrolysis. This model is consistent with recent evidence that the force generation

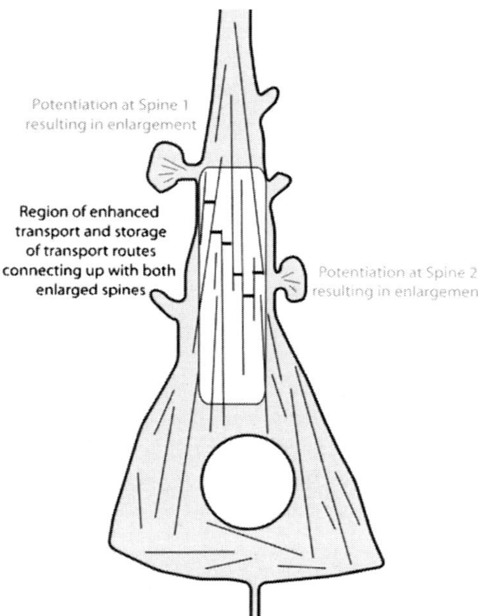

Fig. 4.5. The currently proposed learning and memory model complements prevailing neural network models by adding the notion of intraneuronal connectivity. In the currently proposed model, the neural assembly is altered as in Figure 4.4, but there is also storage within the microtubule matrix stabilized by crossbridges made of MAP2. This restructured matrix provides the neuron with a record of previous transport represented by structural modifications in transport tracks built to accommodate certain patterns of inputs and neural responses to those inputs.

of kinesin (including members of many families) is produced as a result of conformational changes in nine amino acid residues of the N-terminus of the neck linker, cover-neck bundle [212]. Below we summarize the main issues giving rise to the differences between this new torsional spring model and the existing alternative models [211]:

1. This model is a two-dimensional model employing a hand-over-hand movement with respect to the surface of the microtubule.
2. The heads of kinesin are viewed and treated as extended objects in two dimensions, having at least two mass points to define both position and orientation of each kinesin head.
3. The binding of the kinesin heads onto the microtubule depends on a correct orientational fit, and it has to be directed in a vectorial fashion, with scalar coordinates being insufficient to describe it properly.
4. The symmetrical arrangement of the heads in the relaxed state of the kinesin molecule, which has been demonstrated experimentally, is critically incorporated into the model.

5. During simultaneous binding of both heads to the microtubule surface, the kinesin molecule generates internal elastic stress making this conformation energetically unfavorable leading to the subsequent release of one head.
6. To accurately account for the sequence of conformational transformations there must exist at least two torsional springs, which are alternately loaded with elastic energy during the kinesin walk process.
7. Consistent with experimental results for plus-end processive kinesins, the model predicts that if only one head of a processive kinesin is bound, the other unbound head points in the walking direction.
8. According to the model, the internal conformational structure of the kinesin molecule defines the walking direction; however, a change of the conformational structure is able to reverse the walking direction. For example, it is sufficient to change the angle between the head groups in the relaxed state, thus, changing the direction of walking. This raises the possibility of biochemical nanoengineering of the structure of motor proteins, especially the linker domains to achieve a desired direction and speed to processive motion.
9. Consistent with experimental evidence, the directionality-defining springs are predicted to be located in the neck linker region connecting each kinesin head to the dimerization domain.
10. The resultant force-velocity relationship has the usual form found in various experiments supporting the accuracy of this torsional spring model.

Although conventional kinesin is well known to progress approximately 100 steps before it dissociates from the microtubule, it may maintain a "quasi-bound" state. Thus, the internal structure of kinesin (i.e., twisted, relaxed, or transitional) is likely to govern walking along the microtubule as proposed in the model of Bolterauer et al. The following assumptions were made in the derivation of the torsional spring model for kinesin walking [211]:

- The biochemical cycle and mechanical movement are independent.
- There are alternating chemical and mechanical transformations.
- There are combined chemical and mechanical transformations.

As in many chemical reaction-based models, it is common to start with the assumption that for every mechanical step, there is a force-dependent reaction rate. Solving the stationary reaction equations leads to the Michaelis-Menton form for velocity:

$$v = \delta k_{ATPase} = \delta k_{cat} \frac{[ATP]}{K_M + [ATP]} \tag{4.3}$$

where δ is the step size,$[ATP]$ represents the ATP concentration, K_M is the Michaelis-Menten constant, and the k denotes the corresponding reaction rate.

The hand-over-hand model for two-headed kinesin yield results in agreement with expected behavior [213]. The model for the hand-over-hand motion of kinesin is based on the scheme derived from [214]:

$$\mathrm{K} \underset{k_{-1}}{\overset{k_{+1}}{\rightleftharpoons}} \mathrm{KT} \longrightarrow^{k_2} \mathrm{KP} \longrightarrow^{k_3} \mathrm{P} \tag{4.4}$$

with reaction rates denoted by $k_{+1}[ATP]$, k_{-1}, k_2, and k_3, respectively and an Arrhenius assumption for the reaction rates. (The propagation velocity is given by the previous equation.)

As illustrated in Figure 4.6, the torsional spring model depicts a cycle that includes an initial step (State 0) and 4 additional steps (States 1-4) [211]. Initially, kinesin is unbound and the neck region is relaxed. In step 1, kinesin twists as head A turns in a clockwise direction to reach the next binding site on the microtubule. It is during step 2 that the energy from ATP hydrolysis is used to release either head A or B. If head A is released, the cycle goes back to step 1, but if head B is released it is possible for it to swing around head A making an 8-nm step along the microtubule. In step 3, the binding energy is used to load spring B as it turns counterclockwise. The end result is that spring B is loaded and spring A is unloaded. Finally, in step 4, head A is released and swings around head B.

The torsional spring model proposed in [211] has potential applications to nanotechnology and nanomedicine. It provides some of the necessary information to enable the biochemical construction of linker domains of motor proteins having different propagation directions and velocities. These designer motor proteins could be used as components in more elaborate "nanofactories" or in biomedical nanodevices.

Other biophysical models have been applied to the problem of kinesin motor protein transport along microtubules. Kinesin has been modeled as a Brownian stepper whose energetics predominantly produces forward movement along the microtubule, with an occasional backstep [218]. A Brownian motion model predicated on Langevin dynamics has also been applied to specifically to the single-headed kinesin, KIF1A, describing its motive mechanism [219]. Computer simulations suggest that KIF1A makes lane changes onto different microtubule protofilaments that are dependent on local concentration of motor proteins and availability of ATP [220].

Biophysical models have also attempted to solve the problem of how multiple kinesin motors coordinately operate during transport. Kinetic modeling of collective transport among multiple kinesin motors suggests that loose mechanical coupling may enable their ability to transport efficiently [221]. Monte Carlo simulations have reliably predicted how cargos are transported by multiple kinesin motor proteins as opposed to one, the result being two motors can move cargos longer distances, but with more stalling [222]. Compared to experimentally measured parameters of microtubule transport along kinesin-coated surfaces, Monte Carlo simulations yield similar results, indicating such calculations could be useful in rational design of individual or network nanotransport systems [223].

There is also considerable interest in the biophysics of transport in networks of actin filaments and microtubules. Actin filaments have been modeled

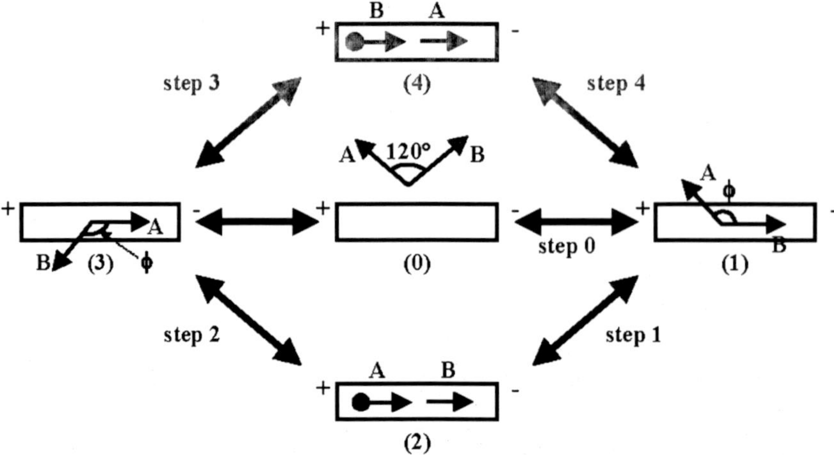

Fig. 4.6. The torsional spring model for kinesin walking. Adapted from [211].

as out-of-equilibrium dynamic networks exhibiting noisy fluctuations governed by stochastic nanoscale motions deriving from chemical interactions between surrounding molecules [224]. A mixed elastic model of myosin and kinesin-mediated transport has been able to efficiently generate saddle points, transition paths, potentials of mean force, and partition functions [225]. Network models of motor protein transport along actin filaments and microtubules are capable of accounting for large-scale collective behavior among nanoscale structures.

4.8 Bioengineering of Transport Molecules and Hybrid Biological Devices

Bioengineers have recently been successful in terms of constructing and modifying transport motors capable of carrying both biological and non-biological materials. Moreover, the next few decades promise to bring more sophisticated devices and advances in design. Practical applications for these nanoscale technologies include nanoanalytical systems, adaptive (self-healing) materials, and directed molecular assembly [226]. Microtubules can also serve as templates to be used in the mass production of more complex nanostructures.

Nanofluidics and microfluidics are emerging analytical technologies having multifold potential applications, including the assay of very small quantities of enzymes, DNA, toxins, and pathogens [227, 228]. Cytoskeletal tracks and motor proteins are being utilized in these and related nanotechnologies. Motor proteins possess qualities making them ideal nanoactuators capable of transporting materials for millimeter distances with nanometer precision [229]. Kinesin and dynein motor proteins have been employed, for example, as nanoactuators in microfluidic chambers to regulate nanotransport [230]. DNA-conjugated microtubules have also been constructed that are capable of loading and unloading quantum dot nanoparticle cargo and transporting this cargo along kinesin-coated surfaces [231]. This multi-operational nanoscale apparatus has potential applications in nanofluidics, nanobiosensors, and as a component in more complex nanoscale devices.

Kinesin motors are not only able to carry nanoparticles (e.g., quantum dots), they also provide sufficient force to transport much large microparticles. According to one experimental design, microtubules conjugated to streptavidin and biotinylated antibodies, which were sandwiched with fluorescent-labeled antibodies, still demonstrated the capacity to travel along kinesin-coated surfaces [232]. Such assemblies constructed from nanoparticle-cytoskeletal protein conjugates can be employed as nanobiosensors by incorporating selective antibodies capable of detecting specific protein molecules that bind to the antibody sandwich. One issue that remains unresolved for some uses of nanoparticle assemblies is whether transport rate is affected by the attachment of the nanoparticle. While one study found that certain fluorescent-labeled conjugates attached to microtubules only slightly reduced their transport rates [233], another study evaluating both metal and semiconductor nanoparticles determined that the size of the particle and other physical parameters contributed to alterations in transport rate [234]. The effects that individual nanoparticle properties have on transport need to be fully evaluated for various nanoparticles under the relevant conditions for each suggested use.

Another major challenge in nanotransport design is to provide a mechanism for external control over the transport function. Possible solutions include using electric fields or functionalizing nanostructures so that they respond to magnetic fields or to the chemical environment. The application of electric fields has been found to exert control over the gliding of microtubules on kinesin-coated surfaces and the bending of microtubules in a length-dependent manner [235, 236]. This sensitivity to electric fields could be exploited in the context of constructing nanotransport systems. Alternatively, functionalized microtubules coated with cobalt ferrite ($CoFe_2O_4$) nanoparticles have been shown to respond to externally applied magnetic field, enabling experimenter control over the orientation of these microtubules [237]. Other nanoparticles having pH sensitivity and capability to bind myosin are suitable as "smart" drug delivery devices that can be controlled by the chemical environment [238].

The earliest eukaryotes already possessed multiple motor proteins and higher eukaryotes have elaborated on those basic motors through gene duplication, alternative splicing, and subunit additions [239]. Nanotechnology allows for experimenter-manipulated modifications of motor proteins. Biological kinesin motors can be modified through genetic engineering, chemical treatments, or nanomanufacturing to better suit a particular need. In many respects, native kinesin is an ideal nanoactuators; nonetheless, it can be modified for a particular task. Controllable switches can be built into the kinesin molecule, thereby enabling researchers to alter its behavior, including velocity of transport. One group of researchers succeeded at producing a genetic mutation in kinesin that altered its neck-linker region to include a binding site for Zn^{2+} [240]. This genetic mutation enabled an ATP-independent manipulation of kinesin by Zn^{2+}.

Another way to modify transport is to stabilize cytoskeletal protein tracks. The microtubule tracks, upon which kinesin travels, have fast turnover rates and half-lives in the order of minutes *in vitro*. These half-lives can be extended up to the order of days with certain modifications [241]. In addition to taxol, which is well known to increase the stability of microtubules, various crosslinking agents – glutaraldehyde, sebacic acid bis(succinimidyl) ester, and ethylene glycol bis(succinic acid N-hydroxysuccinimide ester) – increase the stability of microtubules many times over. These crosslinkers decrease flexibility through intramonomer, intradimer, intraprotofilament, and interprotofilament bonds. Ca^{2+} ordinarily depolymerizes microtubules; however, crosslinking agents that bind to amine residues decrease this depolymerizing effect.

Yet another option is to use nanofabricated materials that simulate critical properties of motor proteins or cytoskeletal protein tracks. Single-walled carbon nanotubes (SWNT) and multiwalled carbon nanotubes (MWNT), which possess structural similarities to microtubules, have been used to transport and deliver materials in living cells and in silico. Functionalized SWNT are capable of delivering small interfering RNAs that silence genetic expression [242]. SWNT can furthermore be controlled by transverse electric fields, which polarize the spins of electrons and initiate their coherent propagation along distances of 0.2 - 1.5 m [243]. MWNT have been used to deliver oligodeoxynucleotide anti-cancer therapy, and fluorescent quantum dots have proved capable of efficient delivery to the nucleus of the cell [244].

More elaborate biologically inspired nanodevices are possible, using natural biomotors such as kinesin and dynein, as well as cytoskeletal protein tracks, as prototypes. One such biologically inspired device is a laser-operated locomotor, which converts optical energy into mechanical energy, and is capable of faster transport rates and moves heavier cargo than its biological counterparts [245]. The application of nanotechnology to transport, in particular to the issue of neural transport, is just beginning to be realized.

References

1. Setou M, Hayasaka T, Yao I. Axonal transport versus dendritic transport. J Neurobiol. 2004 Feb 5;58(2):201-6.
2. Rao GN, Kulkarni SS, Koushika SP, Rau KR. In vivo nanosecond laser axotomy: cavitation dynamics and vesicle transport. Opt Express. 2008 Jun 23;16(13):9884-94.
3. Muthukrishnan G, Hutchins BM, Williams ME, Hancock WO. Transport of semiconductor nanocrystals by kinesin molecular motors. Small. 2006 May;2(5):626-30.
4. Cui B, Wu C, Chen L, Ramirez A, Bearer EL, Li WP, Mobley WC, Chu S. One at a time, live tracking of NGF axonal transport using quantum dots. Proc Natl Acad Sci U S A. 2007 Aug 21;104(34):13666-71.
5. Echarte MM, Bruno L, Arndt-Jovin DJ, Jovin TM, Pietrasanta LI. Quantitative single particle tracking of NGF-receptor complexes: transport is bidirectional but biased by longer retrograde run lengths. FEBS Lett. 2007 Jun 26;581(16):2905-13.
6. Bouzigues C, Lévi S, Triller A, Dahan M. Single quantum dot tracking of membrane receptors. Methods Mol Biol. 2007;374:81-91.
7. Vogel PD. Nature's design of nanomotors. Eur J Pharm Biopharm. 2005 Jul;60(2):267-77.
8. Capitanio M, Vanzi F, Broggio C, Cicchi R, Normanno D, Romano G, Sacconi L, Pavone FS. Exploring molecular motors and switches at the single-molecule level. Microsc Res Tech. 2004 Nov;65(4-5):194-204.
9. Mehta AD, Rief M, Spudich JA, Smith DA, Simmons RM. Single-molecule biomechanics with optical methods. Science. 1999 Mar 12;283(5408):1689-95.
10. Hirokawa N, Takemura R. Molecular motors in neuronal development, intracellular transport and diseases. Curr Opin Neurobiol. 2004 Oct;14(5):564-73.
11. Höök P, Vallee RB. The dynein family at a glance. J Cell Sci. 2006 Nov 1;119(Pt 21):4369-71.
12. Brown ME, Bridgman PC. Myosin function in nervous and sensory systems. J Neurobiol. 2004 Jan;58(1):118-30.
13. Langford GM. Myosin-V, a versatile motor for short-range vesicle transport. Traffic. 2002 Dec;3(12):859-65.
14. Phelps MA, Foraker AB, Swaan PW. Cytoskeletal motors and cargo in membrane trafficking: opportunities for high specificity in drug intervention. Drug Discov Today. 2003 Jun 1;8(11):494-502.

15. Lawrence CJ, Dawe RK, Christie KR, Cleveland DW, Dawson SC, Endow SA, Goldstein LS, Goodson HV, Hirokawa N, Howard J, Malmberg RL, McIntosh JR, Miki H, Mitchison TJ, Okada Y, Reddy AS, Saxton WM, Schliwa M, Scholey JM, Vale RD, Walczak CE, Wordeman L. A standardized kinesin nomenclature. J Cell Biol. 2004 Oct 11;167(1):19-22.
16. Ross JL, Ali MY, Warshaw DM. Cargo transport: molecular motors navigate a complex cytoskeleton. Curr Opin Cell Biol. 2008 Feb;20(1):41-7.
17. Hirokawa N. Kinesin and dynein superfamily proteins and the mechanism of organelle transport. Science. 1998 Jan 23;279(5350):519-26.
18. Muresan V. One axon, many kinesins: What's the logic? J Neurocytol. 2000 Nov-Dec;29(11-12):799-818.
19. Gindhart JG. Towards an understanding of kinesin-1 dependent transport pathways through the study of protein-protein interactions. Brief Funct Genomic Proteomic. 2006 Mar;5(1):74-86.
20. Skowronek KJ, Kocik E, Kasprzak AA. Subunits interactions in kinesin motors. Eur J Cell Biol. 2007 Sep;86(9):559-68.
21. Sablin EP, Fletterick RJ. Coordination between motor domains in processive kinesins. J Biol Chem. 2004 Apr 16;279(16):15707-10.
22. Cai D, Verhey KJ, Meyhöfer E. Tracking single kinesin molecules in the cytoplasm of mammalian cells. Biophys J. 2007 Jun 15;92(12):4137-44.
23. Courty S, Luccardini C, Bellaiche Y, Cappello G, Dahan M. Tracking individual kinesin motors in living cells using single quantum-dot imaging. Nano Lett. 2006 Jul;6(7):1491-5.
24. Seitz A, Surrey T. Processive movement of single kinesins on crowded microtubules visualized using quantum dots. EMBO J. 2006 Jan 25;25(2):267-77.
25. Nan X, Sims PA, Xie XS. Organelle tracking in a living cell with microsecond time resolution and nanometer spatial precision. Chemphyschem. 2008 Apr 4;9(5):707-12.
26. Carter NJ, Cross RA. Mechanics of the kinesin step. Nature. 2005 May 19;435(7040):308-12.
27. Korten T, Diez S. Setting up roadblocks for kinesin-1: mechanism for the selective speed control of cargo carrying microtubules. Lab Chip. 2008 Sep;8(9):1441-7.
28. Leduc C, Ruhnow F, Howard J, Diez S. Detection of fractional steps in cargo movement by the collective operation of kinesin-1 motors. Proc Natl Acad Sci U S A. 2007 Jun 26;104(26):10847-52.
29. Kuznetsov SA, Gelfand VI. Bovine brain kinesin is a microtubule-activated ATPase. Proc Natl Acad Sci U S A. 1986 Nov;83(22):8530-4.
30. Endow SA. Kinesin motors as molecular machines. Bioessays. 2003 Dec;25(12):1212-9.
31. Howard, J. (2001) Mechanics of Motor Proteins and the Cytoskeleton. Sinauer Associates.
32. Block SM. Kinesin motor mechanics: binding, stepping, tracking, gating, and limping. Biophys J. 2007 May 1;92(9):2986-95.
33. Ray S, Meyhofer E, Milligan RA, and Howard J. Kinesin follows the microtubule's protofilament axis. J. Cell Biol. 1993. 121:1083-1093.
34. Kikkawa M. The role of microtubules in processive kinesin movement. Trends Cell Biol. 2008 Mar;18(3):128-35.

35. Skiniotis G, Cochran JC, Müller J, Mandelkow E, Gilbert SP, Hoenger A. Modulation of kinesin binding by the C-termini of tubulin. EMBO J. 2004 Mar 10;23(5):989-99.
36. Seitz A, Kojima H, Oiwa K, Mandelkow EM, Song YH, Mandelkow E. Single-molecule investigation of the interference between kinesin, tau and MAP2c. EMBO J. 2002 Sep 16;21(18):4896-905.
37. Morfini G, Pigino G, Mizuno N, Kikkawa M, Brady ST. Tau binding to microtubules does not directly affect microtubule-based vesicle motility. J Neurosci Res. 2007 Sep;85(12):2620-30.
38. Liao G, Gundersen GG. Kinesin is a candidate for cross-bridging microtubules and intermediate filaments. Selective binding of kinesin to detyrosinated tubulin and vimentin. J Biol Chem. 1998 Apr 17;273(16):9797-803.
39. Reed NA, Cai D, Blasius TL, Jih GT, Meyhofer E, Gaertig J, Verhey KJ. Microtubule acetylation promotes kinesin-1 binding and transport. Curr Biol. 2006 Nov 7;16(21):2166-72.
40. Vallee RB, Shpetner HS, Paschal BM. The role of dynein in retrograde axonal transport. Trends Neurosci. 1989 Feb;12(2):66-70.
41. Vallee R. Cytoplasmic dynein: advances in microtubule-based motility. Trends Cell Biol. 1991 Jul;1(1):25-9.
42. Vallee RB, Williams JC, Varma D, Barnhart LE. Dynein: An ancient motor protein involved in multiple modes of transport. J Neurobiol. 2004 Feb 5;58(2):189-200.
43. Reck-Peterson SL, Yildiz A, Carter AP, Gennerich A, Zhang N, Vale RD. Single-molecule analysis of dynein processivity and stepping behavior. Cell. 2006 Jul 28;126(2):335-48.
44. Watanabe TM, Sato T, Gonda K, Higuchi H. Three-dimensional nanometry of vesicle transport in living cells using dual-focus imaging optics. Biochem Biophys Res Commun. 2007 Jul 20;359(1):1-7.
45. Koonce MP, Samsó M. Of rings and levers: the dynein motor comes of age. Trends Cell Biol. 2004 Nov;14(11):612-9.
46. Garces JA, Clark IB, Meyer DI, Vallee RB. Interaction of the p62 subunit of dynactin with Arp1 and the cortical actin cytoskeleton. Curr Biol. 1999 Dec 16-30;9(24):1497-500.
47. Haghnia M, Cavalli V, Shah SB, Schimmelpfeng K, Brusch R, Yang G, Herrera C, Pilling A, Goldstein LS. Dynactin is required for coordinated bidirectional motility, but not for dynein membrane attachment. Mol Biol Cell. 2007 Jun;18(6):2081-9.
48. Magnani E, Fan J, Gasparini L, Golding M, Williams M, Schiavo G, Goedert M, Amos LA, Spillantini MG. Interaction of tau protein with the dynactin complex. EMBO J. 2007 Oct 31;26(21):4546-54.
49. Shah JV, Flanagan LA, Janmey PA, Leterrier JF. Bidirectional translocation of neurofilaments along microtubules mediated in part by dynein/dynactin. Mol Biol Cell. 2000 Oct;11(10):3495-508.
50. Mallik R, Gross SP. Molecular motors: strategies to get along. Curr Biol. 2004 Nov 23;14(22):R971-82.
51. Bridgman PC. Myosin-dependent transport in neurons. J Neurobiol. 2004 Feb 5;58(2):164-74.
52. Tashiro A, Yuste R. Structure and molecular organization of dendritic spines. Histol Histopathol. 2003 Apr;18(2):617-34.

53. Groc L, Lafourcade M, Heine M, Renner M, Racine V, Sibarita JB, Lounis B, Choquet D, Cognet L. Surface trafficking of neurotransmitter receptor: comparison between single-molecule/quantum dot strategies. J Neurosci. 2007 Nov 14;27(46):12433-7.
54. Volpicelli LA, Lah JJ, Fang G, Goldenring JR, Levey AI. Rab11a and myosin Vb regulate recycling of the M4 muscarinic acetylcholine receptor. J Neurosci. 2002 Nov 15;22(22):9776-84.
55. Fettiplace R. Active hair bundle movements in auditory hair cells. J Physiol. 2006 Oct 1;576(Pt 1):29-36.
56. Gillespie PG. Myosin I and adaptation of mechanical transduction by the inner ear. Philos Trans R Soc Lond B Biol Sci. 2004 Dec 29;359(1452):1945-51.
57. Strissel KJ, Arshavsky VY. Myosin III illuminates the mechanism of arrestin translocation. Neuron. 2004 Jul 8;43(1):2-4.
58. Amos LA. Molecular motors: not quite like clockwork. Cell Mol Life Sci. 2008 Feb;65(4):509-15.
59. Komori T, Nishikawa S, Ariga T, Iwane AH, Yanagida T. Measurement system for simultaneous observation of myosin V chemical and mechanical events. Biosystems. 2008 Jul-Aug;93(1-2):48-57.
60. Frank DJ, Noguchi T, Miller KG. Myosin VI: a structural role in actin organization important for protein and organelle localization and trafficking. Curr Opin Cell Biol. 2004 Apr;16(2):189-94.
61. Schliwa M and Woehlke G. Nature 2003. 422, 759-765.
62. Lasek RJ, Garner JA, Brady ST. Axonal transport of the cytoplasmic matrix. J Cell Biol. 1984 Jul;99(1 Pt 2):212s-221s.
63. Nixon RA. The slow axonal transport of cytoskeletal proteins. Curr Opin Cell Biol. 1998 Feb;10(1):87-92.
64. Brown A. Slow axonal transport: stop and go traffic in the axon. Nat Rev Mol Cell Biol. 2000 Nov;1(2):153-6.
65. Nan X, Sims PA, Chen P, Xie XS. Observation of individual microtubule motor steps in living cells with endocytosed quantum dots. J Phys Chem B. 2005 Dec 29;109(51):24220-4.
66. Loubéry S, Wilhelm C, Hurbain I, Neveu S, Louvard D, Coudrier E. Different microtubule motors move early and late endocytic compartments. Traffic. 2008 Apr;9(4):492-509.
67. Suh J, Choy KL, Lai SK, Suk JS, Tang BC, Prabhu S, Hanes J. PEGylation of nanoparticles improves their cytoplasmic transport. Int J Nanomedicine. 2007;2(4):735-41.
68. Dahlström A, Larsson PA, Goldstein M, Lundmark K, Dahllöf AG, Bööj S. Immunocytochemical studies on axonal transport in adrenergic and cholinergic nerves using cytofluorimetric scanning. Med Biol. 1986;64(2-3):49-56.
69. Garzón M, Pickel VM. Dendritic and axonal targeting of the vesicular acetylcholine transporter to membranous cytoplasmic organelles in laterodorsal and pedunculopontine tegmental nuclei. J Comp Neurol. 2000 Mar 27;419(1):32-48.
70. Li JY, Dahlström AM, Hersh LB, Dahlström A. Fast axonal transport of the vesicular acetylcholine transporter (VAChT) in cholinergic neurons in the rat sciatic nerve. Neurochem Int. 1998 May-Jun;32(5-6):457-67.
71. Nirenberg MJ, Liu Y, Peter D, Edwards RH, Pickel VM. The vesicular monoamine transporter 2 is present in small synaptic vesicles and preferentially localizes to large dense core vesicles in rat solitary tract nuclei. Proc Natl Acad Sci U S A. 1995 Sep 12;92(19):8773-7.

72. Huether G. Acute regulation and long-term modulation of presynaptic serotonin output. Adv Exp Med Biol. 1999;467:1-10.
73. Palkovits M. Peptidergic neurotransmitters in the endocrine hypothalamus. Ciba Found Symp. 1992;168:3-10.
74. Van der Gucht E, Jacobs S, Kaneko T, Vandesande F, Arckens L. Distribution and morphological characterization of phosphate-activated glutaminase-immunoreactive neurons in cat visual cortex. Brain Res. 2003 Oct 24;988(1-2):29-42.
75. Holten AT, Gundersen V. Glutamine as a precursor for transmitter glutamate, aspartate and GABA in the cerebellum: a role for phosphate-activated glutaminase. J Neurochem. 2008 Feb;104(4):1032-42.
76. Waagepetersen HS, Qu H, Sonnewald U, Shimamoto K, Schousboe A. Role of glutamine and neuronal glutamate uptake in glutamate homeostasis and synthesis during vesicular release in cultured glutamatergic neurons. Neurochem Int. 2005 Jul;47(1-2):92-102.
77. Hertz L. Glutamate, a neurotransmitter–and so much more. A synopsis of Wierzba III. Neurochem Int. 2006 May-Jun;48(6-7):416-25.
78. Takamori S. VGLUTs: 'exciting' times for glutamatergic research? Neurosci Res. 2006 Aug;55(4):343-51.
79. Hollenbeck PJ, Saxton WM. The axonal transport of mitochondria. J Cell Sci. 2005 Dec 1;118(Pt 23):5411-9.
80. Boldogh IR, Pon LA. Mitochondria on the move. Trends Cell Biol. 2007 Oct;17(10):502-10.
81. Willner I, Basnar B, Willner B. Nanoparticle-enzyme hybrid systems for nanobiotechnology. FEBS J. 2007 Jan;274(2):302-9.
82. D'Souza GG, Cheng SM, Boddapati SV, Horobin RW, Weissig V. Nanocarrier-assisted sub-cellular targeting to the site of mitochondria improves the pro-apoptotic activity of paclitaxel. J Drug Target. 2008 Aug;16(7):578-85.
83. Boddapati SV, D'Souza GG, Erdogan S, Torchilin VP, Weissig V. Organelle-targeted nanocarriers: specific delivery of liposomal ceramide to mitochondria enhances its cytotoxicity in vitro and in vivo. Nano Lett. 2008 Aug;8(8):2559-63.
84. Kogure K, Akita H, Yamada Y, Harashima H. Multifunctional envelope-type nano device (MEND) as a non-viral gene delivery system. Adv Drug Deliv Rev. 2008 Mar 1;60(4-5):559-71.
85. Huang EJ, Reichardt LF. Trk receptors: roles in neuronal signal transduction. Annu Rev Biochem. 2003;72:609-42.
86. Ibáñez CF. Message in a bottle: long-range retrograde signaling in the nervous system. Trends Cell Biol. 2007 Nov;17(11):519-28.
87. Gao X, Wang T, Wu B, Chen J, Chen J, Yue Y, Dai N, Chen H, Jiang X. Quantum dots for tracking cellular transport of lectin-functionalized nanoparticles. Biochem Biophys Res Commun. 2008 Sep 25.
88. Yano H, Lee FS, Kong H, Chuang J, Arevalo J, Perez P, Sung C, Chao MV. Association of Trk neurotrophin receptors with components of the cytoplasmic dynein motor. J Neurosci. 2001 Feb 1;21(3):RC125.
89. Ure DR, Campenot RB. Retrograde transport and steady-state distribution of 125I-nerve growth factor in rat sympathetic neurons in compartmented cultures. J Neurosci. 1997 Feb 15;17(4):1282-90.
90. Sundara Rajan S, Vu TQ. Quantum dots monitor TrkA receptor dynamics in the interior of neural PC12 cells. Nano Lett. 2006 Sep;6(9):2049-59.

91. Weible MW 2nd, Hendry IA. What is the importance of multivesicular bodies in retrograde axonal transport in vivo? J Neurobiol. 2004 Feb 5;58(2):230-43.
92. von Bartheld CS, Wang X, Butowt R. Anterograde axonal transport, transcytosis, and recycling of neurotrophic factors: the concept of trophic currencies in neural networks. Mol Neurobiol. 2001 Aug-Dec;24(1-3):1-28.
93. Wittig JH Jr, Ryan AF, Asbeck PM. A reusable microfluidic plate with alternate-choice architecture for assessing growth preference in tissue culture. J Neurosci Methods. 2005 May 15;144(1):79-89.
94. Du J, Feng L, Zaitsev E, Je HS, Liu XW, Lu B. Regulation of TrkB receptor tyrosine kinase and its internalization by neuronal activity and Ca2+ influx. J Cell Biol. 2003 Oct 27;163(2):385-95.
95. Watson FL, Heerssen HM, Moheban DB, Lin MZ, Sauvageot CM, Bhattacharyya A, Pomeroy SL, Segal RA. Rapid nuclear responses to target-derived neurotrophins require retrograde transport of ligand-receptor complex. J Neurosci. 1999 Sep 15;19(18):7889-900.
96. McAllister AK, Katz LC, Lo DC. Neurotrophin regulation of cortical dendritic growth requires activity. Neuron. 1996 Dec;17(6):1057-64.
97. Magby JP, Bi C, Chen ZY, Lee FS, Plummer MR. Single-cell characterization of retrograde signaling by brain-derived neurotrophic factor. J Neurosci. 2006 Dec 27;26(52):13531-6.
98. Guan CB, Xu HT, Jin M, Yuan XB, Poo MM. Long-range Ca^{2+} signaling from growth cone to soma mediates reversal of neuronal migration induced by slit-2. Cell. 2007 Apr 20;129(2):385-95.
99. Reynolds AR, Tischer C, Verveer PJ, Rocks O, Bastiaens PI. EGFR activation coupled to inhibition of tyrosine phosphatases causes lateral signal propagation. Nat Cell Biol. 2003 May;5(5):447-53.
100. Markevich NI, Tsyganov MA, Hoek JB, Kholodenko BN. Long-range signaling by phosphoprotein waves arising from bistability in protein kinase cascades. Mol Syst Biol. 2006;2:61.
101. Baas PW, Buster DW. Slow axonal transport and the genesis of neuronal morphology. J Neurobiol. 2004 Jan;58(1):3-17.
102. Galbraith JA, Gallant PE. Axonal transport of tubulin and actin. J Neurocytol. 2000 Nov-Dec;29(11-12):889-911.
103. Hoffman PN, Lopata MA, Watson DF, Luduena RF. Axonal transport of class II and III beta-tubulin: evidence that the slow component wave represents the movement of only a small fraction of the tubulin in mature motor axons. J Cell Biol. 1992 Nov;119(3):595-604.
104. Myers KA, He Y, Hasaka TP, Baas PW. Microtubule transport in the axon: Re-thinking a potential role for the actin cytoskeleton. Neuroscientist. 2006 Apr;12(2):107-18.
105. Kimura T, Watanabe H, Iwamatsu A, Kaibuchi K. Tubulin and CRMP-2 complex is transported via Kinesin-1. J Neurochem. 2005 Jun;93(6):1371-82.
106. Motil J, Chan WK, Dubey M, Chaudhury P, Pimenta A, Chylinski TM, Ortiz DT, Shea TB. Dynein mediates retrograde neurofilament transport within axons and anterograde delivery of NFs from perikarya into axons: regulation by multiple phosphorylation events. Cell Motil Cytoskeleton. 2006 May;63(5):266-86.
107. Shea TB, Yabe J. Occam's Razor slices through the mysteries of neurofilament axonal transport: can it really be so simple? Traffic. 2000 Jun;1(6):522-3.

108. Shah JV, Cleveland DW. Slow axonal transport: fast motors in the slow lane. Curr Opin Cell Biol. 2002 Feb;14(1):58-62.
109. Baas PW, Vidya Nadar C, Myers KA. Axonal transport of microtubules: the long and short of it. Traffic. 2006 May;7(5):490-8.
110. Gallant PE. Axonal protein synthesis and transport. J Neurocytol. 2000 Nov-Dec;29(11-12):779-82.
111. Arimura N, Kaibuchi K. Neuronal polarity: from extracellular signals to intracellular mechanisms. Nat Rev Neurosci. 2007 Mar;8(3):194-205.
112. Kanai Y, Hirokawa N. Sorting mechanisms of tau and MAP2 in neurons: suppressed axonal transit of MAP2 and locally regulated microtubule binding. Neuron. 1995 Feb;14(2):421-32.
113. Sheetz MP, Pfister KK, Bulinski JC, Cotman CW. Mechanisms of trafficking in axons and dendrites: implications for development and neurodegeneration. Prog Neurobiol. 1998 Aug;55(6):577-94.
114. Tang BL. Emerging aspects of membrane traffic in neuronal dendrite growth. Biochim Biophys Acta. 2008 Feb;1783(2):169-76.
115. Lai KO, Zhao Y, Ch'ng TH, Martin KC. Importin-mediated retrograde transport of CREB2 from distal processes to the nucleus in neurons. Proc Natl Acad Sci U S A. 2008 Nov 4;105(44):17175-80.
116. Conti F, Weinberg RJ. Shaping excitation at glutamatergic synapses. Trends Neurosci. 1999 Oct;22(10):451-8.
117. Chen L, Tracy T, Nam CI. Dynamics of postsynaptic glutamate receptor targeting. Curr Opin Neurobiol. 2007 Feb;17(1):53-8.
118. Kim E, Sheng M. PDZ domain proteins of synapses. Nat Rev Neurosci. 2004 Oct;5(10):771-81.
119. Setou M, Nakagawa T, Seog DH, Hirokawa N. Kinesin superfamily motor protein KIF17 and mLin-10 in NMDA receptor-containing vesicle transport. Science. 2000 Jun 9;288(5472):1796-802.
120. Guillaud L, Setou M, Hirokawa N. KIF17 dynamics and regulation of NR2B trafficking in hippocampal neurons. J Neurosci. 2003 Jan 1;23(1):131-40.
121. Setou M, Seog DH, Tanaka Y, Kanai Y, Takei Y, Kawagishi M, Hirokawa N. Glutamate-receptor-interacting protein GRIP1 directly steers kinesin to dendrites. Nature. 2002 May 2;417(6884):83-7.
122. Triller A, Choquet D. Surface trafficking of receptors between synaptic and extrasynaptic membranes: and yet they do move! Trends Neurosci. 2005 Mar;28(3):133-9.
123. Hirokawa N, Takemura R. Molecular motors and mechanisms of directional transport in neurons. Nat Rev Neurosci. 2005 Mar;6(3):201-14.
124. Jacob TC, Moss SJ, Jurd R. GABA(A) receptor trafficking and its role in the dynamic modulation of neuronal inhibition. Nat Rev Neurosci. 2008 May;9(5):331-43.
125. Hanus C, Ehrensperger MV, Triller A. Activity-dependent movements of postsynaptic scaffolds at inhibitory synapses. J Neurosci. 2006 Apr 26;26(17):4586-95.
126. Kuriu T, Inoue A, Bito H, Sobue K, Okabe S. Differential control of postsynaptic density scaffolds via actin-dependent and -independent mechanisms. J Neurosci. 2006 Jul 19;26(29):7693-706.
127. El-Husseini Ael-D, Schnell E, Dakoji S, Sweeney N, Zhou Q, Prange O, Gauthier-Campbell C, Aguilera-Moreno A, Nicoll RA, Bredt DS. Synaptic

strength regulated by palmitate cycling on PSD-95. Cell. 2002 Mar 22;108(6):849-63.

128. Blanpied TA, Kerr JM, Ehlers MD. Structural plasticity with preserved topology in the postsynaptic protein network. Proc Natl Acad Sci U S A. 2008 Aug 26;105(34):12587-92.
129. Lidke DS, Nagy P, Arndt-Jovin DJ. In vivo imaging using quantum-dot-conjugated probes. Curr Protoc Cell Biol. 2007 Sep;Chapter 25:Unit 25.1.
130. Bouzigues C, Morel M, Triller A, Dahan M. Asymmetric redistribution of GABA receptors during GABA gradient sensing by nerve growth cones analyzed by single quantum dot imaging. Proc Natl Acad Sci U S A. 2007 Jul 3;104(27):11251-6.
131. Bouzigues C, Dahan M. Transient directed motions of GABA(A) receptors in growth cones detected by a speed correlation index. Biophys J. 2007 Jan 15;92(2):654-60.
132. Steward O, Falk PM. Polyribosomes under developing spine synapses: growth specializations of dendrites at sites of synaptogenesis. J Neurosci Res. 1985;13(1-2):75-88.
133. Davis L, Banker GA, Steward O. Selective dendritic transport of RNA in hippocampal neurons in culture. Nature. 1987 Dec 3-9;330(6147):477-9.
134. Hirokawa N. mRNA transport in dendrites: RNA granules, motors, and tracks. J Neurosci. 2006 Jul 5;26(27):7139-42.
135. Garner CC, Tucker RP, Matus A. Selective localization of messenger RNA for cytoskeletal protein MAP2 in dendrites. Nature. 1988 Dec 15;336(6200):674-7.
136. Bramham CR, Wells DG. Dendritic mRNA: transport, translation and function. Nat Rev Neurosci. 2007 Oct;8(10):776-89.
137. Martin KC, Zukin RS. RNA trafficking and local protein synthesis in dendrites: an overview. J Neurosci. 2006 Jul 5;26(27):7131-4.
138. Lyford GL, Yamagata K, Kaufmann WE, Barnes CA, Sanders LK, Copeland NG, Gilbert DJ, Jenkins NA, Lanahan AA, Worley PF. Arc, a growth factor and activity-regulated gene, encodes a novel cytoskeleton-associated protein that is enriched in neuronal dendrites. Neuron. 1995 Feb;14(2):433-45.
139. Bassell GJ, Zhang H, Byrd AL, Femino AM, Singer RH, Taneja KL, Lifshitz LM, Herman IM, Kosik KS. Sorting of beta-actin mRNA and protein to neurites and growth cones in culture. J Neurosci. 1998 Jan 1;18(1):251-65.
140. Dynes JL, Steward O. Dynamics of bi-directional transport of Arc mRNA in neuronal dendrites. J Comp Neurol. 2007 Jan 20;500(3):433-47.
141. Benson DL. Dendritic compartmentation of NMDA receptor mRNA in cultured hippocampal neurons. Neuroreport. 1997 Mar 3;8(4):823-8.
142. Grooms SY, Noh KM, Regis R, Bassell GJ, Bryan MK, Carroll RC, Zukin RS. Activity bi-directionally regulates AMPA receptor mRNA abundance in dendrites of hippocampal neurons. J Neurosci. 2006 Aug 9;26(32):8339-51.
143. Tongiorgi E, Righi M, Cattaneo A. Activity-dependent dendritic targeting of BDNF and TrkB mRNAs in hippocampal neurons. J Neurosci. 1997 Dec 15;17(24):9492-505.
144. Righi M, Tongiorgi E, Cattaneo A. Brain-derived neurotrophic factor (BDNF) induces dendritic targeting of BDNF and tyrosine kinase B mRNAs in hippocampal neurons through a phosphatidylinositol-3 kinase-dependent pathway. J Neurosci. 2000 May 1;20(9):3165-74.

145. Schratt GM, Nigh EA, Chen WG, Hu L, Greenberg ME. BDNF regulates the translation of a select group of mRNAs by a mammalian target of rapamycin-phosphatidylinositol 3-kinase-dependent pathway during neuronal development. J Neurosci. 2004 Aug 18;24(33):7366-77.
146. Lu B. BDNF and activity-dependent synaptic modulation. Learn Mem. 2003 Mar-Apr;10(2):86-98.
147. Benson DL, Gall CM, Isackson PJ. Dendritic localization of type II calcium calmodulin-dependent protein kinase mRNA in normal and reinnervated rat hippocampus. Neuroscience. 1992;46(4):851-7.
148. Huang YS, Jung MY, Sarkissian M, Richter JD. N-methyl-D-aspartate receptor signaling results in Aurora kinase-catalyzed CPEB phosphorylation and alpha CaMKII mRNA polyadenylation at synapses. EMBO J. 2002 May 1;21(9):2139-48.
149. Muslimov IA, Nimmrich V, Hernandez AI, Tcherepanov A, Sacktor TC, Tiedge H. Dendritic transport and localization of protein kinase Mzeta mRNA: implications for molecular memory consolidation. J Biol Chem. 2004 Dec 10;279(50):52613-22.
150. Sacktor TC. PKMzeta, LTP maintenance, and the dynamic molecular biology of memory storage. Prog Brain Res. 2008;169:27-40.
151. Thoumine O, Ewers H, Heine M, Groc L, Frischknecht R, Giannone G, Poujol C, Legros P, Lounis B, Cognet L, Choquet D. Probing the dynamics of protein-protein interactions at neuronal contacts by optical imaging. Chem Rev. 2008 May;108(5):1565-87.
152. Lin YL, Jen JC, Hsu SH, Chiu IM. Sciatic nerve repair by microgrooved nerve conduits made of chitosan-gold nanocomposites. Surg Neurol. 2008 Apr 25.
153. Suk JS, Suh J, Lai SK, Hanes J. Quantifying the intracellular transport of viral and nonviral gene vectors in primary neurons. Exp Biol Med (Maywood). 2007 Mar;232(3):461-9.
154. Bergen JM, Pun SH. Analysis of the intracellular barriers encountered by non-viral gene carriers in a model of spatially controlled delivery to neurons. J Gene Med. 2008 Feb;10(2):187-97.
155. LoPachin RM, Lehning EJ. Mechanism of calcium entry during axon injury and degeneration. Toxicol Appl Pharmacol. 1997 Apr;143(2):233-44.
156. Glass TF, Reeves B, Sharp FR. Modeling both the mechanical and hypoxic features of traumatic brain injury in vitro in rats. Neurosci Lett. 2002 Aug 9;328(2):133-6.
157. Fitzpatrick MO, Dewar D, Teasdale GM, Graham DI. The neuronal cytoskeleton in acute brain injury. Br J Neurosurg. 1998 Aug;12(4):313-7.
158. Smith DH, Uryu K, Saatman KE, Trojanowski JQ, McIntosh TK. Protein accumulation in traumatic brain injury. Neuromolecular Med. 2003;4(1-2):59-72.
159. Fitzpatrick MO, Maxwell WL, Graham DI. The role of the axolemma in the initiation of traumatically induced axonal injury. J Neurol Neurosurg Psychiatry. 1998 Mar;64(3):285-7.
160. Hanz S, Fainzilber M. Retrograde signaling in injured nerve–the axon reaction revisited. J Neurochem. 2006 Oct;99(1):13-9.
161. Perlson E, Hanz S, Ben-Yaakov K, Segal-Ruder Y, Seger R, Fainzilber M. Vimentin-dependent spatial translocation of an activated MAP kinase in injured nerve. Neuron. 2005 Mar 3;45(5):715-26.

162. Helfand BT, Chou YH, Shumaker DK, Goldman RD. Intermediate filament proteins participate in signal transduction. Trends Cell Biol. 2005 Nov;15(11):568-70.
163. Hoffman PN, Luduena RF. Changes in the isotype composition of beta-tubulin delivered to regenerating sensory axons by slow axonal transport. Brain Res. 1996 Dec 2;742(1-2):329-33.
164. Li GL, Farooque M, Lewen A, Lennmyr F, Holtz A, Olsson Y. MAP2 and neurogranin as markers for dendritic lesions in CNS injury. An immunohistochemical study in the rat. APMIS. 2000 Feb;108(2):98-106.
165. Posmantur RM, Kampfl A, Liu SJ, Heck K, Taft WC, Clifton GL, Hayes RL. Cytoskeletal derangements of cortical neuronal processes three hours after traumatic brain injury in rats: an immunofluorescence study. J Neuropathol Exp Neurol. 1996 Jan;55(1):68-80.
166. Nawashiro H, Shima K, Chigasaki H. Selective vulnerability of hippocampal CA3 neurons to hypoxia after mild concussion in the rat. Neurol Res. 1995 Dec;17(6):455-60.
167. Taft WC, Yang K, Dixon CE, Hayes RL. Microtubule-associated protein 2 levels decrease in hippocampus following traumatic brain injury. J Neurotrauma. 1992 Fall;9(3):281-90.
168. Huh JW, Raghupathi R, Laurer HL, Helfaer MA, Saatman KE. Transient loss of microtubule-associated protein 2 immunoreactivity after moderate brain injury in mice. J Neurotrauma. 2003 Oct;20(10):975-84.
169. Folkerts MM, Berman RF, Muizelaar JP, Rafols JA. Disruption of MAP-2 immunostaining in rat hippocampus after traumatic brain injury. J Neurotrauma. 1998 May;15(5):349-63.
170. Li S, Kuroiwa T, Ishibashi S, Sun L, Endo S, Ohno K. Transient cognitive deficits are associated with the reversible accumulation of amyloid precursor protein after mild traumatic brain injury. Neurosci Lett. 2006 Dec 6;409(3):182-6.
171. Lewén A, Li GL, Olsson Y, Hillered L. Changes in microtubule-associated protein 2 and amyloid precursor protein immunoreactivity following traumatic brain injury in rat: influence of MK-801 treatment. Brain Res. 1996 May 6;719(1-2):161-71.
172. Ost M, Nylén K, Csajbok L, Ohrfelt AO, Tullberg M, Wikkelsö C, Nellgrd P, Rosengren L, Blennow K, Nellgrd B.Initial CSF total tau correlates with 1-year outcome in patients with traumatic brain injury. Neurology. 2006 Nov 14;67(9):1600-4.
173. Smith C, Graham DI, Murray LS, Nicoll JA. Tau immunohistochemistry in acute brain injury. Neuropathol Appl Neurobiol. 2003 Oct;29(5):496-502.
174. Gabbita SP, Scheff SW, Menard RM, Roberts K, Fugaccia I, Zemlan FP. Cleaved-tau: a biomarker of neuronal damage after traumatic brain injury. J Neurotrauma. 2005 Jan;22(1):83-94.
175. McCracken E, Hunter AJ, Patel S, Graham DI, Dewar D. Calpain activation and cytoskeletal protein breakdown in the corpus callosum of head-injured patients. J Neurotrauma. 1999 Sep;16(9):749-61.
176. Haranishi Y, Kawata R, Fukuda S, Kiyoshima T, Morimoto Y, Matsumoto M, Sakabe T. Moderate hypothermia, but not calpain inhibitor 2, attenuates the proteolysis of microtubule-associated protein 2 in the hippocampus following traumatic brain injury in rats. Eur J Anaesthesiol. 2005 Feb;22(2):140-7.

177. Henninger N, Dützmann S, Sicard KM, Kollmar R, Bardutzky J, Schwab S. Impaired spatial learning in a novel rat model of mild cerebral concussion injury. Exp Neurol. 2005 Oct;195(2):447-57.
178. Clausen F, Lewén A, Marklund N, Olsson Y, McArthur DL, Hillered L. Correlation of hippocampal morphological changes and morris water maze performance after cortical contusion injury in rats. Neurosurgery. 2005 Jul;57(1):154-63; discussion 154-63.
179. Selzer ME. Promotion of axonal regeneration in the injured CNS. Lancet Neurol. 2003 Mar;2(3):157-66.
180. Han PJ, Shukla S, Subramanian PS, Hoffman PN. Cyclic AMP elevates tubulin expression without increasing intrinsic axon growth capacity. Exp Neurol. 2004 Oct;189(2):293-302.
181. Liu HH, Brady ST. cAMP, tubulin, axonal transport, and regeneration. Exp Neurol. 2004 Oct;189(2):199-203.
182. Murray M, Kim D, Liu Y, Tobias C, Tessler A, Fischer I. Transplantation of genetically modified cells contributes to repair and recovery from spinal injury. Brain Res Brain Res Rev. 2002 Oct;40(1-3):292-300.
183. Al-Majed AA, Tam SL, Gordon T. Electrical stimulation accelerates and enhances expression of regeneration-associated genes in regenerating rat femoral motoneurons. Cell Mol Neurobiol. 2004 Jun;24(3):379-402.
184. Blizzard CA, Haas MA, Vickers JC, Dickson TC. Cellular dynamics underlying regeneration of damaged axons differs from initial axon development. Eur J Neurosci. 2007 Sep;26(5):1100-8.
185. Bouquet C, Soares S, von Boxberg Y, Ravaille-Veron M, Propst F, Nothias F. Microtubule-associated protein 1B controls directionality of growth cone migration and axonal branching in regeneration of adult dorsal root ganglia neurons. J Neurosci. 2004 Aug 11;24(32):7204-13.
186. Björkblom B, Ostman N, Hongisto V, Komarovski V, Filén JJ, Nyman TA, Kallunki T, Courtney MJ, Coffey ET. Constitutively active cytoplasmic c-Jun N-terminal kinase 1 is a dominant regulator of dendritic architecture: role of microtubule-associated protein 2 as an effector. J Neurosci. 2005 Jul 6;25(27):6350-61.
187. Chang L, Jones Y, Ellisman MH, Goldstein LS, Karin M. JNK1 is required for maintenance of neuronal microtubules and controls phosphorylation of microtubule-associated proteins. Dev Cell. 2003 Apr;4(4):521-33.
188. Sekino Y, Kojima N, Shirao T. Role of actin cytoskeleton in dendritic spine morphogenesis. Neurochem Int. 2007 Jul-Sep;51(2-4):92-104.
189. Halpain S, Hipolito A, Saffer L. Regulation of F-actin stability in dendritic spines by glutamate receptors and calcineurin. J Neurosci. 1998 Dec 1;18(23):9835-44.
190. Ultanir SK, Kim JE, Hall BJ, Deerinck T, Ellisman M, Ghosh A. Regulation of spine morphology and spine density by NMDA receptor signaling in vivo. Proc Natl Acad Sci U S A. 2007 Dec 4;104(49):19553-8.
191. Satoh D, Sato D, Tsuyama T, Saito M, Ohkura H, Rolls MM, Ishikawa F, Uemura T. Spatial control of branching within dendritic arbors by dynein-dependent transport of Rab5-endosomes. Nat Cell Biol. 2008 Oct;10(10):1164-71.
192. Horton AC, Yi JJ, Ehlers MD. Cell type-specific dendritic polarity in the absence of spatially organized external cues. Brain Cell Biol. 2006 Feb;35(1):29-38.

193. Hanus C, Ehlers MD. Secretory outposts for the local processing of membrane cargo in neuronal dendrites. Traffic. 2008 Sep;9(9):1437-45.
194. Horton AC, Rácz B, Monson EE, Lin AL, Weinberg RJ, Ehlers MD. Polarized secretory trafficking directs cargo for asymmetric dendrite growth and morphogenesis. Neuron. 2005 Dec 8;48(5):757-71.
195. Kennedy MJ, Ehlers MD. Organelles and trafficking machinery for postsynaptic plasticity. Annu Rev Neurosci. 2006;29:325-62.
196. Matsuzaki M. Factors critical for the plasticity of dendritic spines and memory storage. Neurosci Res. 2007 Jan;57(1):1-9.
197. Ostroff LE, Fiala JC, Allwardt B, Harris KM. Polyribosomes redistribute from dendritic shafts into spines with enlarged synapses during LTP in developing rat hippocampal slices. Neuron. 2002 Aug 1;35(3):535-45.
198. Ehrlich I, Klein M, Rumpel S, Malinow R. PSD-95 is required for activity-driven synapse stabilization. Proc Natl Acad Sci U S A. 2007 Mar 6;104(10):4176-81.
199. Fukazawa Y, Saitoh Y, Ozawa F, Ohta Y, Mizuno K, Inokuchi K. Hippocampal LTP is accompanied by enhanced F-actin content within the dendritic spine that is essential for late LTP maintenance in vivo. Neuron. 2003 May 8;38(3):447-60.
200. Park M, Salgado JM, Ostroff L, Helton TD, Robinson CG, Harris KM, Ehlers MD. Plasticity-induced growth of dendritic spines by exocytic trafficking from recycling endosomes. Neuron. 2006 Dec 7;52(5):817-30.
201. Wang Z, Edwards JG, Riley N, Provance DW Jr, Karcher R, Li XD, Davison IG, Ikebe M, Mercer JA, Kauer JA, Ehlers MD. Myosin Vb mobilizes recycling endosomes and AMPA receptors for postsynaptic plasticity. Cell. 2008 Oct 31;135(3):535-48.
202. Wong RW, Setou M, Teng J, Takei Y, Hirokawa N. Overexpression of motor protein KIF17 enhances spatial and working memory in transgenic mice. Proc Natl Acad Sci U S A. 2002 Oct 29;99(22):14500-5.
203. Hirokawa N, Takemura R. Kinesin superfamily proteins and their various functions and dynamics. Exp Cell Res. 2004 Nov 15;301(1):50-9.
204. Kopec CD, Real E, Kessels HW, Malinow R. GluR1 links structural and functional plasticity at excitatory synapses. J Neurosci. 2007 Dec 12;27(50):13706-18.
205. Tanaka J, Horiike Y, Matsuzaki M, Miyazaki T, Ellis-Davies GC, Kasai H. Protein synthesis and neurotrophin-dependent structural plasticity of single dendritic spines. Science. 2008 Mar 21;319(5870):1683-7.
206. Chen LY, Rex CS, Casale MS, Gall CM, Lynch G. Changes in synaptic morphology ac company actin signaling during LTP. J Neurosci. 2007 May 16;27(20):5363-72.
207. Sakurai Y. How do cell assemblies encode information in the brain? Neurosci Biobehav Rev. 1999;23(6):785-96.
208. Woolf, NJ. A structural basis for memory storage in mammals. Prog. Neurobiol. 1998, 55, 59-77.
209. Woolf NJ, Zinnerman MD, Johnson GV. Hippocampal microtubule-associated protein-2 alterations with contextual memory. Brain Res. 1999 Mar 6;821(1):241-9.
210. Mershin A, Pavlopoulos E, Fitch O, Braden BC, Nanopoulos DV, Skoulakis EM. Learning and memory deficits upon TAU accumulation in Drosophila mushroom body neurons.Learn Mem. 2004 May-Jun;11(3):277-87.

211. Bolterauer H, Tuszyński JA, Unger E. Directed binding: a novel physical mechanism that describes the directional motion of two-headed kinesin motor proteins. Cell Biochem Biophys. 2005;42(2):95-119.
212. Hwang W, Lang MJ, Karplus M. Force generation in kinesin hinges on cover-neck bundle formation. Structure. 2008 Jan;16(1):62-71. Erratum in: Structure. 2008 Jul;16(7):1147.
213. Munárriz J, Mazo JJ, Falo F. Model for hand-over-hand motion of molecular motors. Phys Rev E Stat Nonlin Soft Matter Phys. 2008 Mar;77(3 Pt 1):031915.
214. Howard, J. (2001) Mechanics of Motor Proteins and the Cytoskeleton, Sinauer Associates Inc., Sunderland, MA.
215. Doering, C. R., Horsthemke, W., and Riordan, J. (1994) Nonequilibrium fluctuation-induced transport. Phys. Rev. Lett. 72, 2984-2987.
216. Astumian, R. D. and Bier, M. (1994) Fluctuation driven ratchets: molecular motors. Phys. Rev. Lett. 72, 1766-1769.
217. Bier, M. and Astumian, R. D. (1996) Biased brownian motion as the operating principle for microscopic engines. Bioelectrochem. Bioenerg. 39, 67-75.
218. Bier M. The energetics, chemistry, and mechanics of a processive motor protein. Biosystems. 2008 Jul-Aug;93(1-2):23-8.
219. Harada T, Sasa S. Fluctuations, responses and energetics of molecular motors. Math Bio sci. 2007 Jun;207(2):365-86.
220. Chowdhury D, Garai A, Wang JS. Traffic of single-headed motor proteins KIF1A: effects of lane changing. Phys Rev E Stat Nonlin Soft Matter Phys. 2008 May;77(5 Pt 1):050902.
221. Bieling P, Telley IA, Piehler J, Surrey T. Processive kinesins require loose mechanical coupling for efficient collective motility. EMBO Rep. 2008 Sep 19.
222. Kunwar A, Vershinin M, Xu J, Gross SP. Stepping, strain gating, and an unexpected force-velocity curve for multiple-motor-based transport. Curr Biol. 2008 Aug 26;18(16):1173-83.
223. Nitta T, Tanahashi A, Hirano M, Hess H. Simulating molecular shuttle movements: towards computer-aided design of nanoscale transport systems. Lab Chip. 2006 Jul;6(7):881-5.
224. Bursac P, Fabry B, Trepat X, Lenormand G, Butler JP, Wang N, Fredberg JJ, An SS. Cytoskeleton dynamics: fluctuations within the network. Biochem Biophys Res Commun. 2007 Apr 6;355(2):324-30.
225. Zheng W, Brooks BR, Hummer G. Protein conformational transitions explored by mixed elastic network models. Proteins. 2007 Oct 1;69(1):43-57.
226. Hess H, Bachand GD, Vogel V. Powering nanodevices with biomolecular motors. Chemistry. 2004 May 3;10(9):2110-6.
227. Han, C.; Jonas, O. T.; Robert, H. A.; Stephen, Y. C. Applied Physics Letters 2002, 81, 3058-3060.
228. Squires, T. M.; Quake, S. R. Reviews of Modern Physics 2005, 77, 977-1026.
229. Yokokawa R, Tarhan MC, Kon T, Fujita H. Simultaneous and bi-directional transport of kinesin-coated microspheres and dynein-coated microspheres on polarity-oriented microtubules. Biotechnol Bioeng. 2008 Sep 1;101(1):1-8.
230. Böhm KJ, Stracke R, Unger E. Motor proteins and kinesin-based nanoactuatoric devices. Tsitol Genet. 2003 Mar-Apr;37(2):11-21.
231. Taira S, Du YZ, Hiratsuka Y, Uyeda TQ, Yumoto N, Kodaka M.Loading and unloading of molecular cargo by DNA-conjugated microtubule. Biotechnol Bioeng. 2008 Feb 15;99(3):734-9.

232. Ramachandran S, Ernst KH, Bachand GD, Vogel V, Hess H. Selective loading of kinesin-powered molecular shuttles with protein cargo and its application to biosensing. Small. 2006 Mar;2(3):330-4.
233. Martin BD, Soto CM, Blum AS, Sapsford KE, Whitley JL, Johnson JE, Chatterji A, Ratna BR. An engineered virus as a bright fluorescent tag and scaffold for cargo proteins–capture and transport by gliding microtubules. J Nanosci Nanotechnol. 2006 Aug;6(8):2451-60.
234. Bachand M, Trent AM, Bunker BC, Bachand GD. Physical factors affecting kinesin-based transport of synthetic nanoparticle cargo. J Nanosci Nanotechnol. 2005 May;5(5):718-22.
235. Stracke, R., K. J. Böhm, L. Wollweber, J. A. Tuszyński, and E. Unger.2002. Analysis of the migration behaviour of single microtubules in electric fields. Biochem. Biophys. Res. Commun. 293:602-609.
236. Kim T, Kao MT, Hasselbrink EF, Meyhöfer E. Nanomechanical model of microtubule translocation in the presence of electric fields. Biophys J. 2008 May 15;94(10):3880-92.
237. Platt M, Muthukrishnan G, Hancock WO, Williams ME. Millimeter scale alignment of magnetic nanoparticle functionalized microtubules in magnetic fields. J Am Chem Soc. 2005 Nov 16;127(45):15686-7.
238. Sawant RM, Hurley JP, Salmaso S, Kale A, Tolcheva E, Levchenko TS, Torchilin VP. "SMART" drug delivery systems: double-targeted pH-responsive pharmaceutical nanocarriers. Bioconjug Chem. 2006 Jul-Aug;17(4):943-9.
239. Vale RD. The molecular motor toolbox for intracellular transport. Cell. 2003 Feb 21;112(4):467-80.
240. Greene AC, Trent AM, Bachand GD. Controlling kinesin motor proteins in nanoengineered systems through a metal-binding on/off switch. Biotechnol Bioeng. 2008 Apr 21.
241. Boal AK, Tellez H, Rivera SB, Miller NE, Bachand GD, Bunker BC. The stability and functionality of chemically crosslinked microtubules. Small. 2006 Jun;2(6):793-803.
242. Krajcik R, Jung A, Hirsch A, Neuhuber W, Zolk O. Functionalization of carbon nanotubes enables non-covalent binding and intracellular delivery of small interfering RNA for efficient knock-down of genes. Biochem Biophys Res Commun. 2008 May 2;369(2):595-602.
243. Son YW, Cohen ML, Louie SG. Electric field effects on spin transport in defective metallic carbon nanotubes. Nano Lett. 2007 Nov;7(11):3518-22
244. Jia N, Lian Q, Shen H, Wang C, Li X, Yang Z. Intracellular delivery of quantum dots tagged antisense oligodeoxynucleotides by functionalized multiwalled carbon nanotubes. Nano Lett. 2007 Oct;7(10):2976-80.
245. Wang Z. Bioinspired laser-operated molecular locomotive. Phys Rev E Stat Nonlin Soft Matter Phys. 2004 Sep;70(3 Pt 1):031903.

5

Nanotechnology, Nanostructure, and Nervous System Disorders

Summary

Nanoscience impacts on nervous system diseases in at least two distinct ways. Nanomechanical structures within neurons are fundamentally impaired in multiple nervous system disorders and nanotechnology is instrumental to the development of novel drug and gene therapies and prosthetic nanodevices. A striking number of neurodevelopmental, neurological, and neuropsychiatric disorders exhibit disruption of the nanomechanical properties of the cytoskeleton, affecting subunit proteins, binding proteins, related signal transduction molecules, or indirectly impairing transport mechanisms. The neurodevelopmental disorders such as the fragile X syndrome, Turner syndrome, Williams syndrome, autism, Rett syndrome, and Down syndrome are associated with abnormalities to dendrites and spines, indicating underlying cytoskeletal involvement. Motor neuron diseases, such as amyotrophic lateral sclerosis, and degenerative neurological disorders, such as Alzheimer's, Parkinson's, and Huntington's disease, present with profound disruption of the neuronal cytoskeleton, as well as compromised axonal transport. There is also evidence of cytoskeletal abnormalities in neuropsychiatric disorders, such as schizophrenia, bipolar disorder, and major depression. Identifying the genetic causes of nervous system disorders leads to new treatment targets. The genetic basis for many neurodevelopmental disorders is known, and in many cases expression of a cytoskeleton-related protein is abnormal. The genetic basis for many neurological and neuropsychiatric disorders remains largely undetermined; however, in those sporadic cases that have a gene locus specified, a deficit in a cytoskeleton-related proteins or impaired transport is often noted. Nanotechnological approaches to neurodevelopmental, neurological, and neuropsychiatric disorders include (1) using nanoparticles or nanocarriers to deliver drug or gene therapies, (2) using nanotechnology to reconstruct, reinforce, and/or stabilize the cytoskeletal matrix, (3) using nanofabrication methods to make biohybrid transport devices, and (4) coating electrodes with nanoparticles. Tangentially related to nanotechnological approaches are rational drug design techniques. High throughput scanning of huge molecular databases can be used to identify potential drugs that will target specific proteins in damaged neurons in an effort to restore nanomechanical function.

5.1 Identifying Nanomechanical Dysfunction in Nervous System Disorders

There are three levels of evidence linking nanomechanical dysfunction (i.e., cytoskeletal protein disruption or impaired transport) to a wide range of nervous system disorders:

- Microscopic studies reveal cellular changes that reflect alterations in the cytoskeleton in numerous diseases of the nervous system. These include distortions in neuronal size and shape, dendrite number and length, and spine densities - changes that imply cytoskeletal involvement due to the fact that these parameters are directly determined by cytoskeletal proteins. Microscopic analyses also reveal signs of degeneration, and some of these degenerative changes are directly related to the neuronal cytoskeleton.
- Biochemical and immunohistochemical analyses of cytoskeletal proteins, cytoskeleton-stabilizing proteins (e.g., microtubule-associated proteins: MAPs), motor proteins, scaffolding proteins, and related signal transduction molecules reveal their involvement in a wide variety of nervous system disorders.
- Genetic linkage studies demonstrate that many nervous system diseases are due to genetic deletions or single nucleotide polymorphisms affecting cytoskeletal proteins, MAPs, motor proteins, scaffolding proteins, and related signal transduction molecules.

5.2 Neurodevelopmental Disorders: Cytoskeletal Protein Abnormalities and Impaired Transport

Neurodevelopmental disorders resulting in varying degrees of cognitive impairment (ranging from severe mental retardation to minor problems with specific cognitive tasks) include fragile X syndrome, Turner syndrome, Williams syndrome, autism, Rett syndrome, and Down syndrome. Each of these neurodevelopmental disorders has a unique genetic basis (see Table 5.1). In fragile X syndrome and Williams syndrome, deletion of particular genes and their proteins represents the underlying cause of the disorder. In Turner syndrome and Down syndrome, chromosomes are either missing or duplicated resulting in chromosome X monosomy or trisomy 21, respectively. The cause of autism is polygenetic and still not completely understood. Despite all the differences in etiology, the cytoskeleton of neurons is affected in each of these disorders. Dendritic abnormalities, in particular, decreased numbers of branches and reduced spine densities, occur in most neurodevelopmental disorders, and furthermore correlate with degree of cognitive impairment [1].

Table 5.1. Genetic bases of the major neurodevelopmental syndromes and their effects on the cytoskeleton*.

Neurodevelopmental syndrome	Associated genetic links	Effect on cytoskeleton
Fragile X syndrome	FMRP deletion	Decreased repression of MAP1B expression; disrupted actin dynamics subsequent to down-regulation of ADF/cofilin
Turner syndrome	Monosomy of chromosome X	Decreased dendrite pruning
Williams syndrome	LIMK1 and CLIP115 deletion	Disrupted actin dynamics subsequent to decreased phosphorylation of ADF/cofilin; upregulation of CLIP170 and dynactin
Autism	MARK1 single nucleotide polymorphism (occurring in 1% of autistics)	Altered phosphorylation of MAPs and disrupted dendrite length and transport
Rett syndrome	MECP2 deletion	Downstream effects on MAP2
Down syndrome	Trisomy of chromosome 21	Decreased drebrin, centractin-α, F-actin capping protein, Arp2/3, coronin-like p57, and moesin leading to disrupted actin dynamics; MAP alterations during neurodegeneration

*See text for more details.

5.2.1 Fragile X Syndrome: Impaired mRNA Transport

Fragile X syndrome is the most common of inherited causes of mental retardation affecting 1 in 2000 male offspring and 1 in 4000 female offspring [2]. A lack of expression of the fragile X mental retardation protein (FMRP) is the underlying cause of the disorder [3]. FMRP is an mRNA-binding protein that is responsible for attaching ribosomes to microtubules, and then transporting these complexes or granules to dendrites and spines as part of synaptogenesis and synaptic maintenance [4]. Local protein synthesis in the spine is regulated by FMRP [5].

Experiments tracking green fluorescent protein (GFP)-labeled FMRP granules demonstrated that FMRP is capable of moving both anterogradely and bi-directionally at an average speed of 0.19 μm/s and a maximum speed of 0.71 μm/s [6]. The motor responsible for transporting FMRP along microtubule tracks was shown to be KIF3C [7]. FMRP has multiple domains capable of binding diverse mRNAs, and it also works with various RNA transport factors [8].

Using high-resolution fluorescent imaging methods, FMRP granules have been found in actin-rich compartments of developing neurons, particularly in filopodia, spines, and growth cones [9]. Given its normal distribution, an absence of FMRP would be expected to impair neuronal development, particularly causing a delay in the maturation of dendrites and dendritic spines. Misregulation of the microtubule-actin crosslinker, MAP1B, also occurs in fragile X syndrome, and this appears to be related to the deletion of FMRP [10]. FMRP usually represses the local synthesis of MAP1B, so its absence leads to an elevation of MAP1B, which in turn increases microtubule stability and decreases microtubule dynamics. Although more than 400 mRNAs associate with FMRP, MAP1B is particularly critical because of its role in early neuronal development. Of the MAPs, MAP1B is expressed first in the developmental sequence.

FMRP exerts control over actin dynamics, in part, by regulating profilin, as has been demonstrated in Drosophila [11]. It has also been demonstrated in Drosophila that FRMP knockout leads to down-regulation of actin depolymerizing factor (ADF)/cofilin, which in turn influences Rac1 signaling, and alters the balance between actin polymerization and depolymerization [12].

Other signal transduction molecules associated with FMRP underscore its role in neural plasticity. FMRP granules contain Ca^{2+}/calmodulin-dependent kinase II (CaMK II), staufen, and cytoplasmic polyadenylation element binding protein (CPEB) [13]. CaMK II is known to phosphorylate MAPs and plays a role in synaptic plasticity. The metabotropic glutamate receptor (mGlu) activates the movement of FMRP-mRNA complexes into spines where they colocalize with postsynaptic scaffolding proteins, such as PSD95 and Shank, and become translationally active. It is likely that interruption of these functions of FMRP is responsible for the cognitive impairments and mental retardation attendant with fragile X syndrome.

5.2.2 Turner Syndrome: Failure of Dendrite Pruning

Turner syndrome occurs in 1:2500 female births and is caused by complete or partial monosomy of the X chromosome (in which an individual only has one full X chromosome rather than the usual two) [14]. Although Turner syndrome is not associated with mental retardation, patients demonstrate specific cognitive impairments on visuospatial and mathematical tasks, and there are often deficits in executive functions, such as working memory.

The cognitive symptoms of Turner syndrome have been attributed to a failure of dendrite pruning or loss of neuronal integrity [15, 16]. These hypotheses are based on magnetic resonance imaging (MRI) measurements of cortical volumes and subsequent magnetic resonance spectroscopic assays of the chemical composition of brain regions showing altered volumes on MRI scans. More precise demonstration of altered dendrite pruning, which would clearly implicate the neuronal cytoskeleton, awaits confirmation at the histological level.

5.2.3 Williams Syndrome: Deletions of Cytoskeleton-Related Proteins

Williams syndrome is associated with an elfin facial appearance and mental retardation; it occurs at a rate of 1:7500 live births [17]. The cause of Williams syndrome is a heterozygous deletion of chromosome 7q11.23, containing some 20 genes, two of which code for proteins associated with the cytoskeleton [18]. Lin-11/Isl-1/Mec-3-domain-containing protein kinase-1 (LIMK1) and cytoplasmic linker protein of 115 kD (CLIP115) are the proteins deleted in Williams syndrome having cytoskeletal function and implicated in the cognitive impairments associated with the disorder. Accordingly, mice in which the genes for LIMK1 and CLIP115 are deleted provide viable animal models for the disorder, enabling a way to assess its molecular underpinnings.

Knockout mice null for LIMK1 have cortical and hippocampal dendrites with abnormal spines and show frequency-dependent enhancement or impairment of long-term potentiation (LTP) compared to that of wild type mice [19]. LTP stimulated by high frequencies (50 - 100 Hz) is enhanced in LIMK1 knockouts, whereas lower frequency stimuli (5 - 10 Hz) produced no LTP in the knockout mice. Actin misregulation underlies these spine abnormalities and the resultant physiological changes. Ordinarily, LIMK1 regulates the phosphorylative state of ADF/cofilin. Since ADF/cofilin is responsible for depolymerizing actin, deficits in LIMK1 impair the balance between polymerized and depolymerized actin, a dynamic necessary for the proper regulation of spine function. As a result, there is an abnormal distribution of actin and related signal transduction molecules in the absence of LIMK1. To illustrate, dendrite shafts in LIMK1-knockout mice accumulate abnormal clusters of cofilin and actin. Moreover, spine shapes are altered in these mutant mice from the normal thin-neck, mushroom-head shape to a thick-neck, small-head shape. These atypical distributions and shape abnormalities reflect deficits in actin depolymerization-polymerization dynamics and transport.

Knockout mice haploinsufficient for the protein CLIP115 result from the mutation of the gene Cyln2 and also duplicate some of the biochemical manifestations of Williams syndrome [20]. CLIP115 is a microtubule-binding protein that preferentially associates with the growing (plus) ends of microtubules. In CLIP115-knockout mice, other cytoskeletal linkers, such as

CLIP170 and dynactin, are upregulated. The end result is altered function for microtubules and actin filaments.

Cytoskeletal abnormalities in Williams syndrome contribute to the structural alterations noted in that disorder. Although overall hippocampal size is normal in Williams syndrome, the shape of the hippocampus is altered as compared to control brain [21]. These alterations in hippocampal shape were found to be accompanied by decreased blood flow at rest (in the hippocampus) and lessened responses to visual stimuli as shown in PET and fMRI scans.

5.2.4 Autism Spectrum Disorder: Disruptions of MAPs Due to Deletions of MAP Kinase and Reelin Genes

Autism spectrum disorder is a neurodevelopment disorder characterized by mental retardation (both verbal and non-verbal), social communication deficits, and repetitive behaviors, and having a prevalence of 1:150 [22]. Although there is some degree of co-morbidity, in which autism co-occurs with fragile X syndrome or Rett syndromes, each of these syndromes represent a small fraction (1% or less) of autistic cases [23]. Autism spectrum disorder represents a genetically heterogeneous group, whose characteristics are beginning to be better understood. Diverse, polygenetic causes are suspected.

Single nucleotide polymorphisms (SNPs) occur in at least 1% of the population and are linked to vulnerability to certain disorders and diseases [24]. An SNP on chromosome 1q41 - q42 that is associated with MAP/microtubule affinity-regulating kinase 1 (MARK1) has been found in some patients with autism [25]. MARK1 is also selectively overexpressed in the prefrontal cortex of autistic brain, a region known to participate in higher cognitive function, such as those compromised in the disorder. The normal role of MARK1 is to phosphorylate MAPs and to regulate trafficking along microtubules. Overexpression or deletion of MARK1 leads to altered dendritic length and transport rates for mitochondria traveling along dendrite tracks.

Consistent with these genetic perturbations in kinases affecting MAPs, there are findings of reduced MAPs in autistic brain. One study of post-mortem samples obtained from two adults with autism revealed reduced MAP2 immunohistochemistry in the dorsolateral part of the prefrontal cortex [26]. Although cell density in the prefrontal cortex appeared normal, cell organization was disrupted and dendrite number was decreased. These kinds of changes are consistent with cytoskeletal dysfunction.

There is also evidence of a weak association between the gene coding for the protein reelin and some individual cases of autism, insofar as larger than normal reelin genes (having more CGG repeats) may contribute to some subtypes of the disorder [27]. In a study comparing post-mortem brain samples from autistic patients to those of controls, researchers found reductions in reelin, the mRNA for reelin, and β-actin in the frontal cortex and cerebellum [28]. Reelin is a protein that regulates synaptic plasticity. One of its downstream effects is the activation of glycogen synthase kinase-3β (GSK3β), which in turn is responsible for phosphorylating the MAP, tau. To the extent then that reelin expression is disrupted in select cases of autism, one would expect microtubule function, as regulated by tau, to be similarly impaired.

5.2.5 Rett Syndrome: Decreases in MAP2 Possibly Linked to Mutations of the MCEP2 Gene

Rett syndrome is a neurodevelopmental disorder, associated with autistic-like symptoms and having a prevalence of 1:10,000 female offspring [29]. The mutation responsible for Rett syndrome is X-linked and lethal if transmitted to potential male offspring. The majority of those diagnosed with Rett syndrome have a deletion to the gene coding for methyl CpG-binding protein-2 (MECP2). MECP2 is a transcription repressor, responsible for silencing other genes. MECP2 is found in both MAP2-positive and MAP2-negative cells [30]. The absence of MECP2 in Rett syndrome may underlie alterations in MAP2, abnormal dendrite morphology, and profound changes in overall brain size [31].

Girls with Rett syndrome begin to demonstrate cognitive and social deficits at 6 months to 1 years of age, which parallel the time at which the brain begins to fail to grow at the appropriate rate [32]. The brain weight of children with Rett syndrome is on average half that of normal controls. Gross brain atrophy is particularly distinct in brain regions associated with higher cognitive functions, including prefrontal, posterior frontal, and anterior temporal cortex. Within the most dramatically affected areas of cortex, individual neurons have markedly diminished dendrite arbors. Rett syndrome brains also exhibit a dramatic loss of spines [33].

Immunohistochemical studies of brain tissue samples from different cortical areas of Rett syndrome patients show decreased MAP2 staining, especially in pyramidal cells, as compared to controls [34]. Conversely, samples from some of the cortical areas in Rett syndrome brains showed enhanced staining for non-phosphorylated neurofilament protein. This pattern of change is consistent with a defect in dendrite branching, but not in dendrite stability. Dendritic branching abnormalities found in Rett syndrome brain have been noted in both apical and basilar dendrites, although these abnormalities varied across individual regions of cortex and differ from those found in Down syndrome brain [35].

5.2.6 Down Syndrome: Early and Late Defects in the Microtubule and Actin Cytoskeleton

Down syndrome (also called trisomy 21) is a major cause of mental retardation, caused by an extra copy of all or part of chromosome 21 and having a prevalence of 1 in 800 births [36]. While what causes failure of chromosome non-disjunction is still incompletely understood, some have speculated that subtle forms of maternal microtubule dysfunction may predispose to meiotic or mitotic errors resulting in hyperploidy, or extra copies of chromosomes [37]. Microtubule hyperstabilizing and destabilizing agents, taxol and vincristine, are known to induce hyperploidy [38]. Moreover, older women (who are generally at increased risk for having a Down syndrome child) and young women who have had a Down syndrome offspring have microtubules that are less sensitive to the microtubule polymerization inhibitor, colchicine [39].

Cytoskeletal abnormalities have been noted in Down syndrome patients at different points along the lifespan. As detailed below, deficits in actin-related proteins are present during early development, whereas abnormalities among MAPs accumulate in association with neural degeneration during middle adulthood to old age. These changes have been further correlated with spine and dendrite development and degeneration.

Down syndrome infants have deficiencies in dendrites and spines, which appear to result from prenatal deficits in the spine-specific protein drebrin evident during the early part of the second trimester [40]. In addition to reductions in drebrin, the fetal Down syndrome brain shows decreased expression of centractin-α and the F-actin capping protein [41]. These decreases contrast with normal levels of dynein and kinesin. Centractin-α and the F-actin capping protein are part of the dynactin complex, which links microtubules to spectrin in spines. Also compromised in fetal Down syndrome brain is actin-related protein complex (Arp2/3) and coronin-like p57 [42]. These proteins are responsible for crosslinking, branching, nucleating, and capping actin. Moesin, a protein responsible for crosslinking actin to the neuronal membrane, is also significantly decreased in fetal Down syndrome brain [43]. Although these reductions in actin-related proteins are most often discussed in the context of disrupted spine function, reductions in these actin-related proteins also contribute to deficient axonal transport and mitotic failure.

Besides developmental delays and lifelong mental retardation, Down syndrome patients show a markedly increased risk of acquiring a dementia similar to that of Alzheimer's disease, except that the dementia begins much earlier and has a slightly different neuropathological profile [44]. Individual assessments of neuropathology in 10 Down syndrome and Alzheimer's disease patients found senile plaques in all cases of Down syndrome, including those in their late teens and twenties, whereas senile plaques and neurofibrillary tangles were only found in the older Down syndrome and Alzheimer's brains. These neurofibrillary tangles reacted with antibodies to tau and MAP2.

It is possible that a deficit in tubulin or tau underlies a predisposition to dementia in adult Down syndrome brain. There are reduced levels of the βIII-tubulin isotype and neurofilament proteins in Down syndrome brains, whereas Alzheimer's disease brains only show decreased levels of neurofilaments [45]. Neurofibrillary tangles in Down syndrome brain also associate more frequently with β-tubulin more than do neurofibrillary tangles in Alzheimer's disease brain [46]. An insoluble pool of highly phosphorylated (hyperphosphorylated) tau is found in adult Down syndrome brains and in Alzheimer's disease brains, but not in fetal Down syndrome brains [47]. Taken together, these results suggest that while abnormalities in actin-related proteins associated with spines produce dramatic early effects, MAPs may also be altered in Down syndrome, but the functional consequences are not fully expressed until a later time.

5.3 Neurological Disorders Involving Nanomechanical Dysfunction

Many neurological disorders, including amyotrophic lateral sclerosis, and Alzheimer's, Parkinson's, and Huntington's disease, share the common feature of having some kind of nanomechanical dysfunction in terms of a cytoskeletal impairment. At least three solid lines of evidence exist in support of this relationship. (1) Deficits in axonal transport occur across these disorders [48, 49]. (2) There are pathological profiles that include cytoskeletal protein components in these disorders [50, 51]. (3) Established genetic markers associated with sporadic or specific familial forms of these neurological disorders can be linked to cytoskeletal dysfunction [52]-[59]. Genetic links, along with their particular impact on the cytoskeleton, are summarized in Table 5.2.

5.3.1 Neuromuscular Disorders and Disrupted Axonal Transport

Amyotrophic lateral sclerosis (also known as Lou Gehrig's disease) is a neuromuscular disorder characterized by symptoms of progressive muscle weakness, spasticity, and atrophy, and having a prevalence of 6 persons in 100,000 [66]. Most cases are sporadic; only 5 - 10% of cases have a familial origin. Mutations in genes for copper-zinc superoxide dismutase (SOD1), the cytoplasmic dynein/dynactin complex, and possibly the intermediate filament NF-L contribute to a small percentage of familial cases, as well as to a presently unknown number of sporadic cases of amyotrophic lateral sclerosis (see Table 5.2).

There are over 90 mutations to SOD1 linked to amyotrophic lateral sclerosis, each of which commonly result in misfolding of the SOD1 protein altering its normal interactions with dynein [67]. This in turn affects axonal transport of the complex; the end result being that SOD1 protein complexes associated with dynein and other MAPs form large insoluble aggregates that interfere with nanomechanical function and eventually cause motor neuron degeneration and death.

A limited number of both familial and sporadic cases of amyotrophic lateral sclerosis have been linked to mutations to the gene for dynactin protein DCTN1 [68]. The relationship between the mutation of the DCNT1 gene and amyotrophic lateral sclerosis is only certain in the familial cases; more evidence is needed to determine a meaningful link in the sporadic cases. Several other genes have been implicated as contributing to sporadic amyotrophic lateral sclerosis, among them the gene for the intermediate filament NF-H.

Table 5.2. Genetic bases of neurological condition having an effect on the cytoskeleton.

Neurological condition	Associated genetic links	Effect on cytoskeleton
Neuromuscular disorders		
Amyotrophic lateral sclerosis	Copper-zinc superoxide dismutase (SOD1) gene mutations (20% of familial and 1% of sporadic cases)	Decreased binding of MAP1A, MAP2, and tau; tau hyperphosphorylation; impaired axonal transport
	DCTN1 deletion (familial and sporadic)	Depleted cytoplasmic dynein and dynactin; impaired axonal transport
	Intermediate filaments (NF-H)	Filamentous aggregates; impaired axonal transport
Charcot-Marie-Tooth disease	Intermediate filaments (NF-L)	Filamentous aggregates; impaired axonal transport
	Kinesin-3 (KIF1Bβ) gene deletion	Impaired synaptic vesicle transport
Hereditary spastic paraplegia (SPG10)	Kinesin-1 (KIF5A) gene deletion	Impaired axonal transport
Neurodegenerative diseases		
Alzheimer's disease (early-onset)	Amyloid precursor protein (APP), presenilin-1 (PSEN1), and PSEN2 mutations (10% of cases)	Tau hyperphosphorylation; impaired axonal transport
Alzheimer's disease (late-onset)	Apolipoprotein E (ApoE) (risk factor in 50% of cases)	Tau hyperphosphorylation; impaired axonal transport
Frontotemporal dementia with Parkinsonism linked to chromosome 17 (FTDP-17)	Tau mutation	Tau hyperphosphorylation; impaired axonal transport
Parkinson's disease	PARK1-13 (5 - 10% of familial and sporadic cases)	Deficits in α-synuclein (PARK1/4) and parkin (PARK3) leading to impaired axonal transport
	Expression of 4-repeat isoform of tau	Impaired tau function
Huntington's disease	CAG repeats in huntingtin gene	Polyglutamine repeats in huntington protein disrupts its binding to microtubules and leads to abnormal aggregation and impaired axonal transport

Adapted from [52]-[65]. See text for details.

There is some doubt, however, whether NF-H mutations are causally associated with amyotrophic lateral sclerosis, in contrast to Charcot-Marie-Tooth disease, which seems to be definitively associated with mutations to NF-L.

Dynactin activates dynein; the two proteins work together as a complex. Among its many cargoes, dynein is responsible for transporting neurofilament proteins along microtubule tracks. Dynein depletion, but not that of kinesin, induces aggregation of neurofilament proteins in axons, and this might explain how axons become disrupted in amyotrophic lateral sclerosis [69]. Mutations in kinesin genes have also been detected in other neuromuscular disorders, such as Charcot-Marie-Tooth disorder and hereditary spastic paraplegia, but it is the dynein-dynactin complex that is particularly targeted in amyotrophic lateral sclerosis (see Table 5.2).

The role of dynein in transport and axonal integrity is shown in multiple mouse transgenic models. Transgenic mice expressing less dynein and dynactin due to overexpression of dynamitin exhibited abnormal microtubule-mediated transport and progressive motor neuron degeneration [70]. Other transgenic mice, in which the dynactin $p150^{Glued}$ subunit was deleted, demonstrated deficits in axonal transport, specifically of synaptic vesicles [71]. According to some researchers, amyotrophic lateral sclerosis results primarily from deficits in motor neuron axonal transport, whether it is due to dynactin gene mutations that target the motor neuron transport specifically or SOD1 gene mutations that target the motor neuron, the muscle, and surrounding glial cells [72].

Links between amyotrophic lateral sclerosis and microtubule-associated proteins also exist. Highly phosphorylated tau protein has been detected in patients with amyotrophic lateral sclerosis and cognitive impairment [73]. This correlates with upregulated glucagon synthase kinase-3β (GSK3β). Levels of GSK3β were elevated in patients with amyotrophic lateral sclerosis with and without cognitive impairment, possibly contributing to increased risk of tau hyperphosphorylation. Stable-tubule-only-polypeptide (STOP), a microtubule-stabilizing protein, also binds to neurofilament aggregates in amyotrophic lateral sclerosis [74].

5.3.2 Nanomechanical Dysfunction in Alzheimer's Disease: Tauopathies and Impaired Transport

Alzheimer's disease has as its chief symptom dementia, and it currently affects more than 24 million people worldwide [75]. It is an aging-related disorder that preferentially targets individuals over 65 years of age. Diagnosing "probable" Alzheimer's disease first involves documenting a persistent and progressive cognitive impairment using a neuropsychological test for cognitive decline, such as the Mini Mental Status Exam (MMSE) or the Alzheimer's Disease Assessment Scale (ADAS) [76]. Other diagnostic tests can be used to corroborate a diagnosis of Alzheimer's disease, such as low levels of β-amyloid, high levels of tau, and highly phosphorylated tau in the cerebrospinal fluid.

Still, the most definitive diagnosis for Alzheimer's disease is the presence of senile plaques and neurofibrillary tangles in the hippocampus and cerebral cortex, usually detected in post-mortem tissue samples [77]. The nature of these pathological profiles suggests Alzheimer's disease is a disorder of the neuronal cytoskeleton.

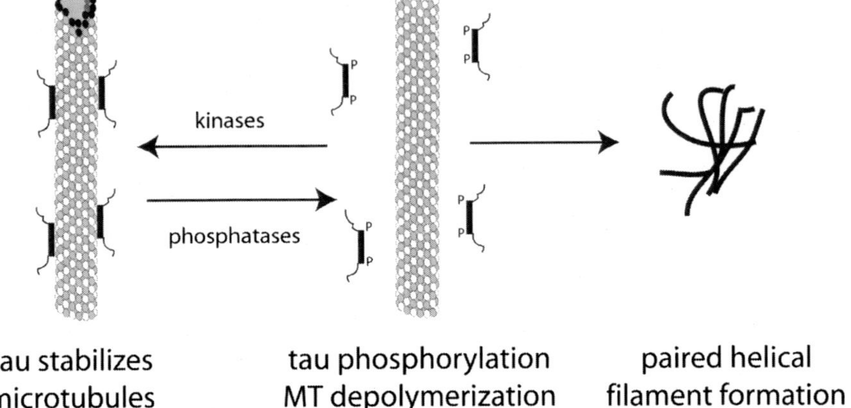

Fig. 5.1. Formation of neurofibrillary tangles begins with hyperphosphorylation of tau. One of the consequences is depolymerization of microtubules.

Hyperphosphorylated Tau and Its Relationship to Microtubules

As depicted in Figure 5.1, the neuropathological cascade for neurofibrillary tangle formation in Alzheimer's disease starts off with the tau protein becoming highly phosphorylated, or what is called hyperphosphorylated [78, 79]. This in turn leads to aggregation of tau, along with MAP2 and other microtubule-associated proteins into insoluble protein masses. The tau aggregates further coil into paired helical filaments, at which point the tau protein component adopts a β-conformation, and the entire mass ultimately transforms into neurofibrillary tangle material [80]. Because tau is central to their formation, neurofibrillary tangles are also known as tauopathies. An unresolved question is whether a different primary event (such as amyloid deposition) triggers tau hyperphosphorylation or if tau hyperphosphorylation is the initiator of subsequent neuronal damage. There is experimental evidence consistent with both possibilities.

Hyperphosphorylated tau can dramatically alter microtubule dynamics. Assembly of new microtubules was blocked when researchers added hyperphosphorylated tau isolated from Alzheimer's disease brain (AD P-tau) to mouse embryonic fibroblasts, which had their interiors replaced with adult rat brain cytosol [81]. AD P-tau at concentrations greater than 50μg/ml not only completely prevented the assembly of new microtubules, but AD P-tau destroyed pre-existing microtubules. These results suggest that tau hyperphosphorylation is responsible for the microtubule dysfunction in Alzheimer's disease. Another possibility, however, is that AD P-tau exacerbates microtubule defects already present in Alzheimer's disease brain. This is consistent with certain studies indicating that microtubule dysfunction is a primary problem in Alzheimer's disease and not merely a result of neurofibrillary tangles compromising neuronal integrity.

When neurons in Alzheimer's disease brain were evaluated for microtubule abnormalities, even healthy neurons (i.e., those lacking neurofibrillary tangles or tau filaments) exhibited microtubule defects. Researchers demonstrated this by estimating the microtubule number and total microtubule length in cortical pyramidal cells from Alzheimer's disease and control brains [82]. As expected, microtubule numbers and total lengths were decreased in tau filament-containing neurons in Alzheimer's disease brains as compared to those from normal neurons in control brains. What was unexpected was that microtubule numbers and total microtubule lengths for normal neurons from Alzheimer's disease brains were also decreased compared to those in neurons from controls. While the reason for a primary defect in microtubules remains unknown, one possible explanation is that microtubules may be a common target in cellular aging [83].

Another possible explanation for tau hyperphosphorylation is that trafficking of MAPs may be abnormal in Alzheimer's disease due to imbalances in the levels of different MAPs. Overexpressing tau in hippocampal neurons results in increased levels of tau entering into dendrites and a subsequent loss of spines [84]. As tau accumulates in dendrites, it is vulnerable to hyperphosphorylation and it competes with other MAPs affecting microtubule dynamics. Gene expression for MAP2 is decreased in Alzheimer's disease brain, while levels of tau expression are normal [85]. Moreover, MAP2 immunostaining is negatively correlated with neurofibrillary tangle load [86]. This latter result suggests an inverse relationship between MAP2 levels and presence of tangles, but does not indicate which of the two might cause the other, or if both pathological profiles simply co-occur.

Amyloid-β Peptide, Amyloid Precursor Protein, and Axonal Transport

Whereas neurofibrillary tangles outstrip senile plaques in the early stages of Alzheimer's disease, both neurofibrillary tangles and senile plaques are abundant in more advanced stages of Alzheimer's disease [77]. Senile plaques are extracellular accumulations of the β-amyloid protein, which in turn attract dystrophic neurites, glia, and immunoglobulins [87, 88]. Of particular interest is the highly toxic amyloid-β peptide, thought to be key in the etiology of Alzheimer's disease [89]. That β-amyloid plays a primary role in Alzheimer's disease dovetails with amyloid precursor protein (APP) and presenilin (PSEN) mutations being detected in some familial cases of Alzheimer's disease; however, it fails to sufficiently address the relationship between tauopathies and the deposition of amyloid [90].

Axonal transport is a possible function in which both amyloid-related proteins and tau participate. There is evidence that APP binds directly to kinesin and is thereby fundamental to axonal transport. APP is the receptor protein responsible for binding β-secretase and PSEN1 to kinesin-1 [91]. During normal transport this complex is transported intact; however, stalling of transport under pathological conditions might lead to the cleavage of amyloid-β peptide by both β- and γ-secretase. Under such conditions, the highly toxic amyloid-β peptide would be released into the axon and subsequently cause neural damage.

Other data implicate impaired axonal transport mechanisms in Alzheimer's disease. Transgenic mouse models of Alzheimer's disease, such as those increasing expression of tau, APP, or cholesterol-transporting apoliprotein E (ApoE), are all associated with defects in axonal transport [92]. Tau ordinarily regulates axonal transport. When tau is in a pathological hyperphosphorylated state, axons swell and transport is impaired. Disrupted anterograde transport is suspected to result in increased production of amyloid-β peptide and its delivery to senile plaques. Amyloid-β peptide and ApoE also have effects on tau hyperphosphorylation; however, many details of the dynamic relationships between tau, amyloid-β peptide, ApoE, and axonal transport remain to be worked out. Although massive disruption of axonal transport is evident in late-stage Alzheimer's disease, it is unclear how much of a role impaired axonal transport plays early in the course of Alzheimer's disease [93].

Neurotrophins and Early Deficits in Axonal Transport

In addition to neurofibrillary tangles and senile plaques, there is marked degeneration of cholinergic basal forebrain neurons in Alzheimer's disease; this includes neurons in the nucleus basalis of Meynert, which project to the entire neocortex, and neurons in the medial septal nucleus, which project to the hippocampus and limbic cortex [94, 95]. This cholinergic degeneration is more than simply a neurochemical deficit. Cholinergic axons that show abnormal swellings before the onset of dementia preferentially undergo degeneration in Alzheimer's disease [96]. While the intrinsic mechanism for cholinergic neuron degeneration remains unknown, the cytoskeleton remains a possible determinant.

The cholinergic deficit in Alzheimer's disease appears to be related to neurotrophic factors, with either insufficient or overabundant levels of these factors potentially contributing to cytoskeletal derangement [97, 98]. Cholinergic basal forebrain neurons are sensitive to the neurotrophin NGF, relying upon optimal levels of NGF to be retrogradely transported to the cell body in order to sustain cell survival [99]. Excess of NGF in the target regions has been interpreted as evidence that retrograde axonal transport may be deficient in these neurons, resulting in insufficient numbers of NGF reaching the cell body where it is able to influence transcription. Besides NGF, other neurotrophins that are affected in Alzheimer's disease include BDNF, NT3, and NT4 [100]. These additional neurotrophins are taken up and retrogradely transported by cortical and hippocampal neurons. The dysregulation of all these neurotrophins in Alzheimer's disease is consistent with an impaired axonal transport. Precursors for these neurotrophins (e.g., pro-NGF, pro-BDNF, pro-NT3, and pro-NT4) bind to a common receptor $p75^{NTR}$, which triggers cell death. Given impairment to retrograde transport in Alzheimer's disease, precursors accumulate and activate $p75^{NTR}$ receptors at an elevated rate. Other factors exacerbate the deterioration. The β-amyloid protein also binds to $p75^{NTR}$, thereby increasing levels of cholinergic cell death [101].

Deficits in retrograde transport in the cholinergic basal forebrain as well as in the cerebral cortex suggest significant impairment of microtubules in Alzheimer's disease. This would be consistent with evidence of abnormal sprouting in both groups of these neurons in Alzheimer's disease brain, a phenomenon that is mediated by cytoskeletal proteins [102]-[105]. To the extent that microtubules are deficient in Alzheimer's disease, potential electric signaling along microtubules or the generation of phosphoprotein waves, as postulated in Chapters 3, would likely be impaired. As further discussed in Chapter 4, such mechanisms may be crucial to the retrograde transport of neurotrophins.

5.3.3 Nanomechanical Dysfunction in Parkinson's Disease: Microtubule Instability and Synucleinopathies

Parkinson's disease cases total over 4 million worldwide and that number is rapidly growing as the aged population continues to increase [106]. The symptoms of Parkinson's disease include tremor at rest, bradykinesia, rigidity, loss of postural control, and dyskinesia; moreover, dementia occurs in a significant number of cases [107]. The hallmark pathological marker of Parkinson's disease is a marked degeneration of dopamine-containing neurons in the substantia nigra, the neurons that provide dopamine to the striatum [108]. As shown in Table 5.2, the genetic mutations that have been linked to Parkinson's disease include mutations in PARK1-13 genes. The gene PARK1/4 codes for α-synuclein and PARK3 codes for parkin, with mutations in both contributing to early-onset Parkinson's disease and a small number of sporadic Parkinson's disease cases [109, 110]. Also, the expression of the 4-repeat isoform of the MAP tau has been linked to both familial and sporadic Parkinson's disease [64]. Ranking high among the other suspected causative factors for Parkinson's disease are environmental toxins. Evidence supporting a role for neurotoxins comes from laboratory experiments and clinical case studies. The exposure to the pesticide rotenone causes parkinsonism in rats, for example

[111]. The accidental ingestion of the compound 1-methy-4-phenyl-1,2,3,6-tetrahydropyridine (MPTP) caused an immediate onset of parkinsonism in several drug users who administered a narcotic drug laced with the toxin [112]. MPTP has subsequently been used to produce an animal model of Parkinson's disease, an animal model that reliably replicates the dopamine degeneration associated with the disease, but not other aspects of the disorder.

Genetic and Environmental Causes of Parkinson's Disease Affect Microtubules

The two main putative causes of Parkinson's disease, genetic and environmental toxins, have a common denominator – an effect on brain microtubules [113]. It has been argued that mutations to parkin or exposure to toxins, such as rotenone or MPTP, compromise microtubule stability, and that with such increases in microtubule depolymerization, there is an increased need to clear toxic excess tubulins. Impaired microtubule stability further leads to decreased transport of dopamine-containing vesicles and subsequent leakage of dopamine into the cytoplasm. Dopamine that is leaked into the cell becomes highly toxic following oxidation by monoamine oxidase, which leads to increased oxidative stress and neuronal death [114].

Genetic mutations to parkin would also be expected to affect tubulin clearance. Parkin is an E3-ubiquitin ligase responsible for ubiquinating tubulin dimers and then mobilizing them to the 23S proteasome for degradation. This pathway is responsible for removing highly toxic misfolded tubulins from the neuron [115]. Misfolded proteins may indeed lie at the core of many degenerative diseases [116], thereby implicating effects on tubulin and microtubules in the onset of these disorders.

Synucleinopathies Resemble Tauopathies

Synucleinopathies found in Parkinson's disease are related to the cytoskeleton, particularly to microtubules. Fundamental similarities exist between α-synuclein and tau: (1) both are functional MAPs and (2) both contribute to pathologies – synucleinopathies and tauopathies, respectively [117]. Moreover, the presence of tauopathies in Parkinson's disease with dementia suggests convergence of pathways affecting both MAPs [118]. Tubulin binds α-synuclein during the initial formation of α-synuclein fibrils, or what are called Lewy bodies in Parkinson's disease brain [119]. Subsequent to initiation, α-synuclein fibrils coil into protofilaments, which adopt a β-sheet conformation that resembles that of paired helical filaments made from tau. It is possible that excess α-synuclein triggers the initiation of synucleinopathies. Experiments consistent with this notion show that the overexpression of α-synuclein impairs trafficking along microtubules, which then leads to aggregation of α-synuclein on microtubules and finally to microtubule degeneration [120, 121].

A direct relationship exists between rotenone and microtubules. Rotenone binds to tubulin at the colchicine-binding site; moreover, taxol, a microtubule-stabilizing compound, decreases the toxicity of rotenone [122]. In addition to poisoning dopamine neurons, rotenone exerts a toxic effect on serotonergic neurons by increasing microtubule polymerization [123]. The resulting effect on serotonin may be of relevance to depression that is diagnosed in some Parkinson's disease cases.

Conversely, the relationship between MPTP and microtubules is indirect. Once MPTP gains entry, monoamine oxidase in glia cells metabolizes MPTP to the free radical 1-methyl-4-phenylpyridinium (MPP+), and it is this free radical that interferes with complex I of the mitochondrial electron transport system. As a result of toxic damage to dopamine neurons and greater breakdown of tubulin, the need to ubiquinate and clear those damaged proteins increases [113]. Such a notion is supported by experimental evidence. When the free radical MPP+ is administered to neuron-like PC12 cells, this decreases microtubule polymerization and increases levels of unbound α-tubulin [124].

5.3.4 Huntington's Disease: Involvement of Microtubules and Axonal Transport of Neurotrophin Receptors

Huntington disease is a debilitating, autosomal dominant disease that affects over 1 in 10,000 people [125]. Symptoms include uncontrollable spastic and ballistic movements, along with psychiatric disturbances and dementia. The disease has an onset in middle adulthood and is fatal, leading to death in an average of 10 - 20 years. The genetic basis of Huntington's disease is an excess number of polyglutamine repeats (> 37 CAG repeats) leading to a mutated huntingtin protein [65]. Mutated huntingtin causes cellular degeneration that is particularly severe among GABAergic neurons in the striatum. Huntingtin is a protein that ordinarily binds to huntingtin-associated protein-1 (HAP1), huntingtin-interacting protein-1 (HIP1), HIP2, and microtubules. The excess in polyglutamine repeats in the huntingtin protein interferes with its normal function and interaction with the cytoskeleton, as demonstrated by depletion of α-tubulin, MAP2, and phosphorylated MAP2 early in the course of Huntington's disease [126]. Microtubule instability further contributes to the toxicity of mutated huntingtin protein [127].

Accumulation of abnormal huntingtin protein leads to dysfunctional axonal transport [128]. Ordinarily the microtubule motors dynein and dynactin associate with huntingtin protein; the complex additionally binds HAP1 and is responsible for transporting vesicles down the axon [129]. Excess polyglutamine repeats in huntingtin interfere with transport of vesicles. HAP1 binds to kinesin light chain, and linkages between huntingtin and HAP1 similarly impair transport mediated by this motor protein [130]. In addition to impairing transport of vesicles, accumulating N-terminal fragments of huntingtin proteolytic cleavage products, which result from misfolding and aggregation, disrupt mitochondrial transport [131].

In neurons, over half the huntingtin protein and HAP1 is found in association with microtubules [132, 133]. Moreover, the binding of these proteins to microtubules is affected by posttranslational modifications to tubulin. The acetylation of α-tubulin (which increases microtubule stability) enables axonal transport of vesicles to overcome impairments induced by abnormal huntingtin protein [134]. Additionally, phosphorylation of HAP1 affects its binding to dynactin and kinesin and may therefore be relevant to the deleterious effects that abnormal huntingtin has on HAP1-mediated trafficking of TrkA receptor and neurite growth [135].

5.4 Neuropsychiatric Disorders and Nanomechanical Dysfunction

Major forms of mental illness, including schizophrenia, bipolar disorder, and major depression, demonstrate some degree of nanomechanical dysfunction in terms of involvement of the cytoskeleton, particularly that of microtubules. As listed in Table 5.3, genetic markers having the strongest linkages to these psychiatric illnesses code for proteins that interact with microtubules. There are also structural abnormalities in neurons of persons having schizophrenia and bipolar disorder that can be attributed to dysfunctional organization among cytoskeletal proteins.

5.4.1 Schizophrenia: Genetic Mutations of Proteins Linked to Microtubules and Abnormal Neuron Morphology

Schizophrenia affects approximately 1% of the population and has as its chief symptoms hallucinations, delusions, and significant social and occupational impairment [136]. The concordance rate of schizophrenia among identical twins is 60-84%, meaning both genetic and environmental factors influence the expression of the disorder.

Certain genes are associated with increased susceptibility to developing schizophrenia in a small number of individuals; however, most cases of schizophrenia are idiopathic. Disrupted-in-Schizophrenia-1 (DISC1) is a candidate gene increasing vulnerability to schizophrenia, with a truncated mutant form of the protein occurring only in a very few select families [137]-[143]. Intact DISC1 protein binds to axonal microtubules, the presynaptic membrane, postsynaptic membrane, the centrosome, and the nucleus of the cell. Its association with microtubules enables a role in axonal transport, primarily through DISC1 binding to the microtubule motor protein kinesin KIF5B. Mutated and truncated DISC1 leads to impaired transport, as well as to impaired neurite growth because of its reduced binding to microtubules and motor proteins. Over time, decreased neurite growth leads to reduction in neuronal volume,

Table 5.3. Genetic linkages associated with neuropsychiatric disorders and their effects on the cytoskeleton.

Neuropsychiatric disorder	Associated genetic link	Effect on cytoskeleton
Schizophrenia	Disrupted in schizophrenia-1 (DISC1)	Impaired axonal transport
	Dysbindin	Impaired neurogenesis
	Neuregulin-1	
Bipolar disorder	Disrupted in schizophrenia-1 (DISC1)	Impaired axonal transport
	Dysbindin	Impaired neurogenesis
	Neuregulin-1	Neurite growth
	BDNF	Phosphorylates tau
	Glycogen synthase kinase-3β and (GSK3β)	
Major depression	Disrupted in schizophrenia-1 (DISC1)	Impaired axonal transport

Adapted from [137]-[145]. See text for details.

particularly in prefrontal cortex and hippocampus. The loss of dorsolateral prefrontal cortex volume is a hallmark of schizophrenia and is likely to be responsible for the deficits in working memory and executive function that accompany the illness [144].

Two additional gene candidates contributing to schizophrenia in a small number of select families include those coding for neuregulin-1 and dysbindin [141]-[144]. Dysbindin binds microtubules (as well as actin filaments and neurofilaments) and is associated with vesicles and postsynaptic densities [145]. Neuregulin plays a role in neurogenesis, which relies on cytoskeletal rearrangement attendant with neuron growth [146].

The genetic mutations discussed above appear to only account for a small number of cases of schizophrenia. Nonetheless, there is evidence that microtubule proteins may be abnormally expressed in general populations of schizophrenics. STOP isoform 2, or what is also called MAP6, is up regulated in prefrontal cortex of schizophrenic patients [147]. Conversely, mice deficient for STOP exhibit schizophrenic-like behaviors [148]. Microtubule-stabilizing drugs are able to reverse some of the schizophrenic-like behaviors in mice with a genetic deletion of the STOP protein. In a healthy brain, STOP facilitates transport of synaptic vesicles to axonal terminals along microtubules. Thus, it would appear that both excesses and deficits of this protein are damaging to both transport and cognitive function.

A number of studies also report MAP2 alterations in schizophrenic brains, but there are inconsistencies among these reports. Increases in immunohistochemical staining for MAP2 in neurons of the hippocampus, subiculum, and the white matter of the frontal cortex have been reported [149, 150]. However, other studies have shown decreases in MAP2 immunostaining in subiculum and entorhinal cortex [151, 152]. Chronically institutionalized schizophrenics, in particular, showed decreased MAP2. Inconsistencies across these studies may be attributable to different post-mortem delays, differences in antibodies and staining protocols applied, and variance among patient populations. MAP2 staining alterations do, however, accurately pinpoint those neuronal populations that additionally show structural abnormalities in schizophrenia. Disorganized neurons having variably oriented apical dendrites that are not properly aligned have been noted in the hippocampus of schizophrenics at much higher rates than in normal brains, although this has not been observed uniformly across studies [153]. The initial observation of neuronal disarray at the CA1/prosubiculum and CA1/CA2 borders suggested a possible flaw in neural migration during early neural development [154]. Subsequent studies have shown a strong tendency for neural disarray to parallel the severity of psychosis in schizophrenia, but differences did not reach significance [155]. If schizophrenia is a disorder fundamentally affecting microtubule functions such as transport (as suggested here), then it would be expected that some schizophrenic patients would demonstrate morphological abnormalities attendant with those microtubule defects, whereas other patients would not demonstrate structural abnormalities. In those schizophrenics without morphological deformations, the basis of microtubule dysfunction would be indirectly due to defects in other proteins that normally interact with microtubules.

5.4.2 Affective Disorders and Microtubules

Affective disorders, also called mood disorders, include bipolar disorder and major depression. The two major types of bipolar disorder, termed type I and type II, affect approximately 3.9% of the population and are characterized by extreme emotional highs and lows, known respectively as mania (or if present to a lesser degree, hypomania) and depression [156]. Major depression is much more common, exhibiting lifetime prevalence rates as high as 20% and 30% for males and females, respectively [157].

There are multiple gene candidates of potential interest in bipolar disorder, and although many of these are also candidate genes for schizophrenia, bipolar disorder is distinct from schizophrenia in many respects [158]. DISC1, which is a candidate gene for schizophrenia, is associated with bipolar disorder in both Finnish and London cohorts [159]. Additional gene candidates commonly implicated in both schizophrenia and bipolar disorder include neuregulin-1 and dysbindin [160]-[162]. The genes for BDNF and GSK3β are also cited as candidate genes for bipolar disorder, among several others [143]. Major depression is a familial illness, in which environmental factors markedly contribute to its expression [163]. Among the gene candidates associated with schizophrenia and bipolar disorder, DISC1 also demonstrates an association with major depression [139]. Apart from gene mutations, levels of gene expression are modified in patients with major depression compared to controls. One study identified 71 genes that were differentially expressed in major depression, with many of the affected genes participating in cell proliferation and plasticity [164]. Given the presumed polygenic nature of affective disorders, more research is needed to clarify how candidate genes interact with each other and with environmental triggers. To illustrate the complexity of polygenetic contributions, one simulated model of bipolar disorder arbitrarily posits 30 (100%) candidate genes, under which circumstances an affected individual would carry a minimum of 22 (66%) alleles, in contrast to an unaffected individual who would carry on average 15 (50%) alleles [165]. This model does not take into account the varying degrees of penetrance associated with different genes.

The various genes mentioned above have an interaction with the neuronal cytoskeleton, axon transport, and neurite outgrowth (see Table 5.3). The interactions of DISC1, neuregulin-1 and dysbindin with the cytoskeleton were discussed in the previous section. The manner in which BDNF affects the cytoskeleton during the regulation of neurite growth was discussed in Chapter 4. GSK3β is responsible for phosphorylating tau and MAP1B, and is decreased by the mood stabilizers, lithium and valproic acid, which are used to treat bipolar disorder [166, 167].

There is additional evidence that cytoskeletal proteins are abnormally expressed in depression. As compared to that of control brain, immunostaining for both MAP2 and MAP1B is decreased in the anterior cingulate cortex of patients with bipolar disorder [168]. Microtubules are also affected in animal models of depression. One such model, called learned helplessness, uses inescapable footshock to induce a depressed-like state in laboratory rats. Inescapable shock led to decreases down to 50% of normal levels for neurofilament proteins and MAP2 in parts of the hippocampus [169]. Restraint stress, which also leads to behavioral depression, was shown to result in decreased levels of tyrosinated-tubulin, while increasing acetylated-tubulin [170]. These results suggest plasticity is a key component to behavioral depression, operationally defined as the animal learning to inhibit behavioral responses. This is consistent with results showing several plasticity-related proteins, synapsin-1, growth-related protein-43, and MAP2, being modified at 1 and 8 days after learned helplessness training [171]. Antidepressant drugs may act at the level of microtubules to interfere with the learning of these depressed responses [172].

5.5 Nanotechnological Approaches to Nervous System Disorders

Nanotechnology is opening the door to new treatment options for patients having neurological disorders [173]. Recent developments in nanotechnology include major advances in diagnostics, drug delivery, gene delivery, and nanofabrication of biohybrid devices. Progress along these various lines is expected to continue to increase over the next few decades, particularly with regard to nanoparticle systems that are capable of crossing barriers, targeting specific cells or cell compartments, and delivering drugs, proteins, genes, or interference RNA [174, 175, 176].

As described throughout this chapter, a common etiological factor spanning several neurodevelopmental, neurological, and neuropsychiatric disorders is a disruption of nanomechanical function induced by a defective or inoperative cytoskeleton. In many nervous system disorders there is a deficiency in microtubule stability or impaired transport, suggesting that a potential nanotechnological treatment strategy could be aimed at introducing stabilized microtubules into deficient neurons [177]. There are at least two ways that structurally enhanced microtubules could be incorporated into specific parts of the nervous system. One way would be to use gene therapy along with the creation of DNA constructs for tubulin isotypes known to produce more stable microtubules. Another way would be to introduce biohybrid microtubules – consisting of nanocoated microtubules or nanoconstructed synthetic materials – into impaired neurons. Similar approaches could be used to introduce other cytoskeletal proteins, receptor proteins, or signal transduction molecules into neurons.

Nanotechnology also stands to improve deep brain stimulation techniques through use of nanoparticle-coating materials that can be applied to electrodes, decreasing site damage. There is the additional possibility that nanoparticles and nanodevices can assist in guiding neural plasticity in an effort to repair nervous system damage and degeneration.

5.5.1 Nanotechnology and Diagnosis of Nervous System Disorders

A specific category of nanomedicine has been dubbed "nanoneurology" and its capabilities include ultrasensitive imaging resulting in a more accurate diagnosis of neurological disorders [178]. Magnetic nanoparticles and gadofullerenes have recently been identified as potential contrast agents for MRI [179, 180].

In addition to being able to more precisely image sites of neural damage, nanotechnology has advanced detection of minute quantities of neurochemicals. Optical detection of dopamine using gold nanoparticles is a method having possible applications to the diagnosis of Parkinson's disease [181]. An ultrasensitive DNA-barcode method enables the detection of amyloid-β-derived diffusible ligands in the cerebrospinal fluid of Alzheimer's disease patients [182]. Another method being developed to detect tau protein present in Alzheimer's disease cerebrospinal fluid is a surface plasmon-resonance immunochip utilizing a gold-nanoparticle [183]. Also in Alzheimer's disease brain, there is an excessive accumulation of nanoparticles (smaller than 20 nm in diameter) consisting of iron oxide and magnetite, which could be exploited to develop an MRI-based diagnostic measure of the disorder [184].

5.5.2 Nanotechnology Advances in Drug and Gene Delivery

There are multiple ways in which nanotechnology can result in improved drug and gene delivery to the central nervous system, with permeability to the blood-brain barrier being a key issue [185]. Drugs can be modified by nanotechnological approaches to enhance their lipid solubility or otherwise gain access to cross the blood-brain barrier. Nanoparticles can be used as transporters or carriers, or they can be used to mask those properties of the drug that make them impermeable. An active drug can be attached to a "Trojan horse" vector that gains access across the blood-brain barrier, a chimeric peptide can couple a non-soluble drug to a transportable vector, or a monoclonal antibody fusion protein can conjugate a drug to a transport vector that is permeable. Many of these approaches have potential application for treating diseases of the nervous system.

Lipid nanoparticles and gold nanoparticles conjugated to liposomes have proved to be useful as drug and gene delivery systems [186, 187, 188]. Degradable polymeric nanoparticles, which tend not to react with biological fluids, are capable of sustained release of drugs [189], and have even been used to delivery of genes to stem cells [190]. Nanoparticle drug delivery appears to hold promise for treating several nervous system disorders, in particular Parkinson's and Alzheimer's disease.

Research done on experimental animal models of neurodegenerative disease has generated positive results for nanoparticles. A phospholipid nanoparticle has been shown to have neuroprotective effects in an animal model of Parkinson's disease [191]. The dopamine agonist, bromocriptine, has been administered to parkinsonian rats using solid lipid nanoparticles as a delivery system [192]. Also, a nanocarrier made from neurotensin polyplex has proved successful at transfecting dopamine neurons with reporter genes for glial cell line-derived neurotrophic factor and reversing symptoms in parkinsonian rats [193].

Nanoparticle-based drug delivery treatments for Alzheimer's disease have addressed three main issues: increased permeability across the blood-brain barrier, extended release, and selective targeting (e.g., certain nanoparticles have a specific affinity for amyloid protein) [194]. Nanoparticles coated with the surfactant polysorbate 80, for example, were shown to facilitate the uptake of the acetylcholinesterase drug rivastigmine into the brain [195]. Nanoparticles capable of passing through the blood-brain barrier have also been conjugated to iron chelators that in turn are able to solubilize amyloid deposits in Alzheimer's disease brain [196].

Nanoparticles also make it possible to deliver drugs to particular subcellular compartments. The microtubule stabilizing drug, paclitaxel, has been delivered specifically to mitochondria using nanocarriers [196]. Although amyloid deposition is a well-characterized problem associated with Alzheimer's disease, amyloid nanocomplexes have also been found to be excellent drug carriers, which enable sustained drug release because of their extended binding [197].

Gene Therapies Aimed at Restoring Nanomechanical Function

Gene therapies have already been implemented for Alzheimer's and Parkinson's disease, but no gene therapy has yet attempted to fortify an intracellular structure such as the microtubule, despite the preponderance of evidence that these structures are severely compromised in these disorders. Current evidence suggests that gene therapy is nonetheless safe and compares favorably to other treatments for these neurological disorders.

Gene therapies using *ex vivo* and adeno-associated virus (AVV) vectors to introduce genes for NGF have been administered to Alzheimer's disease patients with the goal of rescuing cholinergic basal forebrain neurons that degenerate in that disorder [199]. These NGF gene therapies have met with some degree of success; phase I clinical trials indicate a slowing of cognitive decline among participants by approximately 22 months [200]. Nonetheless, NGF gene therapies do not reverse or halt the downward progression of Alzheimer's disease much the same as anticholinesterase drugs do not reverse or halt the progression of the disease. Despite their prevalent usage, anticholinesterase drugs tend to produce only mild improvements in cognitive status [201]. Given the somewhat disappointing clinical responses to current pharmacological therapies, some Alzheimer's disease researchers have argued for treatment strategies addressing cytoskeletal defects, for example, interventions aimed at preventing the hyperphosphorylation of tau [202, 203].

Gene therapies are also being tried on Parkinson's disease patients. A phase I clinical trial is assessing Parkinson's patients receiving gene therapy boosting neurturin in the striatum, and another clinical trial is investigating the increase of aromatic L-amino acid decarboxylase [204]. Neurturin is known to rescue degenerating dopaminergic neurons in animal models of parkinsonism [205]. The aromatic L-amino acid decarboxylase is the synthetic enzyme responsible for manufacturing dopamine, so gene therapies increasing its expression stand to increase dopamine. Although initial observations suggest these gene therapies are promising, any long-term benefits remain to be established. To the extent that dysfunctional transport mechanisms lead to toxic accumulation of dopamine in Parkinson's disease patients, one might expect only a few years of symptom relief unless the underlying cytoskeletal defect is addressed. Parkinson's disease is currently most often treated with pharmacological replacement of dopamine (L-dopa therapy), and in rarer instances, neurosurgical ablation or brain deep stimulation [206]. L-dopa therapy has a therapeutic window – a limited period of a few years – during which the drug is effective. Given the arguments presented here, a strategy aimed at enhancing the stability of microtubules using a nanotechnological intervention seems worth pursuing as a course of action with possible long-term benefits.

Some researchers have suggested that microtubule-stabilizing drugs might be useful in treating mental disorders like schizophrenia; the major obstacle is that these drugs are highly toxic [207]. Gene therapies that introduce stable microtubules may be a viable alternative treatment strategy for schizophrenia, the challenge being to deliver these genes to the appropriate brain sites. Similar strategies might be applied to bipolar disorder and major depression, with modifications made to the brain sites into which genes would be delivered.

To date there have not been any gene therapies that attempt to introduce stable microtubules, or other cytoskeletal proteins for that matter, that might in turn reorganize the architecture of the neuron directly. This untapped strategy may provide what is needed to restore fundamental neural operations, such as axonal transport, neurite growth, and structural reorganization. Such an approach is viable and exploratory studies in animal models are warranted. Current biotechnologies are perfectly capable of creating novel DNA constructs [208], and these techniques are suitable for making DNA constructs of microtubule-forming tubulins or other cytoskeletal proteins associated with them. Creating DNA constructs for tubulins that assemble into more stable and damage-resistant microtubules should proceed along the following steps:

1. Designing DNA constructs for tubulins that will successfully assemble into microtubules in living cells.
2. Identifying specific point mutations or amino acid substitutions that would produce a more stable (or, in some cases, a more dynamic) microtubule.
3. Assessing what degree enhancement of microtubule stability is desirable to ameliorate a particular disorder.
4. Investigating ways to introduce newly designed genes into the specific regions of the nervous system or into specific neuronal compartments.

Similar strategies could be applied to actin and motor proteins. Table 5.4 provides a list of potential therapies aimed at ameliorating cytoskeletal dysfunction through genetic and nanoengineering means.

5.5.3 Stabilizing the Nanomechanical Machinery in Neurons

Many of the diseases discussed in this chapter involve not only cytoskeletal abnormalities, but also accelerated cell death, especially in select cell populations – dopamine neurons in Parkinson's disease, cortical neurons in Alzheimer's disease, and motor neurons in amyotrophic lateral sclerosis, for example [209]. Both over-stabilized and under-stabilized microtubules contribute to cell death [210], and stabilizing hyperdynamic microtubules appears

Table 5.4. Potential approaches to address nanomechanical dysfunction in neurological and neuropsychiatric disease.

Approach	Effect
Genetically engineered tubulin	Specific mutations that translate into microtubules with specific mechanical, electronic and dynamic properties.
Genetically engineered microtubules	DNA sequences with more sites for stabilizing posttranslational modifications (e.g., acetylation or detyrosination) DNA sequences resulting in microtubules able to bind more avidly to metals or to semi-conductor materials.
Genetically engineered actin	DNA sequences providing more sites for stabilizing posttranslational modifications. DNA sequences resulting in actin filaments that are able to bind more avidly to metals or to semiconductor materials.
Nanoengineered cytoskeletal filaments	Could stabilize existing cytoskeletal networks in neurons.
Nanoengineered scaffolds	Provide structural support to existing cytoskeletal networks in neurons (e.g., synthetic MAPs).
DNA construct approaches	Enhance delivery of materials to neuronal compartments (synapses, axon terminals, etc.) Genetically modify kinesin, dynein, or myosin to form attachments with drugs or nanodevices. Genetically modify kinesin, dynein, or myosin to transport more rapidly or efficiently
Nanoengineered transport devices	
Biologically-inspired implantable hardware	Serves as neural prosthetic to restore memory and higher cognitive functions (e.g., restore hippocampal function)
Silence gene responsible for β-III tubulin	Reduce drug resistance and damage repair in brain cancer.

to counter neuron dysfunction and disease [211]. Thus, a critical balance between stability and instability is necessary for proper microtubule function [212]. This means that for every neurodevelopmental, neurological, or neuropsychiatric disorder, a different imbalance among stable and labile microtubules may be present, and modified microtubules (or genes that encode them) will need to be delivered to different brain sites, depending on where neuron dysfunction predominates in that particular disorder.

A more thorough understanding of the biophysical basis of microtubule stability should provide the framework from which nanoneuroscientists can build more stable and disease-resistant microtubules. Microtubule stability is secondary to the primary amino acid sequence of tubulin, meaning that any stabilizing effect demonstrated in the laboratory can, in principle, be permanently built into a microtubule. Such a conclusion is consistent with the computational modeling of 475 isotypes of tubulin, each having a distinct amino acid sequence, which revealed marked differences in the stability of microtubules assembled from those subunits [213]. There are several factors known to affect microtubule stability, including the ambient temperature, pH and ion concentration levels, the binding of certain drugs, posttranslational modifications to tubulin, and MAPs binding to tubulin.

Drug Binding Affects Microtubule Stability

A number of drugs induce conformational changes to tubulin and microtubules and alter microtubule stability; these include paclitaxel, colchicine, and vinblastine [214, 215]. Different drugs bind to different sites on the tubulin molecule and alter microtubule stability in subtly different ways. Paclitaxel stabilizes the M loop of tubulin, a site responsible for lateral bonding between tubulin dimers [216]. Vinblastine binds to a hydrophobic groove on tubulin, increasing the curvature of individual protofilaments, and subsequently facilitating microtubule depolymerization [217]. The extent to which such known facts could be exploited to design more disease-resistant microtubules is only beginning to be appreciated. Living cells often acquire resistance to anti-cancer drugs; paclitaxel and epothilone, for example, induce point mutations, or amino acid substitutions, in tubulins that affect microtubule stability [218]-[220]. While these examples are of mutations leading to greater vulnerability to disease, there is no reason why the basic logic could not be reversed – in other words, inducing amino acid substitutions in tubulins in order to decrease a disease state.

Posttranslational Modifications Affect Microtubule Stability

Microtubule stability is regulated by posttranslational modifications to tubulin [221]. Posttranslational modifications of tubulin include the presence or absence of tyrosine at the C-terminus (tyrosination or detyrosination), deglutamylation, polyglutamylation, acetylation, phosphorylation, and polyglycylation. An accepted indicator of microtubule stability is a high concentration of detyrosinated and acetylated microtubules, such as is found in the most stable portion of axons [222]. Consistent with the notion that diseased microtubules are not as stable as those in healthy control brains, acetylated tubulin is decreased in Alzheimer's disease brain [223]. Since different amino acid sequences for tubulin correspond with different possibilities for posttranslational modification (such as inclusion of amino acids possessing sites for phosphorylation or acetylation), increased probabilities for various modifications can be built into DNA constructs or biohybrid microtubules.

MAPs Stabilize Microtubules

Upon binding to tubulins the stabilizing MAPs, such as MAP1B, MAP2, and tau, further facilitate assembly and favor the polymerized state of microtubules (see Chapter 2). Early theories suggested that MAPs provided structural support by forming crossbridges between adjacent microtubules (or between microtubules and actin filaments or neurofilaments), and although it is possible that such crossbridges contribute to microtubule stabilization, it is probable that much of the stabilizing influence is due to MAP interactions with the surrounding medium and to MAP binding-induced conformational changes among tubulins constituting the microtubule [224]-[226]. When tau binds to tubulin at its binding site near helices 11 and 12 of α-tubulin, the lateral bonds within the microtubule are strengthened. MAP2c would be expected to also strengthen lateral bonds within the microtubule since it similarly binds at a site near helices 11 and 12. The binding of MAP2c to microtubules also strengthens longitudinal bonds, leading to a decrease in flexibility of microtubules and increase in their stability. These conformational changes to tubulin protein and microtubule stability induced by binding to MAPs have heuristic value in designing more stable and disease-resistant DNA constructs and biohybrid microtubules.

Nanofabrication and Delivery of Biohybrid Nanomechanical Structures

Another approach to repairing defective microtubules, actin filaments, or neurofilaments in the diseased brain is to use nanofabrication techniques to manufacture biohybrid materials capable of carrying out ordinary functions of those cytoskeletal components. A number of nanotechnological approaches have been applied to biohybrid microtubule " nanowires". Biohybrid microtubules can be used to serve their typical biological roles in neurons or be coaxed to take on modified roles. It is possible, for example, to load DNA cargo onto microtubules and coax them into moving along kinesin molecules attached to a Petri dish [230, 231].

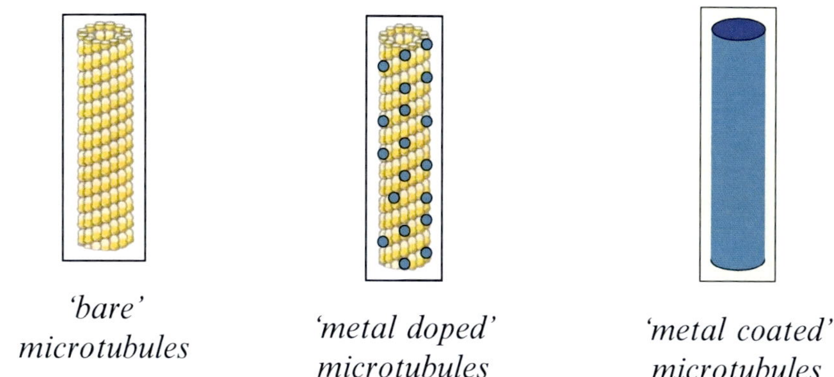

Fig. 5.2. Microtubules can be "metal doped" or "metal coated" altering the basic physical characteristics of the biological polymers. This figure has been obtained courtesy of Dr. Silke Behrens.

An effective way to reinforce the stability or to alter the function of microtubules is to coat them with metals such as silver, gold, and cobalt ferrite [227]-[229] or decorate them with biotin/streptavidin complexes, quantum dots, and dendrimers [232]-[234]. Models for microtubule protein-based nanowires are depicted in Figure 5.2. Issues that remain unresolved include biocompatibility of functionalized microtubules, how such functionalized microtubules might operate *in vivo*, and how to introduce them into specific parts of the nervous system.

An interesting feature of metalized microtubules is that they can be guided by externally generated electric or magnetic fields [227, 235]. Therefore, at least in principle, metalized microtubules (or metalized nanocarriers attached to microtubules, biohybrids, or DNA constructs) could be positioned in a particular brain region using externally applied magnetic or electrical fields. Microtubules also have intrinsic properties that could be exploited when attempting to position them accurately in a particular brain area. It has been previously demonstrated that microtubules generate endogenous electric fields relevant to transport function [236]. Mitochondria also generate electric fields inside cells [237].

As illustrated in Figure 5.3, microtubule protofilaments can serve as nanoscale building blocks that can be assembled into a diversity of shapes including fibers, hoops, ribbons, curled ribbons, sheets, heaped sheets, macrotubules, and paracrystalline formations [238]. The nanotechnological development of diverse microtubule structures could prove useful in constructing more complex biohybrid devices with tunable biophysical properties.

Nanoneuroscientists have the capabilities needed to begin serious consideration of designing DNA constructs, nanodevices, nanocarriers, and biohybrids that could ameliorate cytoskeletal deficits and treat the underlying cause of debilitating diseases such as Alzheimer's and Parkinson's disease, schizophrenia,

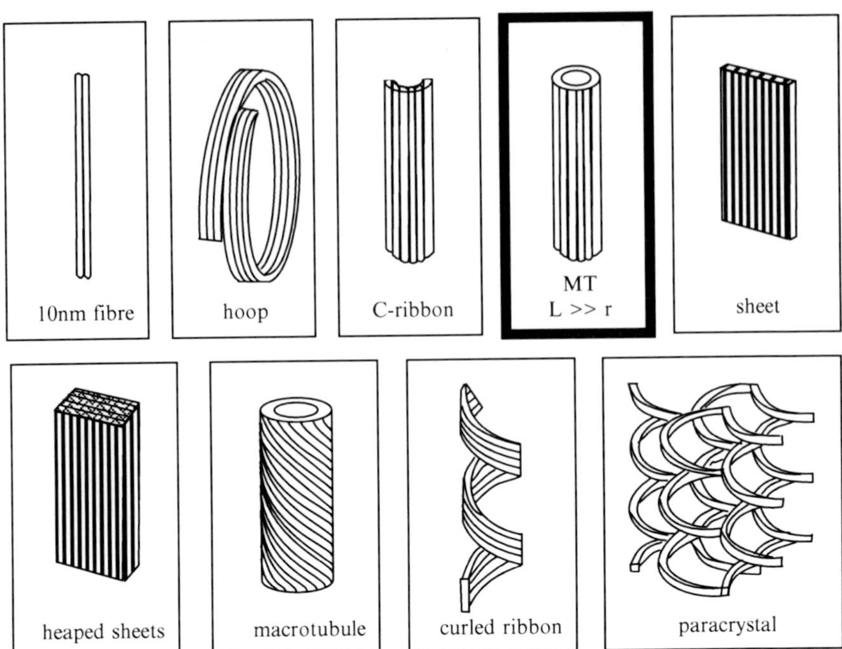

Fig. 5.3. Microtubule protofilaments can assemble into a variety of shapes having potentially diverse functional applications. Adapted from [238].

and bipolar disorder. Past treatments for these disorders have focused on neurotransmitter agonists and antagonists. A new treatment approach focused on repairing damaged microtubules or introducing more stable disease-resistant microtubules (or other cytoskeletal proteins) seems to be both timely and worthwhile in addition to addressing the problem at its most fundamental level.

Table 5.5. Nanostructures and Their Biological Counterparts in Biohybrid Designs

Nanostructure	Biological protein/structure
Nanowires/transistors	Microtubules
Nanocables	Actin filaments
Nanoswitches	Ion channels, ion gates
Nanosensors	Receptors
Nanostorage devices	Vesicles
Nanorobots	Motor proteins

5.5.4 Nanotechnology, Deep Brain Stimulation, and Neural Plasticity

Deep brain stimulation has been used successfully to treat movement disorders, such as Parkinson's disease, and treatment-resistant depression [239, 240, 241]. Although the optimal locus for deep brain stimulation for Parkinson's disease is located in the internal segment of the globus pallidus or the subthalamic nucleus, the optimal site for deep brain stimulation in clinical depression is less clear. Current models of clinical depression suggest that a network of limbic-related structures (i.e., the cingulate cortex, orbitofrontal cortex, and nucleus accumbens) underlie the disorder; however, individual differences among patients and other unknowns have precluded the identification of a consistently effective site that invariably produces positive results. Deep brain stimulation also has the drawback of being an invasive treatment and is thereby reserved as a last resort after medication has proved ineffective.

Nanotechnology could significantly improve deep brain stimulation therapy, with the potential of making the procedure safer and less invasive. Nanoparticle coatings, such as laminin, have been shown to make stimulating (and recording) probes less damaging to surrounding tissue [242]. As described in Chapter 1, stimulating devices can be driven in accordance with online feedback. As an example, deep stimulation in the globus pallidus could be regulated by activity or dopamine levels assayed in the striatum.

To the extent that nanoparticles could potentially be administered to patients and guided to neurons with selective vectors or by using externally applied electromagnetic fields, the possibility exists for treating neurological and neuropsychiatric disorders using nanoparticles localized to specific brain regions. According to this proposed scenario, magnetic or metalized nanoparticles once guided to neurons would enable researchers or clinicians to more precisely (1) record neural activity-related electromagnetic fields amplified by these nanoparticles, and then (2) apply electric or magnetic fields to counter those electromagnetic patterns correlating with a particular disorder (see Figure 5.4). Biohybrid microtubules, conjugated to metals or semiconductors or genetically modified to more readily bind to metals or semiconductor materials, provide a means to confine metallic nanoparticles to neurons for longer time intervals, especially since certain microtubules, such as those at the axon hillock, are stable for extended periods of time.

As such technologies evolve, it is conceivable that individual patients could be given specific "prescriptions" delineating precisely defined patterns of electromagnetic stimulation that counteract activity in the specific neuroanatomical network underlying that patient's clinical condition (e.g., major depression). For purposes of illustration, this "prescription" could be realized as a customized pattern of surface electrodes woven into a cloth cap. Electrode bearing caps are currently used for EEG recording and transcranial magnetic stimulation. Customized electrode-bearing caps coupled with nanotechnologies (e.g., use of magnetic or metalized nanoparticles) could become an alternative or adjunct to drug therapies such as antidepressants. An electromagnetic animation cap has been designed to treat depression [243]; moreover, low-frequency repetitive transcranial magnetic stimulation decreases symptoms of depression and panic disorder [244]; nonetheless, electromagnetic stimulation has yet to be combined with nanotechnologies such as administration of magnetic or metalized nanoparticles. The incorporation of magnetic or metalized nanoparticles into neurons would be expected to enable externally applied electromagnetic fields to exert much more potent effects on neural activity and function and to reach neurons in deeper layers of the cerebral cortex or buried structures like the hippocampus. Gold nanocap particles have been shown to enhance applied electric fields more than 30,000-fold [245].

Another futuristic application of nanoparticles to nervous system disorders lies in precision sculpting of neural plasticity. Researchers have been successful using magnetic fields to direct cell movements following endosomal uptake of magnetic nanoparticles [246]. Although a number of technical difficulties would need to be addressed, it is conceivable that eventually cortical connectivity could be altered using externally applied magnetic fields to manipulate encapsulated magnetic nanoparticles. As an example, lateral connectivity in select portions of the frontal, parietal, or temporal lobe, could be enhanced or diminished. Given the potential to use magnetic nanoparticles to detect specific patterns of abnormal activity, neural plasticity could be induced to remedy patient-specific abnormalities in neural networks, thereby offering potentially long-term cures for a variety of nervous system disorders. Yet another manner in which neuronal growth can be directed by nanotechnology is through the use of microfluidic chips that release small quantities of neurotrophic factors [247].

5.6 Bioinformatics and Rational Drug Design

Bioinformatics and rational drug design draw from a wide range of methods – X-ray crystallography to determine 3-dimensional protein structure, NMR spectroscopy to explore protein-ligand interactions, computation chemistry, molecular modeling, high throughput screening, and virtual screening [248]. By combining methods in molecular biology, genomics, and proteomics, it is possible to determine or predict the 3-dimensional protein conformation as well as the nature and strength of interactions involving a ligand-protein complex leading to optimized and site-specific drug design. In cases where direct experimental data are lacking, computer-aided drug design often utilizes sequence analysis and homology modeling to predict 3-dimensional protein conformation and ligand-protein binding. Virtual high throughput screening enables several millions of compounds to be rapidly evaluated, selecting a limited number of promising candidates having the appropriate binding characteristics for further evaluation. High throughput screening of genes contributing

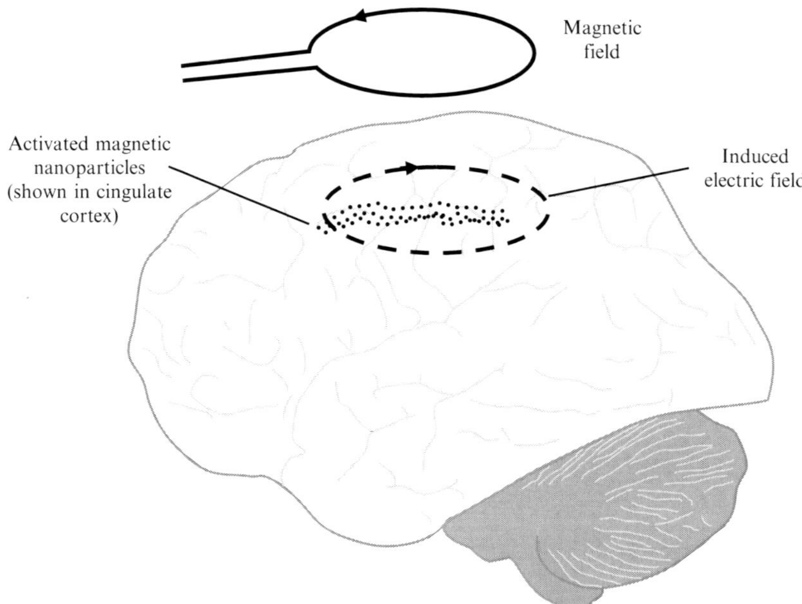

Fig. 5.4. Externally applied magnetic fields induce electric fields in the brain, which have the potential to interact with magnetic or metalized nanoparticles positioned in precise brain regions, such as the cingulate cortex (which is implicated in clinical depression).

to specific disorders can also be used to identify protein targets for potential drugs.

Rational drug design technologies have enabled the screening of numerous potential drugs for treating Alzheimer's disease. Acetylcholinesterase inhibitors – the number one drug type used to treat the cognitive impairments with Alzheimer's disease – have been screened in an effort to identify those inhibitors that additionally interfere with β-amyloid aggregation [249]. Potential inhibitors of β-secretase, an enzyme involved in the neuropathological cascade in Alzheimer's disease, have been deduced from docking simulations based on both the geometric and electrostatic parameters essential to enzymatic inhibition [250, 251]. One laboratory screened genes that rescued cells from amyloid-β peptide toxicity and discovered a peptide named Humanin as having neuroprotective properties [252]. Another approach used a fluorescence resonance energy transfer (FRET)-based screen to identify potential inhibitors of amyloid-β-derived diffusible ligands [253]. Rational drug design has also identified potential vaccines for Alzheimer's disease. Despite some complications with earlier vaccines directed at amyloid-β peptide, new vaccines appear to hold promise [254].

High (and medium) throughput screening has been applied to other nervous system disorders, such as Parkinson's disease, amyotrophic lateral sclerosis, and neuropsychiatric disorders [255]-[257]. Another strategy is to screen drugs targeting a protein that is commonly affected across nervous system disorder, such as a cytoskeletal protein. Combining NMR spectroscopy and computational analysis has revealed the binding mode for tubulin-stabilizing drug, epothilone A [258]. Bioinformatics and rational drug design strategies could be applied to other drugs that bind tubulin or other cytoskeleton-related proteins.

References

1. Kaufmann WE, Moser HW. Dendritic anomalies in disorders associated with mental retardation. Cereb Cortex. 2000 Oct;10(10):981-91.
2. Kooy RF, Willemsen R, Oostra BA. Fragile X syndrome at the turn of the century. Mol Med Today. 2000 May;6(5):193-8.
3. O'Donnell W. T., Warren S. T. A decade of molecular studies of fragile x syndrome. Annu. Rev. Neurosci. 2002;25:315-338.
4. Wang H, Dictenberg JB, Ku L, Li W, Bassell GJ, Feng Y. Dynamic association of the fragile X mental retardation protein as a messenger ribonucleoprotein between microtubules and polyribosomes. Mol Biol Cell. 2008 Jan;19(1):105-14
5. Grossman AW, Aldridge GM, Weiler IJ, Greenough WT. Local protein synthesis and spine morphogenesis: Fragile X syndrome and beyond. J Neurosci. 2006 Jul 5;26(27):7151-5.
6. De Diego Otero Y, Severijnen LA, van Cappellen G, Schrier M, Oostra B, Willemsen R. Transport of fragile X mental retardation protein via granules in neurites of PC12 cells. Mol Cell Biol. 2002 Dec;22(23):8332-41.
7. Davidovic L, Jaglin XH, Lepagnol-Bestel AM, Tremblay S, Simonneau M, Bardoni B, Khandjian EW. The fragile X mental retardation protein is a molecular adaptor between the neurospecific KIF3C kinesin and dendritic RNA granules. Hum Mol Genet. 2007 Dec 15;16(24):3047-58
8. Rackham O, Brown CM. Visualization of RNA-protein interactions in living cells: FMRP and IMP1 interact on mRNAs. EMBO J. 2004 Aug 18;23(16):3346-55.
9. Antar LN, Dictenberg JB, Plociniak M, Afroz R, Bassell GJ. Localization of FMRP-associate mRNA granules and requirement of microtubules for activity-dependent trafficking in hippocampal neurons. Genes Brain Behav. 2005 Aug;4(6):350-9.
10. Lu R, Wang H, Liang Z, Ku L, O'donnell WT, Li W, Warren ST, Feng Y. The fragile X protein controls microtubule-associated protein 1B translation and microtubule stability in brain neuron development. Proc Natl Acad Sci U S A. 2004 Oct 19;101(42):15201-6.
11. Reeve SP, Bassetto L, Genova GK, Kleyner Y, Leyssen M, Jackson FR, Hassan BA. The Drosophila fragile X mental retardation protein controls actin dynamics by directly regulating profilin in the brain. Curr Biol. 2005 Jun 21;15(12):1156-63.

12. Castets M, Schaeffer C, Bechara E, Schenck A, Khandjian EW, Luche S, Moine H, Rabilloud T, Mandel JL, Bardoni B. FMRP interferes with the Rac1 pathway and controls actin cytoskeleton dynamics in murine fibroblasts. Hum Mol Genet. 2005 Mar 15;14(6):835-44.
13. Ferrari F, Mercaldo V, Piccoli G, Sala C, Cannata S, Achsel T, Bagni C. The fragile X mental retardation protein-RNP granules show an mGluR-dependent localization in the post-synaptic spines. Mol Cell Neurosci. 2007 Mar;34(3):343-54
14. Kesler SR. Turner syndrome. Child Adolesc Psychiatr Clin N Am. 2007 Jul;16(3):709-22.
15. Rae C, Joy P, Harasty J, Kemp A, Kuan S, Christodoulou J, Cowell CT, Coltheart M. Enlarged temporal lobes in Turner syndrome: an X-chromosome effect? Cereb Cortex. 2004 Feb;14(2):156-64.
16. Cutter WJ, Daly EM, Robertson DM, Chitnis XA, van Amelsvoort TA, Simmons A, Ng VW, Williams BS, Shaw P, Conway GS, Skuse DH, Collier DA, Craig M, Murphy DG. Influence of X chromosome and hormones on human brain development: a magnetic resonance imaging and proton magnetic resonance spectroscopy study of Turner syndrome. Biol Psychiatry. 2006 Feb 1;59(3):273-83.
17. Stromme P, Bjornstad PG, Ramstad K (2002). Prevalence estimation of Williams syndrome. Journal of child neurology 17 (4): 269-71.
18. Hoogenraad CC, Akhmanova A, Galjart N, De Zeeuw CI. LIMK1 and CLIP-115: linking cytoskeletal defects to Williams syndrome. Bioessays. 2004 Feb;26(2):141-50.
19. Meng Y, Zhang Y, Tregoubov V, Janus C, Cruz L, Jackson M, Lu WY, MacDonald JF, Wang JY, Falls DL, Jia Z. Abnormal spine morphology and enhanced LTP in LIMK-1 knockout mice. Neuron. 2002 Jul 3;35(1):121-33.
20. Hoogenraad CC, Koekkoek B, Akhmanova A, Krugers H, Dortland B, Miedema M, van Alphen A, Kistler WM, Jaegle M, Koutsourakis M, Van Camp N, Verhoye M, van der Linden A, Kaverina I, Grosveld F, De Zeeuw CI, Galjart N. Targeted mutation of Cyln2 in the Williams syndrome critical region links CLIP-115 haploinsufficiency to neurodevelopmental abnormalities in mice. Nat Genet. 2002 Sep;32(1):116-27.
21. Meyer-Lindenberg A, Mervis CB, Sarpal D, Koch P, Steele S, Kohn P, Marenco S, Morris CA, Das S, Kippenhan S, Mattay VS, Weinberger DR, Berman KF. Functional, structural, and metabolic abnormalities of the hippocampal formation in Williams syndrome. J Clin Invest. 2005 Jul;115(7):1888-95. Epub 2005 Jun 9.
22. Amaral DG, Schumann CM, Nordahl CW. Neuroanatomy of autism. Trends Neurosci. 2008 Mar;31(3):137-45.
23. Abrahams BS, Geschwind DH. Advances in autism genetics: on the threshold of a new neurobiology. Nat Rev Genet. 2008 May;9(5):341-55.
24. Pennisi E. Breakthrough of the year. Human genetic variation. Science. 2007 Dec 21;318(5858):1842-3.
25. Maussion G, Carayol J, Lepagnol-Bestel AM, Tores F, Loe-Mie Y, Milbreta U, Rousseau F, Fontaine K, Renaud J, Moalic JM, Philippi A, Chedotal A, Gorwood P, Ramoz N, Hager J, Simonneau M. Convergent evidence identifying MAP/microtubule affinity-regulating kinase 1 (MARK1) as a susceptibility gene for autism. Hum Mol Genet. 2008 May 20.

26. Mukaetova-Ladinska EB, Arnold H, Jaros E, Perry R, Perry E. Depletion of MAP2 expression and laminar cytoarchitectonic changes in dorsolateral prefrontal cortex in adult autistic individuals. Neuropathol Appl Neurobiol. 2004 Dec;30(6):615-23.
27. Zhang H, Liu X, Zhang C, Mundo E, Macciardi F, Grayson DR, Guidotti AR, Holden JJ. Reelin gene alleles and susceptibility to autism spectrum disorders. Mol Psychiatry. 2002;7(9):1012-7.
28. Fatemi SH, Snow AV, Stary JM, Araghi-Niknam M, Reutiman TJ, Lee S, Brooks AI, Pearce DA. Reelin signaling is impaired in autism. Biol Psychiatry. 2005 Apr 1;57(7):777-87.
29. Percy AK, Lane JB, Childers J, Skinner S, Annese F, Barrish J, Caeg E, Glaze DG, MacLeod P. Rett syndrome: North American database. J Child Neurol. 2007 Dec;22(12):1338-41.
30. Nagai K, Miyake K, Kubota T. A transcriptional repressor MeCP2 causing Rett syndrome is expressed in embryonic non-neuronal cells and controls their growth. Brain Res Dev Brain Res. 2005 Jun 9;157(1):103-6.
31. Armstrong DD. Can we relate MeCP2 deficiency to the structural and chemical abnormalities in the Rett brain? Brain Dev. 2005 Nov;27 Suppl 1:S72-S76.
32. Armstrong DD. Rett syndrome neuropathology review 2000. Brain Dev. 2001 Dec;23 Suppl 1:S72-6.
33. Belichenko PV, Dahlström A. Studies on the 3-dimensional architecture of dendritic spines and varicosities in human cortex by confocal laser scanning microscopy and Lucifer yellowmicroinjections. J Neurosci Methods. 1995 Mar;57(1):55-61.
34. Kaufmann WE, MacDonald SM, Altamura CR. Dendritic cytoskeletal protein expression in mental retardation: an immunohistochemical study of the neocortex in Rett syndrome. Cereb Cortex. 2000 Oct;10(10):992-1004.
35. Armstrong DD, Dunn K, Antalffy B. Decreased dendritic branching in frontal, motor and limbic cortex in Rett syndrome compared with trisomy 21. J Neuropathol Exp Neurol. 1998 Nov;57(11):1013-7.
36. Patterson D, Costa AC. Down syndrome and genetics - a case of linked histories. Nat Rev Genet. 2005 Feb;6(2):137-47.
37. Ford JH. Spindle microtubular dysfunction in mothers of Down syndrome children. Hum Genet. 1984;68(4):295-8.
38. Nitta M, Tsuiki H, Arima Y, Harada K, Nishizaki T, Sasaki K, Mimori T, Ushio Y, Saya H. Hyperploidy induced by drugs that inhibit formation of microtubule promotes chromosome instability. Genes Cells. 2002 Feb;7(2):151-62.
39. Dotan A, Avivi L. Increased spindle resistance to antimicrotubule agents in women of high risk for meiotic nondisjunction. Hum Genet. 1986 Jul;73(3):199-204.
40. Weitzdoerfer R, Dierssen M, Fountoulakis M, Lubec G. Fetal life in Down syndrome starts with normal neuronal density but impaired dendritic spines and synaptosomal structure. J Neural Transm Suppl. 2001;(61):59-70.
41. Gulesserian T, Kim SH, Fountoulakis M, Lubec G. Aberrant expression of centractin and capping proteins, integral constituents of the dynactin complex, in fetal down syndrome brain. Biochem Biophys Res Commun. 2002 Feb 15;291(1):62-7.
42. Weitzdoerfer R, Fountoulakis M, Lubec G. Reduction of actin-related protein complex 2/3 in fetal Down syndrome brain. Biochem Biophys Res Commun. 2002 May 3;293(2):836-41.

43. Lubec B, Weitzdoerfer R, Fountoulakis M. Manifold reduction of moesin in fetal Down syndrome brain. Biochem Biophys Res Commun. 2001 Sep 7;286(5):1191-4.
44. Murphy GM Jr, Eng LF, Ellis WG, Perry G, Meissner LC, Tinklenberg JR. Antigenic profile of plaques and neurofibrillary tangles in the amygdala in Down's syndrome: a comparison with Alzheimer's disease. Brain Res. 1990 Dec 24;537(1-2):102-8.
45. Pollak D, Cairns N, Lubec G. Cytoskeleton derangement in brain of patients with Down syndrome, Alzheimer's disease and Pick's disease. J Neural Transm Suppl. 2003;(67):149-58.
46. Schwab C, McGeer PL. Tubulin immunopositive structures resembling intracellular neurofibrillary tangles. Neurobiol Aging. 1998 Jan-Feb;19(1):41-5.
47. Hanger DP, Brion JP, Gallo JM, Cairns NJ, Luthert PJ, Anderton BH. Tau in Alzheimer's disease and Down's syndrome is insoluble and abnormally phosphorylated. Biochem J. 1991 Apr 1;275 (Pt 1):99-104.
48. Chevalier-Larsen E, Holzbaur EL. Axonal transport and neurodegenerative disease. Biochim Biophys Acta. 2006 Nov-Dec;1762(11-12):1094-108.
49. Roy S, Zhang B, Lee VM, Trojanowski JQ. Axonal transport defects: a common theme in neurodegenerative diseases. Acta Neuropathol. 2005 Jan;109(1):5-13.
50. Cairns NJ, Lee VM, Trojanowski JQ. The cytoskeleton in neurodegenerative diseases. J Pathol. 2004 Nov;204(4):438-49.
51. Goedert M. Filamentous nerve cell inclusions in neurodegenerative diseases: tauopathies and alpha-synucleinopathies. Philos Trans R Soc Lond B Biol Sci. 1999 Jun 29;354(1386):1101-18.
52. Luquin N, Yu B, Trent RJ, Morahan JM, Pamphlett R. An analysis of the entire SOD1 gene in sporadic ALS. Neuromuscul Disord. 2008 May 24.
53. Julien JP, Kriz J. Transgenic mouse models of amyotrophic lateral sclerosis. Biochim Biophys Acta. 2006 Nov-Dec;1762(11-12):1013-24. 51.
54. Farah CA, Nguyen MD, Julien JP, Leclerc N. Altered levels and distribution of microtubule-associated proteins before disease onset in a mouse model of amyotrophic lateral sclerosis. J Neurochem. 2003 Jan;84(1):77-86.
55. Duncan JE, Goldstein LS. The genetics of axonal transport and axonal transport disorders. PLoS Genet. 2006 Sep 29;2(9):e124.
56. Waring SC, Rosenberg RN. Genome-wide association studies in Alzheimer disease. Arch Neurol. 2008 Mar;65(3):329-34.
57. Ertekin-Taner N. Genetics of Alzheimer's disease: a centennial review. Neurol Clin. 2007 Aug;25(3):611-67.
58. Ikeuchi T, Kaneko H, Miyashita A, Nozaki H, Kasuga K, Tsukie T, Tsuchiya M, Imamura T, Ishizu H, Aoki K, Ishikawa A, Onodera O, Kuwano R, Nishizawa M. Mutational analysis in early-onset familial dementia in the Japanese population: the role of PSEN1 and MAPT R406W mutations. Dement Geriatr Cogn Disord. 2008 Jun 28;26(1):43-49.
59. Boutajangout A, Authelet M, Blanchard V, Touchet N, Tremp G, Pradier L, Brion JP. Characterisation of cytoskeletal abnormalities in mice transgenic for wild-type human tau and familial Alzheimer's disease mutants of APP and presenilin-1. Neurobiol Dis. 2004 Feb;15(1):47-60.
60. Adalbert R, Gilley J, Coleman MP. Abeta, tau and ApoE4 in Alzheimer's disease: the axonal connection. Trends Mol Med. 2007 Apr;13(4):135-42.
61. Belin AC, Westerlund M. Parkinson's disease: a genetic perspective. FEBS J. 2008 Apr;275(7):1377-83.

62. Pankratz N, Foroud T. Genetics of Parkinson disease. Genet Med. 2007 Dec;9(12):801-11.
63. Mizuno Y, Hattori N, Kitada T, Matsumine H, Mori H, Shimura H, Kubo S, Kobayashi H, Asakawa S, Minoshima S, Shimizu N. Familial Parkinson's disease. Alpha-synuclein and parkin. Adv Neurol. 2001;86:13-21.
64. Tobin JE, Latourelle JC, Lew MF, Klein C, Suchowersky O, Shill HA, Golbe LI, Mark MH, Growdon JH, Wooten GF, Racette BA, Perlmutter JS, Watts R, Guttman M, Baker KB, Goldwurm S, Pezzoli G, Singer C, Saint-Hilaire MH, Hendricks AE, Williamson S, Nagle MW, Wilk JB, Massood T, Laramie JM, DeStefano AL, Litvan I, Nicholson G, Corbett A, Isaacson S, Burn DJ, Chinnery PF, Pramstaller PP, Sherman S, Al-hinti J, Drasby E, Nance M, Moller AT, Ostergaard K, Roxburgh R, Snow B, Slevin JT, Cambi F, Gusella JF, Myers RH. Haplotypes and gene expression implicate the MAPT region for Parkinson disease: the GenePD Study. Neurology. 2008 Jul 1;71(1):28-34.
65. Walling HW, Baldassare JJ, Westfall TC. Molecular aspects of Huntington's disease. J Neurosci Res. 1998 Nov 1;54(3):301-8.
66. Shoesmith CL, Strong MJ. Amyotrophic lateral sclerosis: update for family physicians. Can Fam Physician. 2006 Dec;52(12):1563-9.
67. Zhang F, Ström AL, Fukada K, Lee S, Hayward LJ, Zhu H. Interaction between familial amyotrophic lateral sclerosis (ALS)-linked SOD1 mutants and the dynein complex. J Biol Chem. 2007 Jun 1;282(22):16691-9.
68. Schymick JC, Talbot K, Traynor BJ. Genetics of sporadic amyotrophic lateral sclerosis. Hum Mol Genet. 2007 Oct 15;16 Spec No. 2:R233-42.
69. Motil J, Dubey M, Chan WK, Shea TB. Inhibition of dynein but not kinesin induces aberrant focal accumulation of neurofilaments within axonal neurites. Brain Res. 2007 Aug 20;1164:125-31.
70. LaMonte BH, Wallace KE, Holloway BA, Shelly SS, Ascaño J, Tokito M, Van Winkle T, Howland DS, Holzbaur EL. Disruption of dynein/dynactin inhibits axonal transport in motor neurons causing late-onset progressive degeneration. Neuron. 2002 May 30;34(5):715-27.
71. Laird FM, Farah MH, Ackerley S, Hoke A, Maragakis N, Rothstein JD, Griffin J, Price DL, Martin LJ, Wong PC. Motor neuron disease occurring in a mutant dynactin mouse model is characterized by defects in vesicular trafficking. J Neurosci. 2008 Feb 27;28(9):1997-2005.
72. Gonzalez de Aguilar JL, Echaniz-Laguna A, Fergani A, René F, Meininger V, Loeffler JP, Dupuis L. Amyotrophic lateral sclerosis: all roads lead to Rome. J Neurochem. 2007 Jun;101(5):1153-60.
73. Yang W, Leystra-Lantz C, Strong MJ. Upregulation of GSK3beta expression in frontal and temporal cortex in ALS with cognitive impairment (ALSci). Brain Res. 2008 Feb 27;1196:131-9.
74. Letournel F, Bocquet A, Dubas F, Barthelaix A, Eyer J. Stable tubule only polypeptides (STOP) proteins co-aggregate with spheroid neurofilaments in amyotrophic lateral sclerosis. J Neuropathol Exp Neurol. 2003 Dec;62(12):1211-9.
75. Ferri CP, Prince M, Brayne C, Brodaty H, Fratiglioni L, Ganguli M, Hall K,Hasegawa K, Hendrie H, Huang Y, Jorm A, Mathers C, Menezes PR, Rimmer E, and Scazufca M; Alzheimer's disease international. Global prevalence of dementia: A Delphi Consensus Study. Lancet. 366, 2112-7 (2005).
76. Dubois B, Feldman HH, Jacova C, Dekosky ST, Barberger-Gateau P, Cummings J, Delacourte A, Galasko D, Gauthier S, Jicha G, Meguro K, O'brien J,

Pasquier F, Robert P, Rossor M, Salloway S, Stern Y, Visser PJ, and Scheltens P. Research criteria for the diagnosis of Alzheimer's disease: revising the NINCDS-ADRDA criteria. Lancet Neurol. 6:734-46 (2007).

77. Braak H and Braak E. Diagnostic criteria for neuropathologic assessment of Alzheimer's disease. Neurobiol Aging. 18:S85-8 (1997).
78. Goedert M, Spillantini MG, Jakes R, Crowther RA, Vanmechelen E, Probst A, Götz J, Bürki K, Cohen P. Molecular dissection of the paired helical filament. Neurobiol Aging. 1995 May-Jun;16(3):325-34
79. Avila J. Tau aggregation into fibrillar polymers: taupathies. FEBS Lett. 2000 Jun 30;476(1-2):89-92.
80. Mandelkow E, von Bergen M, Biernat J, and Mandelkow EM. Structural principles of tau and the paired helical filaments of Alzheimer's disease. Brain Pathol. 17, 83-90 (2007)
81. Li B, Chohan MO, Grundke-Iqbal I, and Iqbal K. Disruption of microtubule network by Alzheimer abnormally hyperphosphorylated tau. Acta Neuropathol. 113, 501-11 (2007).
82. Cash AD, Aliev G, Siedlak SL, Nunomura A, Fujioka H, Zhu X, Raina AK, Vinters HV, Tabaton M, Johnson AB, Paula-Barbosa M, Avla J, Jones PK, Castellani RJ, Smith MA, and Perry G. Microtubule reduction in Alzheimer's disease and aging is independent of tau filament formation. Am J Pathol. 162, 1623-7 (2003).
83. Raes M. Involvement of microtubules in modifications associated with cellular aging. Mutat Res. 256, 149-68 (1991).
84. Thies E and Mandelkow EM. Missorting of tau in neurons causes degeneration of synapses that can be rescued by the kinase MARK2/Par-1. J Neurosci. 27, 2896-907 (2007)
85. Loring JF, Wen X, Lee JM, Seilhamer J, Somogyi R. A gene expression profile of Alzheimer's disease. DNA Cell Biol. 2001 Nov;20(11):683-95. Erratum in: DNA Cell Biol 2002 Mar;21(3):241.
86. Ashford JW, Soultanian NS, Zhang SX, Geddes JW. Neuropil threads are collinear with MAP2 immunostaining in neuronal dendrites of Alzheimer brain. J Neuropathol Exp Neurol. 1998 Oct;57(10):972-8.
87. Armstrong RA. Beta-amyloid plaques: stages in life history or independent origin? Dement Geriatr Cogn Disord. 9, 227-38 (1998).
88. Hardy J, Allsop D. Amyloid deposition as the central event in the aetiology of Alzheimer's disease. Trends Pharmacol Sci. 1991 Oct;12(10):383-8.
89. Selkoe DJ. Toward a comprehensive theory for Alzheimer's disease. Hypothesis: Alzheimer's disease is caused by the cerebral accumulation and cytotoxicity of amyloid beta-protein. Ann N Y Acad Sci. 2000;924:17-25.
90. Hardy J, Duff K, Hardy KG, Perez-Tur J, Hutton M. Genetic dissection of Alzheimer's disease and related dementias: amyloid and its relationship to tau. Nat Neurosci. 1998 Sep;1(5):355-8.
91. Kamal A, Almenar-Queralt A, LeBlanc JF, Roberts EA, Goldstein LS. Kinesin-mediated axonal transport of a membrane compartment containing beta-secretase and presenilin-1 requires APP. Nature. 2001 Dec 6;414(6864):643-8.
92. Adalbert R, Gilley J, Coleman MP. Abeta, tau and ApoE4 in Alzheimer's disease: the axonal connection. Trends Mol Med. 2007 Apr;13(4):135-42.
93. Terwel D, Dewachter I, Van Leuven F. Axonal transport, tau protein, and neurodegeneration in Alzheimer's disease. Neuromolecular Med. 2002;2(2):151-65.

94. Whitehouse PJ, Price DL, Clark AW, Coyle JT, DeLong MR. Alzheimer disease: evidence for selective loss of cholinergic neurons in the nucleus basalis. Ann Neurol. 1981 Aug;10(2):122-6.
95. Woolf NJ. Cholinergic systems in mammalian brain and spinal cord. Prog Neurobiol. 1991;37(6):475-524.
96. Geula C, Nagykery N, Nicholas A, Wu CK. Cholinergic neuronal and axonal abnormalities are present early in aging and in Alzheimer disease. J Neuropathol Exp Neurol. 2008 Apr;67(4):309-18.
97. Butcher LL, Woolf NJ. Neurotrophic agents may exacerbate the pathologic cascade of Alzheimer's disease. Neurobiol Aging. 1989 Sep-Oct;10(5):557-70.
98. Shelanski ML. Cytoskeleton and NGF in the pathogenesis and treatment of Alzheimer's disease. Neurobiol Aging. 1989 Sep-Oct;10(5):577; discussion 588-90.
99. Counts SE, Mufson EJ. The role of nerve growth factor receptors in cholinergic basal forebrain degeneration in prodromal Alzheimer disease. J Neuropathol Exp Neurol. 2005 Apr;64(4):263-72.
100. Schindowski K, Belarbi K, Buée L. Neurotrophic factors in Alzheimer's disease: role of axonal transport. Genes Brain Behav. 2008 Feb;7 Suppl 1:43-56.
101. Rabizadeh S, Bitler CM, Butcher LL, Bredesen DE. Expression of the low-affinity nerve growth factor receptor enhances beta-amyloid peptide toxicity. Proc Natl Acad Sci U S A. 1994 Oct 25;91(22):10703-6.
102. Scheibel AB, Tomiyasu U. Dendritic sprouting in Alzheimer's presenile dementia. Exp Neurol. 1978 May 15;60(1):1-8.
103. Ihara Y. Massive somatodendritic sprouting of cortical neurons in Alzheimer's disease. Brain Res. 1988 Aug 30;459(1):138-44.
104. Arendt T, Brückner MK. Perisomatic sprouts immunoreactive for nerve growth factor receptor and neurofibrillary degeneration affect different neuronal populations in the basal nucleus in patients with Alzheimer's disease. Neurosci Lett. 1992 Dec 14;148(1-2):63-6.
105. McKee AC, Kowall NW, Kosik KS. Microtubular reorganization and dendritic growth response in Alzheimer's disease. Ann Neurol. 1989 Nov;26(5):652-9.
106. Dorsey ER, Constantinescu R, Thompson JP, Biglan KM, Holloway RG, Kieburtz K, Marshall FJ, Ravina BM, Schifitto G, Siderowf A, and Tanner CM. projected number of people with Parkinson disease in the most populous nations, 2005 through 2030. Neurology. 68, 384-6 (2007)
107. Dubois B, Burn D, Goetz C, Aarsland D, Brown RG, Broe GA, Dickson D, Duyckaerts C, Cummings J, Gauthier S, Korczyn A, Lees A, Levy R, Litvan I, Mizuno Y, McKeith IG, Olanow CW, Poewe W, Sampaio C, Tolosa E, and Emre M. Diagnostic procedures for Parkinson's disease dementia: recommendations from the Movement Disorder Society Task Force. Mov Disord. 22, 2314-24 (2007)
108. Hornykiewicz O. Basic research on dopamine in Parkinson's disease and the discovery of the nigrostriatal dopamine pathway: the view of an eyewitness. Neurodegener Dis. 5, 114-7 (2008)
109. Giasson BI, Lee VM. Parkin and the molecular pathways of Parkinson's disease. Neuron. 2001 Sep 27;31(6):885-8.
110. Klein C, Vieregge P, Hagenah J, Sieberer M, Doyle E, Jacobs H, Gasser T, Breakefield XO, Risch NJ, Ozelius LJ. Search for the PARK3 founder haplotype in a large cohort of patients with Parkinson's disease from northern Germany. Ann Hum Genet. 1999 Jul;63(Pt 4):285-91.

111. Biehlmaier O, Alam M, and Schmidt WJ. A rat model of parkinsonism shows depletion of dopamine in the retina. Neurochem Int. 50, 189-95 (2007)
112. Langston JW, Ballard P, Tetrud JW, and Irwin I. chronic parkinsonism in humans due to a product of meperidine-analog synthesis. Science. 219, 979-80 (1983).
113. Feng J. Microtubule: A common target for parkin and Parkinson's disease toxins. Neuroscientist.12, 469-76 (2006)
114. Liu Y and Edwards RH. The role of vesicular transport proteins in synaptic transmission and neural degeneration. Annu Rev Neurosci. 20, 125-56 (1997)
115. Ren Y, Zhao J, and Feng J. Parkin binds to alpha/beta tubulin and increases their ubiquitination and degradation. J Neurosci. 23, 3316-24 (2003)
116. Vendruscolo M, Zurdo J, MacPhee CE, and Dobson CM. Protein folding and misfolding: a paradigm of self-assembly and regulation in complex biological systems. Philos Transact A Math Phys Eng Sci. 361, 1205-22 (2003)
117. Alim MA, Ma QL, Takeda K, Aizawa T, Matsubara M, Nakamura M, Asada A, Saito T, Kaji H, Yoshii M, Hisanaga S, and Uéda K. demonstration of a role for alpha-synuclein as a functional microtubule-associated protein. J Alzheimers Dis. 6, 435-42 (2004)
118. Goris A, Williams-Gray CH, Clark GR, Foltynie T, Lewis SJ, Brown J, Ban M, Spillantini MG, Compston A, Burn DJ, Chinnery PF, Barker RA, Sawcer SJ. Tau and alpha-synuclein in susceptibility to, and dementia in, Parkinson's disease. Ann Neurol. 2007 Aug;62(2):145-53.
119. Kim M, Jung W, Lee IH, Bhak G, Paik SR, and Hahn JS. impairment of microtubule system increases alpha-synuclein aggregation and toxicity. Biochem Biophys Res Commun. 365, 628-35 (2008)
120. Jeannotte AM and Sidhu A. regulation of the norepinephrine transporter by alpha-synuclein-mediated interactions with microtubules. Eur J Neurosci. 26, 1509-20 (2007)
121. Lee HJ, Khoshaghideh F, Lee S, and Lee SJ. impairment of microtubule-dependent trafficking by overexpression of alpha-synuclein. Eur J Neurosci. 24, 3153-62 (2006)
122. Ren Y, Liu W, Jiang H, Jiang Q, and Feng J. Selective vulnerability of dopaminergic neurons to microtubule depolymerization. J Biol Chem. 280, 34105-12 (2005)
123. Ren Y, Feng J. Rotenone selectively kills serotonergic neurons through a microtubule-dependent mechanism. J Neurochem. 2007 Oct;103(1):303-11.
124. Cappelletti G, Maggioni MG, and Maci R. Influence of MPP+ on the state of tubulin polymerisation in NGF-differentiated PC12 cells. J Neurosci Res. 56, 28-35 (1999)
125. Harper PS. The epidemiology of Huntington's disease. Hum Genet. 1992 Jun;89(4):365-76.
126. DiProspero NA, Chen EY, Charles V, Plomann M, Kordower JH, Tagle DA. Early changes in Huntington's disease patient brains involve alterations in cytoskeletal and synaptic elements. J Neurocytol. 2004 Sep;33(5):517-33.
127. Trushina E, Heldebrant MP, Perez-Terzic CM, Bortolon R, Kovtun IV, Badger JD 2nd, Terzic A, Estévez A, Windebank AJ, Dyer RB, Yao J, McMurray CT. Microtubule destabilization and nuclear entry are sequential steps leading to toxicity in Huntington's disease. Proc Natl Acad Sci U S A. 2003 Oct 14;100(21):12171-6.

128. Feany MB, La Spada AR. Polyglutamines stop traffic: axonal transport as a common target in neurodegenerative diseases. Neuron. 2003 Sep 25;40(1):1-2.
129. Caviston JP, Ross JL, Antony SM, Tokito M, Holzbaur EL. Huntingtin facilitates dynein/dynactin-mediated vesicle transport. Proc Natl Acad Sci U S A. 2007 Jun 12;104(24):10045-50.
130. McGuire JR, Rong J, Li SH, Li XJ. Interaction of Huntingtin-associated protein-1 with kinesin light chain: implications in intracellular trafficking in neurons. J Biol Chem. 2006 Feb 10;281(6):3552-9.
131. Orr AL, Li S, Wang CE, Li H, Wang J, Rong J, Xu X, Mastroberardino PG, Greenamyre JT, Li XJ. N-terminal mutant huntingtin associates with mitochondria and impairs mitochondrial trafficking. J Neurosci. 2008 Mar 12;28(11):2783-92.
132. Gutekunst CA, Li SH, Yi H, Ferrante RJ, Li XJ, Hersch SM. The cellular and subcellular localization of huntingtin-associated protein 1 (HAP1): comparison with huntingtin in rat and human. J Neurosci. 1998 Oct 1;18(19):7674-86.
133. Hoffner G, Kahlem P, Djian P. Perinuclear localization of huntingtin as a consequence of its binding to microtubules through an interaction with beta-tubulin: relevance to Huntington's disease. J Cell Sci. 2002 Mar 1;115(Pt 5):941-8.
134. Dompierre JP, Godin JD, Charrin BC, Cordelires FP, King SJ, Humbert S, Saudou F. Histone deacetylase 6 inhibition compensates for the transport deficit in Huntington's disease by increasing tubulin acetylation. J Neurosci. 2007 Mar 28;27(13):3571-83.
135. Rong J, McGuire JR, Fang ZH, Sheng G, Shin JY, Li SH, Li XJ. Regulation of intracellular trafficking of huntingtin-associated protein-1 is critical for TrkA protein levels and neurite outgrowth. J Neurosci. 2006 May 31;26(22):6019-30.
136. Mueser KT, McGurk SR. Schizophrenia. Lancet. 2004 Jun 19;363(9426):2063-72.
137. Harrison PJ. Schizophrenia susceptibility genes and their neurodevelopmental implications: focus on neuregulin 1. Novartis Found Symp. 2007;288:246-55
138. Matsuzaki S and Tohyama M. Molecular mechanism of schizophrenia with reference to Disrupted-In-Schizophrenia 1 (DISC1). Neurochem Int. 51, 165-72 (2007)
139. Chubb JE, Bradshaw NJ, Soares DC, Porteous DJ, and Millar JK. The DISC locus in psychiatric illness. Mol Psychiatry. 13, 36-64 (2008)
140. Ishizuka K, Paek M, Kamiya A, and Sawa A. A review of Disrupted-In-Schizophrenia-1 (DISC1): neurodevelopment, cognition, and mental conditions. Biol Psychiatry. 59, 1189-97 (2006)
141. Camargo LM, Collura V, Rain JC, Mizuguchi K, Hermjakob H, Kerrien S, Bonnert TP, Whiting PJ, and Brandon NJ. Disrupted In Schizophrenia 1 interactome: Evidence for the close connectivity of risk genes and a potential synaptic basis for schizophrenia. Mol Psychiatry. 12, 74-86 (2007)
142. Owen MJ, Craddock N, and Jablensky A. The genetic deconstruction of psychosis. Schizophr Bull. 33, 905-11 (2007)
143. Serretti A, Mandelli L. The genetics of bipolar disorder: genome 'hot regions,' genes, new potential candidates and future directions. Mol Psychiatry. 2008 Aug;13(8):742-71.
144. Cannon TD. Clinical and genetic high-risk strategies in understanding vulnerability to psychosis. Schizophr Res. 79, 35-44 (2005)

145. Talbot K, Cho DS, Ong WY, Benson MA, Han LY, Kazi HA, Kamins J, Hahn CG, Blake DJ, and Arnold SE. Dysbindin-1 is a synaptic and microtubular protein that binds brain snapin. Hum Mol Genet. 15, 3041-54 (2006)
146. Liu Y, Ford BD, Mann MA, and Fischbach GD. Neuregulin-1 increases the proliferation of neuronal progenitors from embryonic neural stem cells. Dev Biol. 283, 437-45 (2005)
147. Shimizu H, Iwayama Y, Yamada K, Toyota T, Minabe Y, Nakamura K, Nakajima M, Hattori E, Mori N, Osumi N, and Yoshikawa T. genetic and expression analyses of the STOP (MAP6) gene in schizophrenia. Schizophr Res. 84, 244-52 (2006)
148. Andrieux A, Salin P, Schweitzer A, Bégou M, Pachoud B, Brun P, Gory-Fauré S, Kujala P, Suaud-Chagny MF, Höfle G, and Job D. Microtubule stabilizer ameliorates synaptic function and behavior in a mouse model for schizophrenia. Biol Psychiatry. 60, 1224-30 (2006)
149. Anderson SA, Volk DW, and Lewis DA. Increased density of microtubule associated protein 2-immunoreactive neurons in the prefrontal white matter of schizophrenic subjects. Schizophr Res. 19, 111-9 (1996)
150. Cotter D, Wilson S, Roberts E, Kerwin R, and Everall IP. Increased dendritic MAP2 expression in the hippocampus in schizophrenia. Schizophr Res. 41, 313-23 (2000)
151. Arnold SE, Lee VM, Gur RE, and Trojanowski JQ. Abnormal expression of two microtubule-associated proteins (MAP2 and MAP5) in specific subfields of the hippocampal formation in schizophrenia. Proc Natl Acad Sci U S A. 88, 10850-4 (1991)
152. Rosoklija G, Keilp JG, Toomayan G, Mancevski B, Haroutunian V, Liu D, Malespina D, Hays AP, Sadiq S, Latov N, Dwork AJ. Altered subicular MAP2 immunoreactivity in schizophrenia. Prilozi. 2005 Dec;26(2):13-34.
153. Casanova MF, Rothberg B. Shape distortion of the hippocampus: a possible explanation of the pyramidal cell disarray reported in schizophrenia. Schizophr Res. 2002 May 1;55(1-2):19-24.
154. Kovelman JA, Scheibel AB. A neurohistological correlate of schizophrenia. Biol Psychiatry. 1984 Dec;19(12):1601-21.
155. Altshuler LL, Conrad A, Kovelman JA, Scheibel A. Hippocampal pyramidal cell orientation in schizophrenia. A controlled neurohistologic study of the Yakovlev collection. Arch Gen Psychiatry. 1987 Dec;44(12):1094-8.
156. Miklowitz DJ and Johnson SL. The psychopathology and treatment of bipolar disorder. Annu Rev Clin Psychol. 2, 199-235 (2006)
157. Kruijshaar ME, Barendregt J, Vos T, de Graaf R, Spijker J, Andrews G. Lifetime prevalence estimates of major depression: an indirect estimation method and a quantification of recall bias. Eur J Epidemiol. 2005;20(1):103-11.
158. Berrettini W. Bipolar disorder and schizophrenia: convergent molecular data. Neuromolecular Med. 5, 109-17 (2004)
159. Hennah W, Thomson P, McQuillin A, Bass N, Loukola A, Anjorin A, Blackwood D, Curtis D, Deary IJ, Harris SE, Isometsä ET, Lawrence J, Lönnqvist J, Muir W, Palotie A, Partonen T, Paunio T, Pylkkö E, Robinson M, Soronen P, Suominen K, Suvisaari J, Thirumalai S, Clair DS, Gurling H, Peltonen L, Porteous D. DISC1 association, heterogeneity and interplay in schizophrenia and bipolar disorder. Mol Psychiatry. 2008 Mar 4.
160. Georgieva L, Dimitrova A, Ivanov D, Nikolov I, Williams NM, Grozeva D, Zaharieva I, Toncheva D, Owen MJ, Kirov G, O'Donovan MC. Support for

Neuregulin 1 as a Susceptibility Gene for Bipolar Disorder and Schizophrenia. Biol Psychiatry. 2008 May 6.
161. Joo EJ, Lee KY, Jeong SH, Chang JS, Ahn YM, Koo YJ, and Kim YS. Dysbindin gene variants are associated with bipolar disorder in a Korean population. Neurosci Lett. 418, 272-5 (2007)
162. Corvin A, Donohoe G, Nangle JM, Schwaiger S, Morris D, Gill M. A dysbindin risk haplotype associated with less severe manic-type symptoms in psychosis. Neurosci Lett. 2008 Jan 31;431(2):146-9.
163. Sullivan PF, Neale MC, Kendler KS. Genetic epidemiology of major depression: review and meta-analysis. Am J Psychiatry. 2000 Oct;157(10):1552-62.
164. Tochigi M, Iwamoto K, Bundo M, Sasaki T, Kato N, Kato T. Gene expression profiling of major depression and suicide in the prefrontal cortex of postmortem brains. Neurosci Res. 2008 Feb;60(2):184-91.
165. Nurnberger JI Jr. A simulated genetic structure for bipolar illness. Am J Med Genet B Neuropsychiatr Genet. 2008 Mar 3.
166. Hong M, Chen DC, Klein PS, Lee VM. Lithium reduces tau phosphorylation by inhibition of glycogen synthase kinase-3. J Biol Chem. 1997 Oct 3;272(40):25326-32.
167. Hall AC, Brennan A, Goold RG, Cleverley K, Lucas FR, Gordon-Weeks PR, Salinas PC Valproate regulates GSK-3-mediated axonal remodeling and synapsin I clustering in developing neurons. Mol Cell Neurosci. 2002 Jun;20(2):257-70.
168. Bouras C, Kövari E, Hof PR, Riederer BM, Giannakopoulos P. Anterior cingulate cortex pathology in schizophrenia and bipolar disorder. Acta Neuropathol. 2001 Oct;102(4):373-9.
169. Reinés A, Cereseto M, Ferrero A, Bonavita C, Wikinski S. Neuronal cytoskeletal alterations in an experimental model of depression. Neuroscience. 2004;129(3):529-38.
170. Bianchi M, Hagan JJ, Heidbreder CA. Neuronal plasticity, stress and depression: involvement of the cytoskeletal microtubular system? Curr Drug Targets CNS Neurol Disord. 2005 Oct;4(5):597-611.
171. Iwata M, Shirayama Y, Ishida H, Kawahara R. Hippocampal synapsin I, growth-associated protein-43, and microtubule-associated protein-2 immunoreactivity in learned helplessness rats and antidepressant-treated rats. Neuroscience. 2006 Sep 1;141(3):1301-13.
172. Bianchi M, Heidbreder C, Crespi F. Cytoskeletal changes in the hippocampus following restraint stress: role of serotonin and microtubules. Synapse. 2003 Sep 1;49(3):188-94.
173. Silva GA. What impact will nanotechnology have on neurology? Nat Clin Pract Neurol. 2007 Apr;3(4):180-1.
174. Tosi G, Costantino L, Ruozi B, Forni F, Vandelli MA. Polymeric nanoparticles for the drug delivery to the central nervous system. Expert Opin Drug Deliv. 2008 Feb;5(2):155-74.
175. Lockman PR, Mumper RJ, Khan MA, Allen DD. Nanoparticle technology for drug delivery across the blood-brain barrier. Drug Dev Ind Pharm. 2002 Jan;28(1):1-13.
176. de Boer AG, Gaillard PJ. Drug targeting to the brain. Annu Rev Pharmacol Toxicol. 2007;47:323-55.
177. Woolf NJ. Bionic microtubules: potential applications to multiple neurological and neuropsychiatric diseases. Journal of Nanoneuroscience 2009, 1:85-94.

178. Ellis-Behnke R. Nano neurology and the four P's of central nervous system regeneration: preserve, permit, promote, plasticity. Med Clin North Am. 2007 Sep;91(5):937-62.
179. Corr SA, Byrne SJ, Tekoriute R, Meledandri CJ, Brougham DF, Lynch M, Kerskens C, O'Dwyer L, Gun'ko YK. Linear assemblies of magnetic nanoparticles as MRI contrast agents. J Am Chem Soc. 2008 Apr 2;130(13):4214-5.
180. Bolskar RD. Gadofullerene MRI contrast agents. Nanomed. 2008 Apr;3(2):201-13.
181. Baron R, Zayats M, Willner I. Dopamine-, L-DOPA-, adrenaline-, and noradrenaline-induced growth of Au nanoparticles: assays for the detection of neurotransmitters and of tyrosinase activity. Anal Chem. 2005 Mar 15;77(6):1566-71.
182. Keating CD. Nanoscience enables ultrasensitive detection of Alzheimer's biomarker. Proc Natl Acad Sci U S A. 2005 Feb 15;102(7):2263-4.
183. Vestergaard M, Kerman K, Kim DK, Ha MH, Tamiya E. Detection of Alzheimer's tau protein using localised surface plasmon resonance-based immunochip. Talanta. 2008 Jan 15;74(4):1038-42.
184. Pankhurst Q, Hautot D, Khan N, Dobson J. Increased levels of magnetic iron compounds in Alzheimer's disease. J Alzheimers Dis. 2008 Feb;13(1):49-52.
185. Jain KK. Nanobiotechnology-based drug delivery to the central nervous system. Neurodegener Dis. 2007;4(4):287-91.
186. Joshi MD, Müller RH. Lipid nanoparticles for parenteral delivery of actives. Eur J Pharm Biopharm. 2008 Sep 13.
187. Rhim WK, Kim JS, Nam JM. Lipid-gold-nanoparticle hybrid-based gene delivery. Small. 2008 Oct;4(10):1651-5.
188. Kojima C, Hirano Y, Yuba E, Harada A, Kono K. Preparation and characterization of complexes of liposomes with gold nanoparticles. Colloids Surf B Biointerfaces. 2008 Oct 15;66(2):246-52.
189. J. Kreuter In: J.B.J. Swarbrick, Editor, Encyclopedia of Pharma. Tech., Marcel Dekker, New York (1994), pp. 165-190.
190. Green JJ, Zhou BY, Mitalipova MM, Beard C, Langer R, Jaenisch R, Anderson DG. Nanoparticles for gene transfer to human embryonic stem cell colonies. Nano Lett. 2008 Oct;8(10):3126-30.
191. Crotty S, Fitzgerald P, Tuohy E, Harris DM, Fisher A, Mandel A, Bolton AE, Sullivan AM, Nolan Y. Neuroprotective effects of novel phosphatidylglycerol-based phospholipids in the 6-hydroxydopamine model of Parkinson's disease. Eur J Neurosci. 2008 Jan;27(2):294-300.
192. Esposito E, Fantin M, Marti M, Drechsler M, Paccamiccio L, Mariani P, Sivieri E, Lain F, Menegatti E, Morari M, Cortesi R. Solid lipid nanoparticles as delivery systems for bromocriptine. Pharm Res. 2008 Jul;25(7):1521-30.
193. Gonzalez-Barrios JA, Lindahl M, Bannon MJ, Anaya-Martínez V, Flores G, Navarro-Quiroga I, Trudeau LE, Aceves J, Martinez-Arguelles DB, Garcia-Villegas R, Jiménez I, Segovia J, Martinez-Fong D. Neurotensin polyplex as an efficient carrier for delivering the human GDNF gene into nigral dopamine neurons of hemiparkinsonian rats. Mol Ther. 2006 Dec;14(6):857-65.
194. Roney C, Kulkarni P, Arora V, Antich P, Bonte F, Wu A, Mallikarjuana NN, Manohar S, Liang HF, Kulkarni AR, Sung HW, Sairam M, Aminabhavi TM. Targeted nanoparticles for drug delivery through the blood-brain barrier for Alzheimer's disease. J Control Release. 2005 Nov 28;108(2-3):193-214.

195. Wilson B, Samanta MK, Santhi K, Kumar KP, Paramakrishnan N, Suresh B. Poly(n-butylcyanoacrylate) nanoparticles coated with polysorbate 80 for the targeted delivery of rivastigmine into the brain to treat Alzheimer's disease. Brain Res. 2008 Mar 20;1200:159-68
196. Liu G, Men P, Harris PL, Rolston RK, Perry G, Smith MA. Nanoparticle iron chelators: a new therapeutic approach in Alzheimer disease and other neurologic disorders associated with trace metal imbalance. Neurosci Lett. 2006 Oct 9;406(3):189-93.
197. D'Souza GG, Cheng SM, Boddapati SV, Horobin RW, Weissig V. Nanocarrier-assisted sub-cellular targeting to the site of mitochondria improves the pro-apoptotic activity of paclitaxel. J Drug Target. 2008 Aug;16(7):578-85.
198. Maji SK, Schubert D, Rivier C, Lee S, Rivier JE, Riek R. Amyloid as a depot for the formulation of long-acting drugs. PLoS Biol. 2008 Feb;6(2):e17.
199. Tuszyński MH. Nerve growth factor gene therapy in Alzheimer disease. Alzheimer Dis Assoc Disord. 21, 179-89 (2007)
200. Tuszyński MH, Thal L, Pay M, Salmon DP, U HS, Bakay R, Patel P, Blesch A, Vahlsing HL, Ho G, Tong G, Potkin SG, Fallon J, Hansen L, Mufson EJ, Kordower JH, Gall C, and Conner J. A Phase 1 clinical trial of nerve growth factor gene therapy for Alzheimer disease. Nat Med. 11, 551-5 (2005)
201. Birks J. Cholinesterase inhibitors for Alzheimer's disease. Cochrane Database Syst Rev. 1, 1-75 (2008)
202. Iqbal K and Grundke-Iqbal I. Alzheimer neurofibrillary degeneration: significance, etiopathogenesis, therapeutics and prevention. J Cell Mol Med. Jan 8 (2008)
203. Iqbal K and Grundke-Iqbal I. Pharmacological approaches of neurofibrillary degeneration. Curr Alzheimer Res. 2, 335-41 (2005)
204. Alexander BL, Ali RR, Alton EW, Bainbridge JW, Braun S, Cheng SH, Flotte TR, Gaspar HB, Grez M, Griesenbach U, Kaplitt MG, Ott MG, Seger R, Simons M, Thrasher AJ, Thrasher AZ, and Ylä-Herttuala S. Progress and prospects: gene therapy clinical trials (Part 1). Gene Ther. 14, 1439-47 (2007)
205. Gasmi M, Brandon EP, Herzog CD, Wilson A, Bishop KM, Hofer EK, Cunningham JJ, Printz MA, Kordower JH, and Bartus RT. AAV2-mediated delivery of human neurturin to the rat nigrostriatal system: long-term efficacy and tolerability of CERE-120 for Parkinson's disease. Neurobiol Dis. 27, 67-76 (2007)
206. Singh N, Pillay V, and Choonara YE. Advances in the treatment of Parkinson's disease. Prog Neurobiol. 81, 29-44 (2007)
207. Andrieux A, Salin P, Schweitzer A, Bégou M, Pachoud B, Brun P, Gory-Fauré S, Kujala P, Suaud-Chagny MF, Höfle G, and Job D. Microtubule stabilizer ameliorates synaptic function and behavior in a mouse model for schizophrenia. Biol Psychiatry. 60, 1224-30 (2006)
208. Villalobos A, Ness JE, Gustafsson C, Minshull J, and Govindarajan S. Gene designer: a synthetic biology tool for constructing artificial DNA segments. BMC Bioinformatics. 7, 285 (2006)
209. Kanazawa I. How do neurons die in neurodegenerative diseases? Trends Mol Med. 2001 Aug;7(8):339-44.
210. Feinstein SC and Wilson L. Inability of tau to properly regulate neuronal microtubule dynamics: a loss-of-function mechanism by which tau might mediate neuronal cell death. Biochim Biophys Acta. 1739, 268-79 (2005).

211. Fanara P, Banerjee J, Hueck RV, Harper MR, Awada M, Turner H, Husted KH, Brandt R, and Hellerstein MK. stabilization of hyperdynamic microtubules is neuroprotective in amyotrophic lateral sclerosis. J Biol Chem. 282, 23465-72 (2007)
212. McLean DR and Graham BP. Stability in a mathematical model of neurite elongation. Math Med Biol. 23, 101-17 (2006)
213. Carpenter EJ, Huzil JT, Ludueña RF, Tuszyński JA. Homology modeling of tubulin: influence predictions for microtubule's biophysical properties. Eur Biophys J. 2006 Dec;36(1):35-43.
214. Wilson L, Panda D, and Jordan MA. Modulation of microtubule dynamics by drugs: a paradigm for the actions of cellular regulators. Cell Struct Funct. 24, 329-35 (1999)
215. Downing KH and Nogales E. Crystallographic structure of tubulin: implications for dynamics and drug binding. Cell Struct Funct. 24, 269-75 (1999)
216. Drabik P, Gusarov S, and Kovalenko A. microtubule stability studied by three-dimensional molecular theory of solvation. Biophys J. 92, 394-403 (2007)
217. Gigant B, Wang C, Ravelli RB, Roussi F, Steinmetz MO, Curmi PA, Sobel A, and Knossow M. Structural basis for the regulation of tubulin by vinblastine. Nature. 435, 519-22 (2005)
218. Hari M, Loganzo F, Annable T, Tan X, Musto S, Morilla DB, Nettles JH, Snyder JP, and Greenberger LM. Paclitaxel-resistant cells have a mutation in the paclitaxel-binding region of beta-tubulin (Asp26Glu) and less stable microtubules. Mol Cancer Ther. 5, 270-8 (2006)
219. He L, Yang CP, and Horwitz SB. Mutations in beta-tubulin map to domains involved in regulation of microtubule stability in epothilone-resistant cell lines. Mol Cancer Ther. 1, 3-10 (2001)
220. Wang Y, O'Brate A, Zhou W, and Giannakakou P. Resistance to microtubule-stabilizing drugs involves two events: beta-tubulin mutation in one allele followed by loss of the second allele. Cell Cycle. 4, 1847-53 (2005)
221. Banerjee A. Coordination of posttranslational modifications of bovine brain alpha-tubulin. polyglycylation of delta2 tubulin. J Biol Chem. 277, 46140-4 (2002)
222. Shea TB. Selective stabilization of microtubules within the proximal region of developing axonal neurites. Brain Res Bull. 48, 255-61 (1999)
223. Hempen B and Brion JP. Reduction of acetylated alpha-tubulin immunoreactivity in neurofibrillary tangle-bearing neurons in Alzheimer's disease. J Neuropathol Exp Neurol. 55, 964-72 (1996)
224. Mukhopadhyay R, Kumar S, Hoh JH. Molecular mechanisms for organizing the neuronal cytoskeleton. Bioessays. 2004 Sep;26(9):1017-25.
225. Santarella RA, Skiniotis G, Goldie KN, Tittmann P, Gross H, Mandelkow EM, Mandelkow E, and Hoenger A. surface-decoration of microtubules by human tau. J Mol Biol. 339, 539-53 (2004)
226. Al-Bassam J, Ozer RS, Safer D, Halpain S, and Milligan RA. MAP2 and tau bind longitudinally along the outer ridges of microtubule protofilaments. J Cell Biol. 157, 1187-96 (2002)
227. Hutchins BM, Platt M, Hancock WO, and Williams ME. Directing transport of CoFe2O4-functionalized microtubules with magnetic fields. Small. 3, 126-31 (2007)

228. Platt M, Muthukrishnan G, Hancock WO, and Williams ME. Millimeter scale alignment of magnetic nanoparticle functionalized microtubules in magnetic fields. J Am Chem. 127, 15686-7 (2005)
229. Bekeredjian R, Behrens S, Ruef J, Dinjus E, Unger E, Baum M, and Kuecherer HF. Potential of gold-bound microtubules as a new ultrasound contrast agent. Ultrasound Med Biol. 28, 691-5 (2002)
230. Suh J, Wirtz D, and Hanes J. Real-time intracellular transport of gene nanocarriers studied by multiple particle tracking. Biotechnol Prog. 20, 598-602 (2004)
231. Taira S, Du YZ, Hiratsuka Y, Uyeda TQ, Yumoto N, and Kodaka M. Loading and unloading of molecular cargo by dna-conjugated microtubule. Biotechnol Bioeng. 99, 734-9 (2008)
232. Ramachandran S, Ernst KH, Bachand GD, Vogel V, and Hess H. Selective loading of kinesin-powered molecular shuttles with protein cargo and its application to biosensing. Small. 2, 330-4 (2006)
233. Hess H, Clemmens J, Brunner C, Doot R, Luna S, Ernst KH, and Vogel V. Molecular self-assembly of “nanowires” and “nanospools” using active transport. Nano Lett. 5, 629-33 (2005)
234. Goldberg M, Langer R, and Jia X. Nanostructured materials for applications in drug delivery and tissue engineering. J Biomater Sci Polym Ed. 18, 241-68 (2007)
235. Van den Heuvel MG, de Graaff MP, and Dekker C. Molecular sorting by electrical steering of microtubules in kinesin-coated channels. Science. 312, 910-4 (2006).
236. Sataric MV, Budinski-Petkovic L, Loncarevic I, Tuszyński JA. Modelling the Role of Intrinsic Electric Fields in Microtubules as an Additional Control Mechanism of Bi-directional Intracellular Transport. Cell Biochem Biophys. 2008;52(2):113-24.
237. Tyner KM, Kopelman R, Philbert MA. “Nanosized voltmeter” enables cellular-wide electric field mapping. Biophys J. 2007 Aug 15;93(4):1163-74.
238. Unger,E., Böhm, K.J. and Vater, W. Structural diversity and dynamics of microtubules and polymorphic tubulin assemblies. Electron Microsc. Rev. 3 (1990), pp. 355-395.
239. Deuschl G, Schade-Brittinger C, Krack P, Volkmann J, Schäfer H, Bötzel K, et al: A randomized trial of deep-brain stimulation for Parkinson's disease. N Engl J Med 355:896-908, 2006
240. Deep Brain Stimulation for Parkinson's Disease Study Group: Deep-brain stimulation of the subthalamic nucleus or the pars interna of the globus pallidus in Parkinson's disease. N Engl J Med 345:956-963, 2001
241. Abosch A, Cosgrove GR. Biological basis for the surgical treatment of depression. Neurosurg Focus. 2008;25(1):E2.
242. He W, McConnell GC, Bellamkonda RV. Nanoscale laminin coating modulates cortical scarring response around implanted silicon microelectrode arrays. J Neural Eng. 2006 Dec;3(4):316-26.
243. W.T. Rogers “Electromagnetic Brain Animation” Generator. Full Utility Patent Pending, USPTO, 2004, published 1 year, final allowance 10,2007. 1 - 37
244. Mantovani A, Lisanby SH, Pieraccini F, Ulivelli M, Castrogiovanni P, Rossi S. Repetitive Transcranial Magnetic Stimulation (rTMS) in the treatment of panic disorder (PD) with comorbid major depression. J Affect Disord. 2007 Sep;102(1-3):277-80.

245. Cankurtaran, B.; Ford, M.J.; Cortie, M., "Local Electromagnetic Fields Surrounding Gold Nano-Cap Particles," Nanoscience and Nanotechnology, 2006. ICONN '06. International Conference on , vol., no., pp.-, 3-7 July 2006
246. Wilhelm C, Gazeau F. Universal cell labelling with anionic magnetic nanoparticles. Biomaterials. 2008 Aug;29(22):3161-74.
247. Gross PG, Kartalov EP, Scherer A, Weiner LP. Applications of microfluidics for neuronal studies. J Neurol Sci. 2007 Jan 31;252(2):135-43.
248. Hubbard, RE. Structure-based Drug Discover: An Overview (2006) RSC Biomolecular Series.
249. García-Palomero E, Muńoz P, Usan P, Garcia P, Delgado E, De Austria C, Valenzuela R, Rubio L, Medina M, Martnez A. Potent beta-amyloid modulators. Neurodegener Dis. 2008;5(3-4):153-6.
250. Limongelli V, Marinelli L, Cosconati S, Braun HA, Schmidt B, Novellino E. Ensemble-docking approach on BACE-1: pharmacophore perception and guidelines for drug design. ChemMedChem. 2007 May;2(5):667-78.
251. Fu H, Li W, Luo J, Lee NT, Li M, Tsim KW, Pang Y, Youdim MB, Han Y. Promising anti-Alzheimer's dimer bis(7)-tacrine reduces beta-amyloid generation by directly inhibiting BACE-1 activity. Biochem Biophys Res Commun. 2008 Feb 15;366(3):631-6.
252. Arakawa T, Kita Y, Niikura T. A rescue factor for Alzheimer's diseases: discovery, activity, structure, and mechanism. Curr Med Chem. 2008;15(21):2086-98.
253. Look GC, Jerecic J, Cherbavaz DB, Pray TR, Breach JC, Crosier WJ, Igoudin L, Hironaka CM, Lowe RM, McEntee M, Ruslim-Litrus L, Wu HM, Zhang S, Catalano SM, Goure WF, Summa D, Krafft GA. Discovery of ADDL–targeting small molecule drugs for Alzheimer's disease. Curr Alzheimer Res. 2007 Dec;4(5):562-7.
254. Okura Y, Matsumoto Y. Development of anti-Abeta vaccination as a promising therapy for Alzheimer's disease. Drug News Perspect. 2007 Jul-Aug;20(6):379-86.
255. Shi M, Caudle WM, Zhang J. Biomarker discovery in neurodegenerative diseases: A proteomic approach. Neurobiol Dis. 2008 Sep 26.
256. Vincent AM, Sakowski SA, Schuyler A, Feldman EL. Strategic approaches to developing drug treatments for ALS. Drug Discov Today. 2008 Jan;13(1-2):67-72.
257. Schrattenholz A, Soskić V. What does systems biology mean for drug development? Curr Med Chem. 2008;15(15):1520-8.
258. Reese M, Sánchez-Pedregal VM, Kubicek K, Meiler J, Blommers MJ, Griesinger C, Carlomagno T. Structural basis of the activity of the microtubule-stabilizing agent epothilone a studied by NMR spectroscopy in solution. Angew Chem Int Ed Engl. 2007;46(11):1864-8.

6

Novel Modes of Neural Computation: From Nanowires to Mind

Summary

The human mind is by far one of the most amazing natural phenomena known to man. It embodies our perception of reality, and is in that respect the ultimate observer. The past century produced monumental discoveries regarding the nature of nerve cells, the anatomical connections between nerve cells, the electrophysiological properties of nerve cells, and the molecular biology of nervous tissue. What remains to be uncovered is that essential something – the fundamental dynamic mechanism by which all these well understood biophysical elements combine to form a mental state. In this chapter, we further develop the concept of an intraneuronal matrix as the basis for autonomous, self-organized neural computing, bearing in mind that at this stage such models are speculative. The intraneuronal matrix – composed of microtubules, actin filaments, and cross-linking, adaptor, and scaffolding proteins – is envisioned to be an intraneuronal computational network, which operates in conjunction with traditional neural membrane computational mechanisms to provide vastly enhanced computational power to individual neurons as well as to larger neural networks. Both classical and quantum mechanical physical principles may contribute to the ability of these matrices of cytoskeletal proteins to perform computations that regulate synaptic efficacy and neural response. A scientifically plausible route for controlling synaptic efficacy is through the regulation of neural transport of synaptic proteins and of mRNA. Operations within the matrix of cytoskeletal proteins that have applications to learning, memory, perception, and consciousness, and conceptual models implementing classical and quantum mechanical physics are discussed. Nanoneuroscience methods are emerging that are capable of testing aspects of these conceptual models, both theoretically and experimentally. Incorporating intra-neuronal biophysical operations into existing theoretical frameworks of single neuron and neural network function stands to enhance existing models of neurocognition.

6.1 Traditional Models of Neural Processing

Traditional models of neural processing rely completely on electrochemical events at the membrane brought about by receptor actions in response to inputs. This response is subsequently transmitted to the axon terminal to initiate neurotransmitter release into the synaptic cleft, which in turn stimulates postsynaptic receptors and begins a reiteration of the same process in the recipient neuron. The integration of inputs to single neurons is quite complex and has been a topic of intense research over the past few decades [1].

Although activity patterns in traditional neural networks and the integration of inputs within single neurons[1] have been repeatedly cited as the underlying mechanisms of neurocognition and of animal behavior, membrane-associated activity alone may not be able to fully address some of the enigmatic aspects of cognitive phenomena. Models in which only membrane-associated currents are computationally relevant depict synaptic activation at a given receptor site producing a flow of current along the neural membrane capable of interacting with "other" synaptic inputs, but the neuron itself lacks a significant intrinsic computational mechanism since these "other" synaptic inputs are largely outside the realm of control of the neural membrane. Regardless of how extensive a neural wiring pattern may be, the lack of multiple intrinsic computational mechanisms renders the overall neural network system passively dependent on extrinsic input – in other words, the neural network as it is presently conceived is a machine lacking intrinsic "intelligence".

6.2 Information Processing in the Intraneuronal Cytoskeletal Matrix

Nanoscience allows for dramatically new perspectives on neural function and enables the conceptualization of novel neurocognitive models that take into consideration the biomolecules of the neuron. Theoretical modeling, the development of early prototypes of biomolecular and quantum computers, and experimental results unveiling the quantum principles governing biomolecules all support the notion that neural computation may be comparable to the futuristic "smart grid" proposed to meet the increased energy demands [3]. A smart grid is adaptable to fluctuations in usage, has built-in capacity for bi-directional flow of electrical current, and possesses multiple microprocessors that direct current to where it is needed. Smart grids intimately link the delivery of electricity with communication and computation. By analogy, the internal matrix of cytoskeletal filaments resembles a smart grid of nanowires by virtue of its capacities for bi-directional flow, distributed activity, parallel processing, multiple-point sensing, autonomous behavior, and self-organization (see Table 6.1).

[1] Interactions between the various compartments of dendrites, for example, increase the computational capacity of individual neurons [3].

Table 6.1. Comparison between traditional neural processing and information processing that additionally involves computation in the intraneuronal matrix.

Traditional neural processing*	Intraneuronal matrix processing
Electrochemical (neural membrane events)	Propagation of electrical current along cytoskeletal filaments (electromagnetic, ferroelectric, and quantum mechanical perturbations among electrons of tubulin dimers)
One-way communication (from point of synaptic contact to axon terminal)	Two-way communication: anterograde and retrograde information flow (between the cell nucleus and axon terminals or dendritic synapses, or between synapses and other synapses)
Few sensors (only synapses serve as sensors and compute inputs)	Multiple sensors (computations performed at multiple sites located throughout the interiors of neurons intrinsically modulate the strength of synaptic inputs)
Centralized organization (localization of function relates to sensory system-specific pathways)	Distributed organization (potential for widespread cortical activity patterns triggered by self-organized networks based on intrinsically modulated synapses)
Radial topology (embedded in the cortical hierarchy)	Network topology (potential for massive parallel processing due to widespread activity)
Passive, input-dependent behavior (nerve cells driven solely by stimuli)	Autonomous, self-organizing behavior (nerve cells capable of generating intrinsic output and regulating strength of input)
Input-dependent restructuring of synapses (capable of accounting for stimulus-driven plasticity, but not necessarily certain categories of higher cognition and related states of consciousness)	Self-organized restructuring of synapses and interior of neurons (could possibly account for certain types of higher cognition, e.g., complex knowledge bases, human creativity, higher intelligence, and consciousness)

*Networks of neurons are capable of more complex functions than single neurons

Throughout this chapter examples will be presented conceptualizing how a matrix of cytoskeletal filaments could supplement and enhance the computational ability of the neural membrane. As was thoroughly discussed in Chapter 3, actin filaments and microtubules possess the capability to transmit, and even amplify electric current [4, 5]. It is also possible that cytoskeletal matrices participate in directing ionic or phosphoprotein waves relevant to neurotrophism and protein transcription, and that these nanowires regulate neural transport (see discussion in Chapter 4). As will be elaborated on in subsequent sections of this chapter, the cytoskeletal matrix may in principle serve either as a classical or quantum mechanical computer enabling these information signaling and transport tracks to act autonomously during the implementation of neurocognitive tasks.

Incorporating intraneuronal matrices in a conceptual scheme or model of neural computing offers the following advantages:

(1) A way to connect what is taking place at the cell membrane to the intrinsic biophysical operations of the neuron.
(2) A way to exponentially increase computational capacity, since each cytoskeletal protein rather than each synapse computes operations. Estimates of computational power (at maximal efficiency) are as follows: For each tubulin dimer there are ~32 possible states (based on 4 C-termini states, 4 electron hopping states, and 2 conformational/GTP states). This translates into 1 GB per neuron, and given 10 billion neurons in the cerebral cortex, results in the information storage capacity of 10^{19} bytes/brain, and assuming nanosecond-range transitions, 10^{28} flops of processing capability.
(3) The potential for novel biophysical properties (e.g., due to the ability for self-assembly, ferroelectric properties, and the presence of electronic double-well potentials) to account for aspects of voluntary action, original thinking, planning, and other types of higher cognition – including those unique to humans (e.g., complex language, higher intelligence, and creativity).

As listed in Table 6.2, there exist a number of models attributing higher cognitive functions, such as learning, memory, perception, and consciousness to operations of microtubules and actin filaments. These models range from studies involving behavior of animals to theoretical models detailing how biomolecules operate in neurons. Each of these models addresses the issues from different levels of physical analysis. First, there are studies demonstrating that microtubules, microtubule-associated proteins (MAPs), and actin filaments are involved in neural plasticity, learning, and memory – all of which provide data strictly at the level of classical physics. Second, there are studies using molecular modeling techniques and experimental manipulations of microtubules and actin filaments demonstrating their basic electromagnetic, ferroelectric, and dipolar characteristics at a level of classical physics, that nonetheless have clear implications for the behavior of these biomolecules

Table 6.2. Description of models of higher cognition involving the cytoskeleton.

Model (level of physics involved)	Description
Classical physics	
Learning critically involves plasticity of microtubule-associated proteins (MAP2 and tau).	MAP2 degradation occurs with classical conditioning [6]-[8]; transgenic mice accumulating tau show impaired learning [9]; and transgenic truncation of MAP2 impairs classical conditioning [10].
Consciousness and learning depend on actin filament plasticity.	Anesthetic compounds impair actin filament motility [11, 12]. Learning and long-term potentiation (LTP) involve actin reorganization in spines [13]-[16]
Classical physics with implications for quantum physics	
Cytoskeletal proteins conduct signals of possible relevance to cognition and consciousness.	Microtubules are capable of propagating kink-like excitations or ferroelectric signals [17, 18]; actin filaments and microtubules are capable of conducting and amplifying electric current [4, 5].
Quantum physics	
Orchestrated objective reduction (Orch OR) accounts for the hard problem of consciousness.	Tubulin subunits of microtubules act as qubits and perform quantum computations that are synchronized with neurophysiological events [19]-[21].
Quantum computing mechanism may be related to the dipole moment of tubulin.	Computer simulations and experimental measurements demonstrate quantum-computing capabilities of tubulin dimers [22, 23].
Microtubules may compute and store information based on quantum coherence of the dipolar vibrational modes of tubulin.	Tubulin has a dipolar vibrational energy in the range that enables quantum coherence [24, 25]

at the quantum mechanical level. Third, there are the completely theoretical models, which propose that quantum mechanical computations in brain microtubules account for higher cognitive functioning, in particular consciousness.

6.2.1 Linking Neural Plasticity to Cognition

A structural-functional organization of the cerebral cortex has long been implicated as being fundamental to higher cognitive activities [26, 27] and it is clear that the cytoskeleton is pivotally responsible for that structure in terms of both neuron shape and neural connections. Among the many neurons in the brain, the large pyramidal neurons in the cerebral cortex and hippocampus are the most likely determinants of complex mental functions such as memory, perception, and consciousness [28, 29]. These neurons have numerous, elaborately branching dendrites, which expand prolifically in the first months of life and then continue to grow moderately throughout the lifespan [30]. The role played by the cytoskeleton in determining neural structure is evident at both small and large scales; cytoskeletal proteins mediate interactions with other biomolecules, chemical side-chains, and ions, as well as define the overall morphology of dendrites and dendritic branches.

Many of the most influential synapses in the mammalian cortex occur on spines, protrusions on dendrites that are rich in actin filaments. In contrast, microtubules are abundant in the underlying dendrite shaft. Within the subsynaptic zone below the synapse, microtubules are in a position to regulate activity in spines, in part by controlling what is delivered to the spine [31]. As a result, microtubules would be expected to determine the strength of synaptic responses as induced by neurotransmitters, which will be further discussed later in this chapter.

Self-Organization of Microtubules, Neural Plasticity, and Cognition

Due to their capacity for self-organization, microtubules possess the ability to exert control over synapses autonomously. Microtubules display dynamic instability, the tendency to switch from polymerization or depolymerization at rates depending on a mix of factors deriving from their internal state and that of the surrounding environment. Ionic conductances and electromagnetic membrane currents appear to influence cytoskeletal dynamics, thereby linking conventional neural computing with potential computational modes served by cytoskeletal proteins.

Dynamic instability of microtubules can be affected by ionic influxes at ion channels associated with the synapse. Typical ionic influxes at synapses include perturbations in intracellular concentrations for Na^+ and K^+. Increases in Na^+ or K^+ concentrations to 150 - 160 mM have been experimentally demonstrated to increase microtubule polymerization, whereas Na^+ or K^+

concentrations over 250 mM decrease microtubule polymerization [32, 33]. Membrane currents also generate electromagnetic fields that influence local microtubules, and magnetic stimulation at very low frequencies enhances microtubule polymerization [34].

MAPs in the surrounding environment can affect microtubule activities; moreover, phosphorylation of these MAPs markedly influences the function of these cross-linking proteins [35]. The phosphorylation state of MAP2 regulates its polymerization and dynamic instability. There are a number of protein kinases that phosphorylate MAP2, resulting in numerous G-protein-linked receptors that activate these protein kinases being in a position to influence these MAPs and the microtubules to which they are bound. While there is a tendency for increased phosphorylation to decrease MAP2 binding to a microtubule, the more than 40 sites of polymerization on MAP2 exert subtly different effects.

Polymerization/depolymerization shifts conceivably represent a computational mode for microtubules enabling them to differentially affect various biomolecules in the subsynaptic zone. Moreover, it is also possible that microtubules additionally store information upon stabilization. The binding of MAP2, post-translational modifications of microtubules, and certain protein conformational changes are capable of rendering microtubules stable, in terms of diminished protein turnover. Acetylated and detyrosinated microtubules are the most stable among microtubules and are concentrated at the initial portion of the axon [36]. In dendrites, microtubules demonstrate a mixture of oppositely polarized plus and minus ends, with plus ends being the primary sites of polymerization and depolymerization [37]. Assuming that dendritic structure is relatively stable, at least some dendritic microtubules must be in a stabilized state, but unlike the accumulation of stable microtubules at the initial segment of the axon, stable microtubules are distributed throughout the dendrite length. Microtubules stabilized by MAPs and post-translational modifications, such as acetylation and detyrosination, could store past histories of synaptic inputs (or more precisely, past histories of microtubule computations relevant to past synaptic inputs), as will be discussed more fully later in this chapter.

Regarding altered MAP2 binding to microtubules, there is evidence that MAP2 mediates restructuring of microtubules in neurons following learning. It has been shown experimentally that MAP2 and tubulin are broken down (or proteolyzed) in the auditory cortex and hippocampus in rats following fear conditioning to tone [6]-[8]. Multiple laboratories have confirmed the participation of MAPs in learning. In one study, accumulation of tau disrupted learning in Drosophila [9]. In another study, impaired fear conditioning was observed in transgenic mice having abnormally truncated MAP2 [10]. Moreover, microtubule polymerization appears to be essential for learning. Learning tasks, such as the Morris water maze and passive avoidance, were impaired by intrahippocampal infusion of colchicine (an anti-microtubule drug causing microtubule depolymerization) even at doses that did not cause neural damage

[38]. Transport along microtubules is also implicated in learning. Transgenic mice overexpressing KIF-17, a kinesin motor protein that attaches to microtubules in hippocampus, showed enhanced spatial and working memory [39].

Evidence that synaptic inputs onto spines affect subsynaptic microtubules and that MAP2 and microtubule restructuring following learning support there being two sets of subsynaptic microtubules: one set in a state of heightened dynamic instability and one set in a relatively stable state. The next issue is how these adjacent microtubules might interact with actin filaments in spines, and what roles those actin filaments might play in higher cognition.

Actin Filaments, Neural Plasticity, and Cognition

The dynamic operations of actin filaments found in dendrite spines underlie functions ranging from neural plasticity to higher cognition. Perhaps most intriguing are reports relating actin dynamics to states of consciousness. Research from the laboratory of Dr. Andrew Matus showed that several volatile anesthetics (chloroform, diethylether, enflurane, halothane, and isoflurane) impair dynamic motility of actin filaments at approximately the same concentrations that produce anesthesia [11]. In a similar study, the anesthetic propofol induced anesthesia and recovery according to the same time course as neurite retraction and recovery; moreover, that plasticity depended on actin and myosin dynamics as well as $GABA_A$ neuroreceptors [12]. These results are some of the most direct and convincing studies that cytoskeletal dynamics and motility are related to level of conscious awareness.

Spine changes are well documented and appear to be related to learning and long-term potentiation (LTP) [13, 14]. According to Matus, large stable spines may be responsible for permanent memory storage, while smaller transient spines may be involved with learning. The size of a spine predicts its response. Stimulation of single spines results in a persistent enlargement of smaller spines, but only a transient enlargement of large mushroom spines [15]. Spine size alterations parallel enhanced alpha-amino-3-hydroxy-5-methyl-4-isoxazolepropionic acid (AMPA) receptor currents, N-methyl-D-aspartate (NMDA) receptor activity, actin polymerization, and depend on transport along microtubule tracks to reach actin-filled spines. NMDA and AMPA glutamate receptors are rapidly transported to newly formed synapses [40], and increased rates of AMPA glutamate receptor transport occur following high frequency stimulation of associational fibers to CA3 hippocampal neurons [41]. Blocking glutamate AMPA receptor transport to synaptic terminals significantly reduces AMPA-mediated responses [42], and NMDA receptor function depends on kinesin-based transport along microtubules [43]. These results, taken together, suggest that there is enhanced transport of subunits of these ionotropic NMDA and AMPA receptors to spines with learning, and that this increased transport sustains the potentiated response. Inside the spine, actin is depolymerized following LTP [44].

Larger spine heads with more ionotropic receptors are capable of larger ionic influxes, thereby generating potentially larger electromagnetic fields, which could be transmitted to the microtubules in the subsynaptic zone. This is consistent with an association between microtubules and actin being critical to maintaining the functioning of synapses [45]. As discussed in Chapter 2, the cytoskeleton-linking protein LIM kinase-1 (LIMK1) links actin filament to the membrane and CLIP-115 links actin to microtubules. LIMK1 knockout mice demonstrate both learning deficits and unusually wide spine necks [46]. Curiously, these LIMK1 knockout mice have enhanced LTP, but impaired spatial learning [47]. While the fine details of these interactions still need to be worked out, the results discussed above are consistent with the linking of actin filaments to both microtubules and the neuronal membrane being essential to neural plasticity and neurocognition.

6.2.2 Novel Electric Signaling Modes for Actin Filaments and Microtubules

Although the roles of microtubules and actin filaments in developmental plasticity and transport are undisputed, these proteins have not been viewed as particularly critical to signal propagation in the neuron until recently. As described fully in Chapter 3, electric signals can be propagated and amplified along actin filaments and microtubules; this has been demonstrated experimentally [4, 5]. Placing greater emphasis on such signal propagation and on the intraneuronal matrix in neural computation is likely to dramatically alter conceptual models of neurocognition.

The physical properties of microtubules that sustain electric signaling modes include their ferroelectric nature, spontaneous emergence of electric dipoles, and the distribution of surface counterions [17, 18]. The microtubule exhibits critical dipole-dipole interactions due to couplings of tubulin to its six nearest neighbors. According to one conceptualization, these dipole-dipole interactions lead to topological modes of “kink-like” excitation that propagate down the microtubule. The kink-like excitation can be conceived as a defect, insofar as it represents a transient switch in the direction of the dipole moment of the tubulin dimer as it passes along the microtubule. Collision-based models of microtubule computing have also been advanced [48, 49]. According to these models, microtubules are visualized as Turing machine-type macromolecular assemblies capable of implementing computational algorithms.

As the biophysical properties of cytoskeletal proteins become increasingly well understood at the level classical physics, conceptual descriptions at the quantum mechanical level (and their potential relevance to quantum computing) become gradually possible. Ferroelectric nanostructures possess interesting features that can be described by both classical and quantum mechanical physical theories [50]. Also, the interiors of tubulin dimers have unique properties that may enable quantum computations [51].

Tubulin is a globular protein, which when folded has an interior cavity. As shown in Figure 6.1, solving the Poisson-Boltzmann equation produced an electrostatic map of tubulin revealing that the interior of each tubulin dimer possesses two relatively large areas of positive charge separated by a negative region creating a double-well potential, through which a mobile electron could tunnel [51]. In principle, surrounding conditions would determine whether such a mobile electron would be able to overcome the potential energy barrier. This double-well electrostatic potential thereby serves as a basis for a cellular automata model based on quantum mechanical events. According to this quantum cellular automata microtubule model, electron position in neighboring tubulins affect one another, creating a computation mechanism that is updated in discrete steps. Computer simulations of the quantum cellular automata microtubule model taken to 250 steps yielded several types of oscillating or static states that may resemble neural network activity patterns. This model is potentially related to the hydrophobic pocket in tubulin, which will be discussed later.

6.2.3 Quantum Computations in Brain Microtubules

Italian physicist Luppachini has noted that a quantum-Turing machine is not incompatible with Turing's own ideas regarding the computational basis of mental processes – in particular those related to the human mind and its capacity for intuition [52]. She further argues that quantum computation results in less uncertainty than does statistical probability because:

> Quantum computation arises from the possibility of exploiting a multiplicity of parallel computational paths in superposition as well as quantum interference to amplify the probability of correct outcomes of computations (p. 44).

A number of theorists, most notably Dr. Stuart Hameroff and Sir Roger Penrose, have proposed that quantum computations are performed in brain microtubules [19]-[21]. These theories have met with some skepticism, especially regarding rapid decoherence of coherent states precluding meaningful quantum computation in brain [53]. Still other critics of the model suggest that some version of this idea might at least be falsifiable, but that much more theoretical and experimental evidence is needed [54, 55]. Quantum mechanical models of neurocognitive states nonetheless remain attractive for many reasons, not least of which is that quantum mechanics deals with the parallel calculations of probability amplitudes similar to the way that cognitive states simultaneously take into account probabilities affiliated with a multitude of past associations and future anticipations. Sir John Eccles, greatly admired for his neuroscientific expertise, but criticized for his seemingly dualist beliefs, speculated that the probability fields of quantum mechanics may enable "mental events" to produce neural events such as neurotransmitter release [56, 57]. It is not necessary to subscribe to any particular philosophical

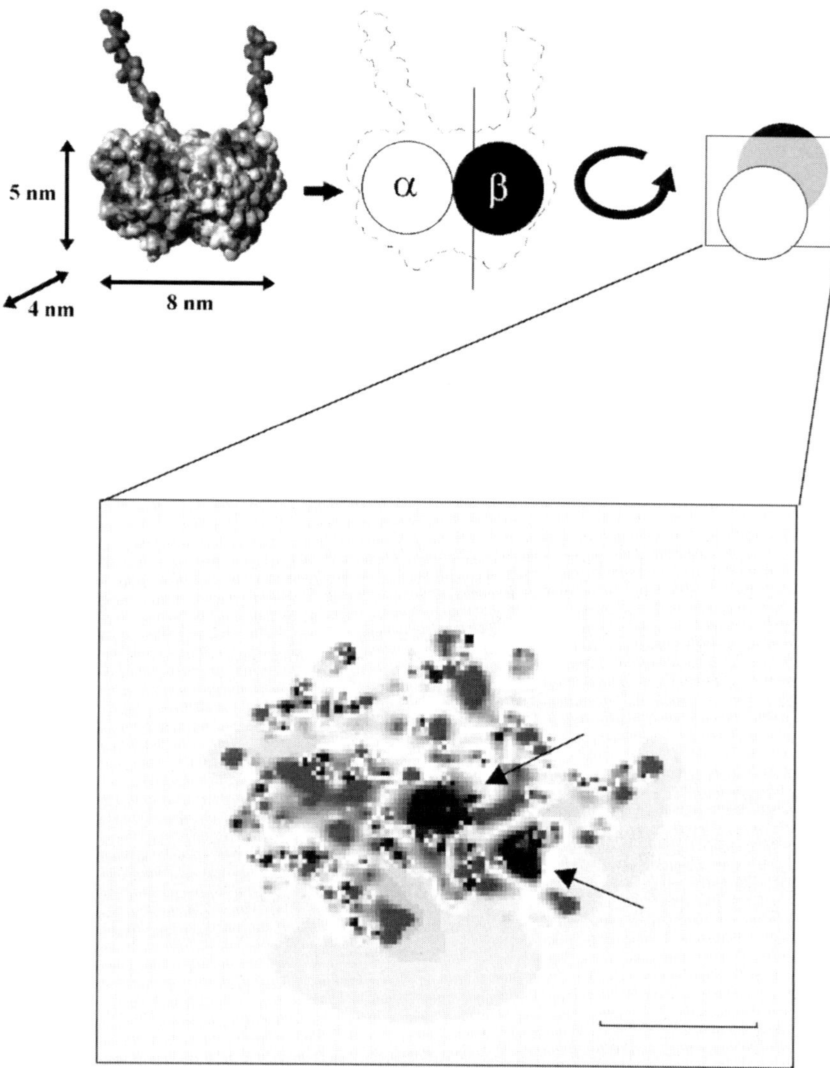

Fig. 6.1. Double-well potentials existing in tubulin dimers, such as illustrated here, enable mobile electrons to tunnel across the energy barrier in accordance with influences exerted by electron positions in surrounding tubulin dimers. Adapted from [51].

argument in favor of quantum models of mind, however. The pragmatic advantage to a quantum model is a higher precision mathematical representation of a cognitive state. Quantum models of mind will ultimately succeed or fail on the basis of their ability to precisely characterize mental states. Accordingly, mathematician Alwyn Scott argued that a neuron might rely on quantum effects in order to attain non-deterministic computational capacities [58]:

> the dynamics of a neuron might be influenced by quantum effects, either in the synapse or the microtubular interior, implying that output signals could be related to input signals only in a probabilistic manner.

While quantum models might be considered exotic compared to deterministic neurocognitive models based strictly on neurophysiological responses, it is particularly those quantum models of mind that incorporate microtubules (or other macromolecules) in neurons that are valuable as they attempt to explain how probabilistic quantum mechanical events on the nanoscale (and below) might produce macroscopic coherent states.

Quantum entanglements (i.e., a connectedness of states, such as electron spins) have been proved theoretically possible for macroscopic systems, even at high temperatures [59, 60]. Moreover, there is theoretical indication the quantum level entanglements may be crucial to macroscopic coherent states in certain solids [61]. Nonetheless, the existence of these macroscopic entanglements remains to be demonstrated in practice. To successfully bridge the gap between nanoscale biomolecular and large-scale neurophysiological events in brain, even conceptually, would provide much needed support for a non-trivial role for quantum mechanics in neurocognition. It would appear, however, that much more theoretical and experimental research needs to be done, given that there is still heated debate as to whether non-trivial quantum effects (e.g., those contributing to information processing) play a role in any aspect of biology [62].[2] In order to place quantum models of mind on firmer ground, both adherents and critics need to remain skeptical and cautious, yet open-minded enough to allow further exploration and testing of these potentially exciting ideas.

There are two broad reasons to continue refining conceptual models of quantum computation in brain and experimental models of classical computation in neuronal microtubules. The first reason is that since all protein structure and function operates according to quantum mechanical principles, only quantum mechanical level analyses of the biophysics of complex neural functions will yield the most accurate depiction. Quantum mechanical and molecular dynamics modeling have been applied to the study of receptor proteins and

[2] Possible quantum mechanical level biological functions include photoisomerization in the retina, energy transfer in photosynthesis, and light-harvesting in bacteria [63].

ion channels, with the consensus being that quantum mechanical properties fundamentally govern how ligands bind to receptors and ion channels operate [64]-[67]. Future studies employing molecular modeling are expected to increase our understanding of how quantum mechanics dictates structure and function of additional classes of proteins found in brain. Since biomolecular activities are core constituents of macroscopic brain behavior, it follows that quantum mechanics will play a role in the latter, and according to some theoretical speculation, a non-trivial role at that [68]. If we presently understood the biophysical underpinnings of neurocognition at the level of classical physics, then quantum mechanical principles might merely refine our existing models. This is not the case, however. Instead, there is a great deal currently unknown regarding how the biophysical properties of macromolecules in the neuron give rise to neurocognition (not merely action potentials), even at the level of classical physics, leaving open the possibility that quantum mechanical principles may be fundamental to the grand scheme of macroscopic brain function. The second reason for pursuing quantum mind models has less to do with elucidating individual quantum mechanical operations (many of which may have negligible impact at the macroscopic level), but instead focusing only on those quantum mechanical properties or effects that contribute to macroscopic coherent states. Macroscopic quantum coherent states have mainly been addressed at the conceptual level [69], and progress regarding the possibility of these states occurring in brain is likely to be limited to conceptual level analyses for some time. However, it is possible that probabilistic quantum mechanical states intrinsic to various biomolecules in neurons contribute to coherent macroscopic states, which are not necessarily quantum in nature. Thus, scientific investigation might be profitably aimed at elucidating potential quantum mechanical operations contributing to microtubule computations, particularly those that affect propagation of signals along microtubules and eventually the macroscopic states deriving from the propagation of those signals.

Macroscopic quantum coherence has been proposed to occur in other cytoskeletal complexes. Matsuno has argued for entangled quantum coherence attaining macroscopic proportions during the slow hydrolysis of ATP, which drives the sliding of actin-myosin complexes with muscle contraction [70]. According to his calculations, actin-myosin complexes (which are also found in nerves) can reach local temperatures as low as 1.6×10^{-3} K. Matsuno speculates that ATP and the actin-myosin complex act as a heat engine/heat sink, which converts heat energy into mechanical energy while maintaining a constant velocity due to quantum mechanical coherence and entanglement.

Explaining the complex basis of macroscopic coherence in brain remains one of the ultimate frontiers in neuroscience, and satisfactorily accomplishing this task will greatly facilitate understanding of higher neurocognition [71]. To the extent that macroscopic coherent states might have a quantum mechanical basis or source, quantum mechanical models of higher cognition have potential relevance to quantum computers and to NEMS devices. Nanoscientists interested in quantum computers can garner inspiration from biomolecular operations in neurons and neuroscientists interested in how the brain computes can find potential mechanisms embedded in select quantum computer designs. In addition to quantum models of mind that directly implicate microtubules, there are quantum mind models that focus on neurotransmitter release from presynaptic terminals based on Heisenberg's uncertainty principle [72] and quantum mind models that claim to better account for neural plasticity than does classical physics [73]. These latter hypotheses indirectly implicate the cytoskeleton, insofar as microtubules and actin filaments are well known to underlie neural plasticity and may also regulate neurotransmitter release [74, 75].

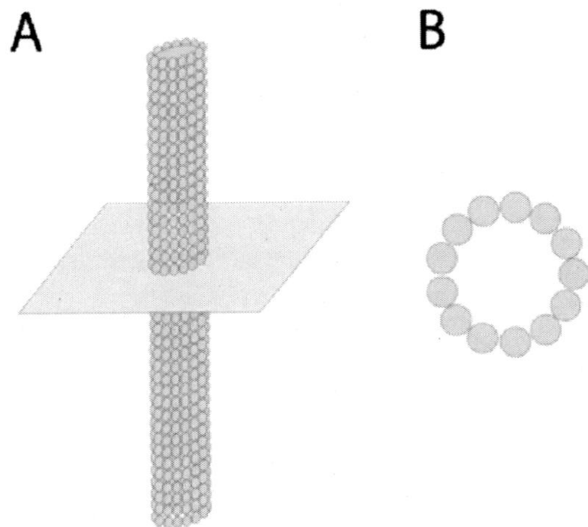

Fig. 6.2. (A) Microtubule cut in cross-section. (B) View of 13 protofilaments arranged around a hollow center.

Are There Quantum Computers That Resemble Microtubules?

As discussed in previous sections, microtubules propagate electric signals consistent with both classical and quantum physical mechanisms and theory. Moreover, the manner in which tubulins are strung together into long polymer chains facilitates individual units acting coherently. In addition to the physical properties of microtubules discussed thus far, electron spins or electric dipole moments in tubulin monomers could participate in neural computation [76]. Each tubulin dimer would be expected to compute in association with its nearest neighbors, and that computation could in turn be further propagated. There are many possible directions for propagation because the structure of microtubules is a lattice, consisting of thirteen individual strands called protofilaments, which surround a hollow center (see Figure 6.2). This architecture might enable quantum computing, provided that a number of stringent criteria are met.

Quantum computers are built to perform computations using qubits, rather than bits. Whereas a bit is either "on" or "off", a qubit can also assume a superposition state of being both on and off simultaneously. This superposition state enables massive parallel computation along with quantum wave interference, which speeds up processing for certain kinds of tasks (e.g., object recognition). Object recognition is a difficult task for conventional digital computers, and it complexly depends on both rate coherence and event coherence [77]. Computational modeling of visual recognition (among other cognitive tasks) represents a large component of neuroscience, and it is likely that an increasing number of computational neuroscientists will model specific brain functions as quantum computations. Modeling neural functions based on quantum algorithms (e.g., those involving large retrieval searches) need not await the availability of quantum computers. Tasks such as high-resolution image analysis can be computed on classical Turing-machine computers implementing quantum logic gates or quantum wave-based holography [78]. In addition to object recognition, quantum computers are expected to be vastly superior at calculating quantum mechanical effects in biomolecules, encryption, and solving certain classes of problems (e.g., optimization) that prove intractable on conventional digital computers. Quantum computers having 100 qubits, for example, should be able to outperform present-day supercomputers at modeling of molecular dynamics [79].

Progress in building potentially viable quantum computers has in some respects been swift, yet in other respects slow – due to instrumental steps not having been realized and significant obstacles restraining this technology. A commercial spin-off company called D-Wave Systems (that originally grew out of research efforts at the University of British Columbia in Vancouver), in a collaborative effort with Google, announced a 28-qubit quantum computer in 2007. This same company has reported future plans for building prototypes for a 512-qubit and a 1024-qubit quantum computer in the near future [80]. All this has transpired only several years after a quantum connection among five photons crossed the threshold needed for distributed quantum information processing [81]. Nonetheless, it may be premature to expect commercially available quantum computers anytime soon. Computer scientist Scott Aaronson claims key breakthroughs in physics are still needed in order to develop practical quantum computers [82].

In addition to increasing the number of qubits, researchers have been working on quantum gates. Solid-state qubits composed of semiconductor quantum dots can be induced to interact via dipole-dipole interactions thereby attaining coherent states [84]. Since quantum dots are capable of self-assembly, therein lies a potential way to manufacture a quantum computational devices deriving from this technology. Theoretical calculations have shown it is possible to polarize coherent spin among quantum-dot/carbon-nanotube assemblies [85]. In one quantum computer design utilizing carbon nanotubes, Buckminster fullerenes are loaded along the longitudinal axis of the slender nanotubes [86, 87]. Another way to align qubits is in an ion trap quantum computer, which consists of strings of ions on linear grids [88, 89]. Single electron spins that have been trapped in quantum dots exhibit long coherence times appropriate for quantum computing [90]. Moreover, information carried by electron spins can be stored more permanently (T $>$ 1 sec) as "quantum memory" by nuclear spins [91].

Biomolecular computing is another approach having potential overlap with quantum computing. Both DNA and DNAzyme computing has been implemented outside the cell [92, 93] and information transfer from DNA to protein is likely to be quantum mechanical [94]. These results indicate a great potential for biomolecular computing, despite obstacles such as the fragile nature of DNA and normal turnover of proteins. A major advantage to biomolecular computers is their capacity for self-assembly. Microtubules rapidly self-assemble, and they are remarkably similar to carbon nanotubes in terms of their biophysical characteristics [95]. This comparability suggests that microtubules could also serve as quantum connects linking longitudinal axes of spin, electron position, or other quantum computational processors, as has been previously proposed [96, 97].

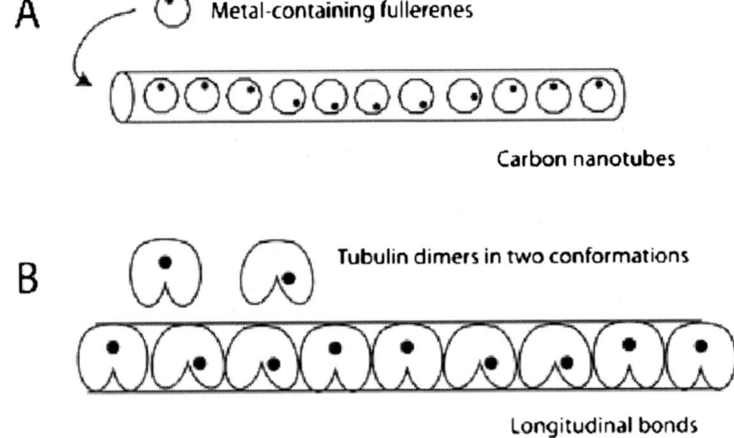

Fig. 6.3. (A) Single-walled carbon nanotubes filled with endohedral fullerenes can embody quantum information in the form of electron spins (adapted from [86, 87]). Small black dots represent cerium atoms, which show irregular collective oscillations. Electrostatic dipole interactions coincide with orientation shifts of the cerium-filled fullerenes. (B) Tubulin dimers align longitudinally in microtubule protofilaments. Two different conformations of tubulin, coinciding with the switching of electric dipole moment, are suggested as the basis of computations. Small black dots represent electrons in a hydrophobic region of the protein, hypothesized to determine protein conformation and electric dipole moment.

Certain recent designs for quantum computers such as Buckminster fullerenes loaded into carbon nanotubes (or strings of ions on linear grids) arguably bear some resemblance to nature's design of a microtubule protofilament consisting of multiple tubulins strung in a row (see Figure 6.3). In this illustrative example, quantum computations depend on electric dipole interactions of Ce atoms embedded in Buckminster fullerenes, with carbon nanotubes serving as extended quantum interconnectors. The function of the carbon nanotube is to carry electrons over long distances without disrupting the direction of magnetic spins among caged atoms. Similar to the way in which the cylindrical carbon nanotube serves as a giant quantum interconnector, the lattice arrangement of tubulins in a microtubule polymer may be capable of facilitating coherent electrostatic coupling between adjacent tubulins [96], and the counterions adhering to the outer surface of microtubules or lining the surface of the interior cavity may further enhance such couplings. The arrangements of tryptophan and histidine amino acids in tubulin give rise to longitudinal and helical conduction pathways, allowing for potential "through bond hopping" that can occur over distances of up to 3 nm [97]. MAP2 bridges or other proteins spanning adjacent microtubules could provide additional means for electrostatic interactions and classical connections between microtubules [98], such that separate microtubules would share the necessary input signals conducive to triggering coherent quantum states. Although highly speculative in nature, quantum interaction at a distance has precedent. Atomic based qubits interact over significant (one-meter) distances without any connectors other than streaming photons [99]. It has previously been proposed that photons might travel down the interior cavities of microtubules [100], but this proposal has received less attention than other models of quantum computing in microtubules.

Microtubules in neighboring neurons have also been proposed to send signals to each other through gap junctions [101]. It remains to be determined whether it is possible to transmit classical or quantum information through gap junctions. What is clear, however, is that microtubule matrices in large ensembles of neurons exhibit collective plasticity insofar as there is simultaneous upregulation or down-regulation of proteolysis (i.e., protein degradation) of MAP2 (see Figure 6.3). This is consistent with microtubule computations being under global control, which could be sustained by both diffusely projecting neural systems regulating attention (e.g., acetylcholine and norepinephrine acting as neuromodulators), as well as the more speculative non-local effects of quantum computations [102]. This collective proteolysis of the cytoskeleton occurs within discrete modules across neocortical and hippocampal fields and it corresponds with recent higher cognitive function. Both classical conditioning and passive avoidance learning leads to this orchestrated proteolytic response as has been demonstrated experimentally in discrete modules of association auditory cortex and in CA1 and CA2 subfields of hippocampus [6]-[8], as well as in other brain areas [103]. As shown in Figure 6.4, enhanced MAP2 immunohistochemical staining indicative of collective proteolysis abruptly stops at the boundaries of cortical modules.

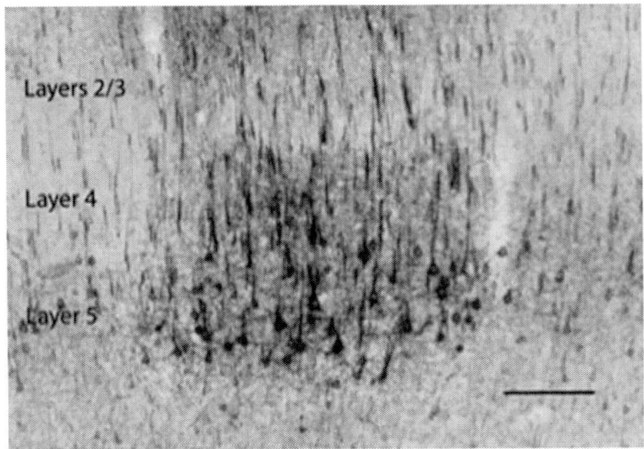

Fig. 6.4. MAP2 proteolysis is upregulated simultaneously in multiple neurons within discrete cortical modules as a result of animal learning. Coherent biomolecular events affecting the entire group of neurons are likely to underlie this large-scale proteolysis as well as the concurrent reorganization of the architecture of the neural assembly. Scale bar equals 100 m. Adapted from [102].

Tubulin as Nature's Qubit?

The proposal that the microtubule serves as a quantum connector only succeeds to the extent that qubits are indeed linked by such a connector. As suggested by Hameroff in his 1987 book Ultimate Computing: Biomolecular Consciousness and Nanotechnology, tubulin itself might act as a qubit underlying biomolecular consciousness [19]. A qubit replaces the bit (or switch) in the classical Turing-machine computer, and is far more powerful than a bit because it can be on, off or both. A qubit in two or more states simultaneously is in a state of superposition, such that superpositions might correspond with computations occurring before a conscious state is decided. Qubits can also interact so strongly as to be considered essentially one – coalesced into a coherent state of entanglement. When a group of entangled qubits is finished computing, the outcome collapses into a single quantum waveform common to all participating qubits. What causes collapse of the quantum waveform remains to be conclusively determined.

Consistent with the notion that biomolecular computations by tubulin underlie consciousness are the findings that the inhaled anesthetic, halothane, binds to both α- and β-isotypes of tubulin [104, 105]. Moreover, exposure to halothane and isoflurane was shown to affect the subsequent expression of tubulin proteins in rat primary cortical neurons, suggestive of a perturbation of those proteins [106]. Based on numerous experimental results from his laboratory, Ekenhoff argues that inhalation anesthetics as a class of drugs most probably exert their effects by occupying hydrophobic pockets measuring ~150 cubic Å, which lie buried inside approximately 10% of soluble proteins in the neuron [107]. When the anesthetic molecule occupies the cavity, this blocks changes in conformational states that are critical to the underlying functions of that protein.

The experimental results from the Ekenhoff laboratory detailed above support the tubulin qubit model of consciousness, although the involvement of other proteins in the mediation of consciousness is not ruled out. A critical aspect of the tubulin qubit model is that a quantum computation determines the conformational state of tubulin. This is consistent with a small part of the tubulin molecule, located in a hydrophobic pocket, controlling its overall 3-dimensional conformation. A hydrophobic pocket has been shown to exist in tubulin near to its taxol-binding site [108]. Taxol, when bound to a microtubule, stabilizes the structure of that microtubule, further demonstrating this hydrophobic region is involved in determining the 3-dimensional conformation of tubulin. Other hydrophobic regions are positioned at the boundaries of different secondary structures of tubulin, suggesting that collective alterations at these hydrophobic regions choreograph the slower, collective movements of the tubulin dimer [109]. At least six sites of mobility are predicted and three kinds of motion. Motions include (1) oppositely directed twisting of the α-tubulin and β-tubulin monomers, (2) bending of the bond between the α–tubulin and β–tubulin monomers, and (3) extending and compressing the length of the tubulin dimer. The binding of MAPs, kinesin and GTP or GDP, as well as lateral and longitudinal contacts between other tubulin dimers, can limit these motions.

More crucially, two stable conformational states hypothesized for tubulin have been validated by direct experimental evidence. Santarella and colleagues showed that tubulin exists in two broadly different conformational states: one that prefers to bind the MAP tau and the other preferably binding kinesin [110]. Since tau binds to the same microtubule site, as does MAP2, it is possible that both tau and MAP2 regulate the conformation of tubulin and that in turn conformation shifts in tubulins determine tau, MAP2, and kinesin binding to the microtubule. Any autonomous or self-organized behavior on the part of microtubules endows the neuron with a capability to regulate its own neural plasticity and transport. Moreover, quantum mechanical events would be expected to contribute a probabilistic element to those functions via their influence on the 3-dimensional conformational state of tubulin.

Decoherence Issues: Are Quantum Mechanical Models of Microtubule Computation Viable?

While there is direct experimental evidence that tubulin can exist in at least two different 3-dimensional conformational states that directly relate to its function and that a quantum computation controlling the conformation of tubulin would be expected to control any function governed by that conformation - this is still not proof that a quantum mechanism in microtubule function contributes to higher cognition. Nonetheless, such a scenario is possible, if not somewhat likely, given that higher cognition depends on neural plasticity and transport. As already discussed, the notion that protein conformations are governed by quantum mechanical properties of the atoms comprising those molecules is experimentally supported. Hence this part of any quantum model that posits 3-dimensional conformational shifts in tubulin sits on firm scientific ground. That any quantum entanglement lasts long enough to contribute to neural function is a different issue, and a major argument against quantum models of mind is that decoherence produced by the brain's environment is sufficient to destroy entanglements that might meaningfully contribute to neural computation [53]. Assuming that the brain consists of one homogenous environment might appear to doom quantum mind theories; however, a number of experimental systems exhibit "quantum decoherence-free subspaces" [111, 112]. The existence of quantum decoherence-free subspaces in and around microtubules could satisfactorily address the decoherence issue. Moreover, calculations put forth to prove quantum mind theories impossible appear to be fundamentally flawed.

Hagan and colleagues point out a number of errors in the calculations given in an attempt to disprove quantum computation in brain microtubules, and further describe various shielding and error correction devices that would be able to isolate microtubules from a noisy environment inside the cell [113]. Typical gel-sol state alterations of the living cell occur due to actin gelation cycles, during which the cytoplasm temporarily alternates between a quantum isolating and a non-isolating state. Also, the counterions on the surface of microtubules would be expected to enhance coherent quantum states. The lattice arrangement of tubulins in microtubules may also allow for topological error correction. There are also the intrinsic properties of tubulin to consider. Based on the computed electric dipole moments of tubulin monomers and dimers, it is reasonable to conclude that quantum computations, by biomolecules such as tubulin, are possible at room temperature under special conditions [23]. This is because microtubule is not a steady-state system; instead biochemical energy is constantly being pumped in and out. Lasers similarly are not steady-state systems and for that reason are capable of quantum coherence at room temperature. Moreover, there is evidence that quantum conductances can occur along carbon nanotubes doped with organic molecules at room temperature [114].

In addition to shielding and isolating mechanisms, something called the quantum Zeno effect enables entanglements to persist when a system is strongly coupled to its environment [62]. Thus, the quantum Zeno effect could account for quantum states of atoms or electrons in tubulins being strongly coupled to their environments extending along microtubule protofilaments, whole microtubules, or even microtubule-MAP matrices. Given possible shielding mechanisms, quantum Zeno effects, and other mechanisms, quantum computations in microtubules at room temperature and entanglements resulting from those computations could exist in time frames relevant to neurocognition. Nonetheless, as long as the matter of how delicate quantum states could avoid rapid decoherence remains unresolved, the burden of proof lies on those who propose quantum mechanical models of higher cognition.

Regarding the underlying mechanism by which information could be stored, it has been recently shown that excitons can be stored on semiconducting quantum dot nanostructures for periods of time exceeding 30 msec at 7 K [115]. This appears to be due to a slow tunneling of the electron away from the electron hole, the separation of which creates an intrinsic dipole. Tunneling of mobile electrons in the interiors of tubulin dimers has been suggested to occur (see Figure 6.1). To the extent that the shielded environments and topological error corrections enable electron tunneling of this type to operate at time intervals relevant to neurocognition remains to be determined [51].

Spin transfer, which involves the angular momentum of an electron changing location, provides another potential means of quantum computation that could be implemented by tubulin proteins. Spin transfer between molecularly bridged quantum dots has been measured at room temperature [116]. While it cannot be ascertained whether these instantaneous entanglements have the potential to affect events at time scales relevant to neurocognition, some researchers have hypothesized a nuclear spin ensemble among neural membrane constituents and proteins as critically underlying neurocognitive states [117].

Although a collapse of the quantum superposition has been suggested to correspond with consciousness [20, 21], it is also conceivable that quantum coherence itself is the closer correlate of the conscious state, particularly the contents of consciousness. This conceptual scheme fits with contents of consciousness "binding" together different sensory elements. Binding is a phenomenon described in the psychology literature that combines different aspects of an object (shape, color, texture) into a singular entity that enables observers to distinguish the object from the background [118].

To the extent that coherence represents a potential correlate of contents of consciousness, collapse (or decoherence) would correspond with the transitions or pauses between those conscious moments. Accordingly, decoherence due to interaction with the environment (a realty that almost all physicists agree upon) becomes an essential ingredient rather than a deterrent to the model. The question remaining is does such an option facilitate the bridging of the gap between quantum mechanical level events and macroscopic coherence? An advantage to considering quantum coherence as a potential correlate of consciousness is that it arguably removes any time scale requirement; the coherent state could be exceedingly brief. Even though longer coherence times are increasingly being measured (e.g., the electrical detection of spin coherence in silicon has recently been measured to be longer than 100 μsec [119]), it remains to be experimentally determined how long coherence can endure in brain – even in protected "decoherence-free subspaces". Many models of consciousness describe this phenomenon as a discontinuous process [120]. Conscious awareness exists as brief discrete events, but there is an illusion of continuity. Accordingly, there is not an accurate estimate for how long individual discrete episodes of phenomenal consciousness (simple experience) last; there is only data pertaining to conscious cycle intervals. It follows that quantum coherence-based phenomenal consciousness could occur on time scales that were significantly briefer than those of neurophysiological events that correlate with conscious cycle intervals.

To illustrate this conceptual scheme consider language processing of morphemes, which are the smallest units of meaning (i.e., prefixes, word roots, and suffixes). At the onset of conscious processing of the word "predisposed" the prefix "pre" activates appropriate interneuronal networks of neurons and their intraneuronal networks of microtubules and tubulin qubits. Presumably towards the end of an information processing interval, quantum coherence is briefly achieved in a requisite number of tubulin qubits, but rather than decohering abruptly (since "pre" means nothing in particular by itself), the coherent state decoheres somewhat gradually as a new set of interneuronal networks of neurons and their intraneuronal networks of microtubules take over processing the second prefix "dis" in its appropriate context. This is repeated (for the word root "pos" and the suffix "ed") until the end of the word is reached, and only then is there a relatively abrupt decoherence.

The point is that a conceptual quantum computing model that incorporates decoherence, rather than seeking to avoid it, is able to utilize subtle variations of decoherence as an operational transition device linking successive states of consciousness in different ways. Although this conceptual scheme lacks an absolute need for a distinct collapse mechanism, an intrinsic collapse mechanism could alternatively work in this model, as long as it could be varied.

6.3 Classical or Quantum Computations As Autonomous Mechanisms Directing Transport

As elaborated upon in the previous section, a classical or quantum computation in a tubulin monomer would be expected to physically alter its 3-dimensional conformational state. Changing the conformation of tubulin alters its binding to MAPs and motor proteins (as well as to its neighboring tubulins). As a result, the output of any classical or quantum computation will in

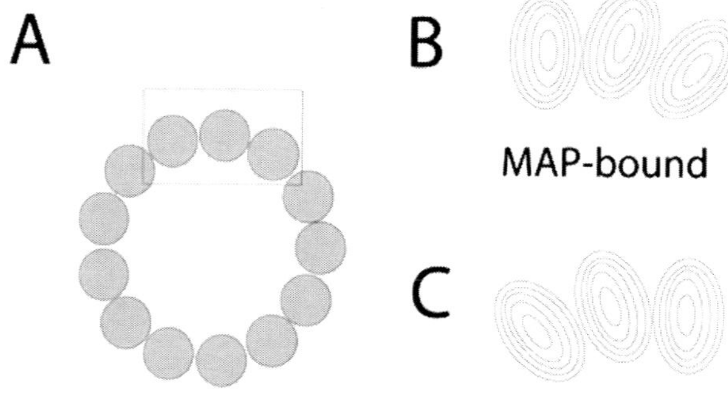

Fig. 6.5. Tubulin dimers in cross-section (A) are found to have a clockwise orientation when bound to MAP (B) and a counterclockwise orientation when bound to kinesin (C). The perspective is the plus-end of the microtubule pointing up. Schematic diagram based on 3-dimensional reconstructions in [110].

turn influence motor protein attachment and progressivity. Controlling transport of synaptic proteins and mRNAs to spines would in turn affect synaptic efficacy and thereby alter subsequent responses to inputs. Accordingly, classical and quantum mechanical computations performed in brain microtubules (and presumably also in actin filaments) have a logical and scientifically supported route to regulate synaptic activity and consequently implement higher cognition.

If we view a microtubule in cross-section (see Figure 6.6), the part of the tubulin that juts out is called the protofilament edge. Studies performed in the Mandelkow laboratory using high-resolution, metal-shadowing, cryoelectron microscopy have shown that the microtubule protofilament edge can lean clockwise or counterclockwise. Different binding sites are exposed depending on the way the tubulin dimer leans. One side of the protofilament binds the stabilizing MAP tau; the other binds the motor protein kinesin. MAPs bind aperiodically at random distances, and the large size and floppiness of tau and MAP2 preclude another protein from binding to sites nearby. Accordingly, the binding of MAPs can under some circumstances interfere with kinesin attachment [121]. Taken together, these results support the notion that quantum computations and their subsequent effects on motor protein binding stand to markedly influence the rate of transport along the microtubule (or actin filament) because transport depends pivotally on motor protein attachment.

There is further experimental evidence suggesting that computations in microtubules would be in a position to fine tune the rate and direction of transport, thereby providing the neuron with an autonomous means to change synaptic efficacy at select sites and alter their responses to future inputs in a very specific way. Transport of NMDA receptors along the microtubules occurs intermittently with cycling of the receptors to the membrane, a process that can be stochastically modeled [122]. In what is called a diffusion-trapping model, the mushroom-shaped spines serve the function to trap receptor molecules; this model allows calculations to be made regarding the timing of transport of receptors [123]. In addition to being transported by cytoskeletal proteins, receptors are also stabilized into clusters in the membrane by these proteins. The NR3A subunit of the NMDA receptor associates with MAP1S, for example, which regulates the transport and function of the NMDA receptor [124]. Rates of glutamate receptor transport have been measured. NMDA receptors travel to nascent synapses at rates of roughly 4 μm/min, and AMPA receptors travel somewhat more slowly [40]. Trafficking of glutamate receptors following LTP depends on both rapid mobilization and delayed protein translation [125], mechanisms that are directly mediated or indirectly modulated by microtubules or actin filaments (see Chapter 4). Kinesin-mediated transport also occurs for synaptic proteins such as post-synaptic density protein-95 (PSD-95) [126], as well as for mRNAs responsible for translating signal transduction molecules critically involved in learning and memory, such as activity-related cytoskeletal protein (Arc), NMDA receptor, Ca^{2+}/calmodulin-dependent kinase II (CaMKII) and MAP2 [127]. Local translation of these proteins within the spine occurs following shifts of polyribosomes from dendrite shafts to spines [128]. Such results show clear-cut nanomechanical rearrangements with learning that could be orchestrated by computation mechanisms intrinsic to cytoskeletal proteins.

There are a number of specific computational mechanisms available to microtubules (and actin filaments) to influence trafficking of synaptic proteins and mRNAs. For one, classical or quantum mechanical computations resulting from alterations in the 3-dimensional conformations of tubulins could alter the timing or direction of transport by way of influencing attachment of motor proteins such as kinesin and dynein – or by modifying their processivity. Second, structural changes to the cytoskeletal matrix would alter the direction of transport. Third, transport could be mediated by changes in C-termini states of the tubulins affecting kinesin-based movements along microtubules [129, 130]. Computations based on classical or quantum mechanical properties of tubulin (or actin) would endow the neuron with the capacity for autonomous control over the way it would respond to future synaptic inputs. The neural response to a familiar pattern of inputs would be a collective pattern of transport – or a "quantum mechanical correlate" of tubulin conformational states-providing a unique signature for a specific cognitive state. Such a quantum mechanical correlate of a cognitive state may give a more accurate representation than a neural correlate provided by a distributed pattern of neurophysiological firing among ensembles of neurons, but arguably both descriptions should ultimately be related to one another.

6.4 Nanoneuroscience and the Theoretical Physical Basis for Mind

The mind as studied by psychologists and neuroscientists has conscious and unconscious components – the subjective nature of consciousness being the most difficult to explain in scientific terms. Nanoneuroscience may be able to provide the necessary tools to describe how individual molecules and atoms in brain can contribute to large mesoscopic or macroscopic coherent states. Whether some of these states can be influenced by quantum mechanics in time frames coinciding with cognition remains to be demonstrated first theoretically, and ultimately proven experimentally using nanoscientific methods.

Many scientists have argued that clear examples are needed that demonstrate quantum mechanics has the potential to better explain human cognition than classical physics, and is thereby likely to play a role in neurocognition, and as a consequence, in animal behavior [68]. At least three such examples exist. (1) There is a lack of a clear time arrow in certain mental activities, which is more comparable to the situation in quantum physics than in classical physics. Only in the imagination or in dreams can one travel both backwards and forwards in time instantaneously. While this may seem a trivial example on the surface, many basic psychological functions (i.e., remembering, anticipating, planning, learning from experience, etc.) are founded on this rather remarkable feature of mind. (2) Behavior cannot be accurately predicted by a simplistic model in which an organism can be expected to exhibit a particular behavior a certain percentage of the time; the present circumstances, prior learning, and future expectations (along with mood, state of arousal, and other internal states) contribute complexly to the resultant behavior. Given that animal behavior is determined by a complex matrix of present inputs, past experiences, and future expectations, the processing of this matrix of information would seemingly be better described by a superposition of multiple probability amplitude vectors, as opposed to simple statistical probabilities. Moreover, quantum interference can amplify correct outcomes, which is consistent with adaptive animal behavior being efficiently and rapidly produced. (3) Similar arguments can be applied to neurocognitive states. Higher neurocognitive states are selected on the basis of present inputs, prior learning, and future expectations; moreover, there are significant effects due to mood, state of arousal, and focusing of attention. Selecting a neurocognitive state would seemingly be accurately described by a superposition of multiple probability amplitude vectors. Moreover, that quantum interference amplifies correct outcomes (and dampens incorrect outcomes) could account for the efficient and rapid determination of a given neurocognitive state.

Nanoneuroscience enables the study of flow and storage of information in the networks or matrices of microtubules and actin filaments (along with MAPs, motors, scaffolding, and adaptor proteins) that span virtually every internal compartment of neurons. The fundamental physical properties of the biomolecules of the intraneuronal matrix operate according to classical and quantum mechanical theories. The big question is: Do we understand these properties well enough to build conceptual models of perception, consciousness, and quantum associative memory?

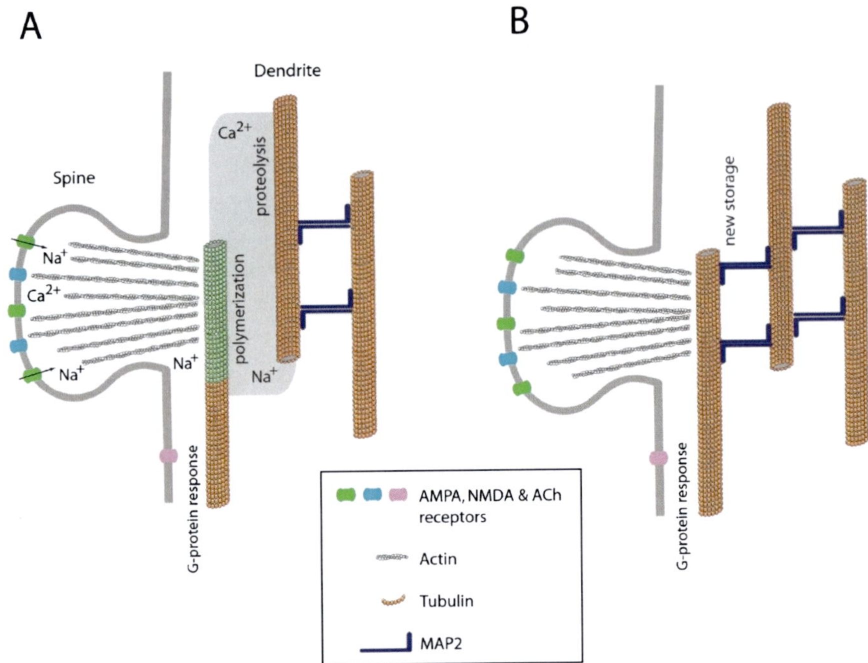

Fig. 6.6. Synaptic input to the spine and to the underlying subsynaptic zone. (A) AMPA, NMDA, and acetylcholine receptor-activation stimulates spine and dendrite membrane. actin filaments connect glutamate receptors with underlying microtubule in an unstable state (i.e., undergoing polymerization/depolymerization). (B) Restructured synapse is stabilized by MAP2 bridges between newly polymerized microtubules and adjacent microtubules.

6.4.1 Putative Steps for Classical and Quantum Information Processing in the IntraNeuronal Matrix

In conventional neural computing models, inputs arising from sensory stimuli impinge upon a neural network that is organized as a system of neurons wired or connected together. Sparsely interconnected networks having excitatory and inhibitory connections can exhibit dynamic oscillatory activity (80c). Nonetheless, the neural network output depends heavily on input patterns and there is no clear mechanism whereby a neural network could self-initiate a particular output pattern *de novo*, even though such a mechanism would seemingly need to exist in order to account for particular higher neurocognitive states, such as creative thinking, planning, decision making, and initiating action.[3] Conventional neural networks typically incorporate neural plasticity (i.e., changes of synaptic strength). Usually, this plasticity is strictly input dependent, with no proposal for an intrinsic mechanism for initiating neural plasticity based on some cellular or organismic motive. Without proposing an intrinsic mechanism for initiating plastic change, it does not matter how many neurons are interconnected to each other, the overall scheme of the system is lacking in a basis for truly autonomous behavior.

In contrast, the intraneuronal matrix model of neural computing integrates synaptic inputs with neural structure and transport of materials to synapses sustaining synaptic strength. This model directly incorporates an intrinsic mechanism accounting for self-organization and autonomy, as well as for storage of spatio-temporal patterns of information in a matrix of cytoskeletal proteins. Conventional models of long-term memory based on enhanced synaptic efficacy (i.e., physiological potentiation) as the primary storage mechanism have a built-in saturation point or ceiling. During the course of learning in behaving animals, synaptic potentiation frequently reaches this ceiling and conversion to structural synaptic reorganization has been hypothesized to enable subsequent learning [132]. Storing information in the intraneuronal matrix (as compared to storing information as potentiated synapses or altered connections) provides an alternative in which synapses need not remain potentiated, but rather could return to baseline conditions following transfer of information to the cytoskeletal matrix. This proposal is consistent with reports of potentiated synapses returning to baseline within days after learning [133, 134] and the dendritic structure of neurons becoming more and more elaborate with learning and experience [135]. Moreover, the present proposal is consistent with evidence of more limited potential for restructuring of cortical circuits among primates compared to rodents [136], whereas cognitive aptitudes are clearly the opposite (i.e., human and non-human primates can acquire more complex information than can rodents). Human pyramidal neurons are larger, have more extensive dendrite arbors, and possess greater spine densities compared to mouse pyramidal neurons [137]. Thus, pyramidal neurons of the human cerebral cortex possess extensive intraneuronal cytoskeletal matrices capable of information processing and storage. It is possible that this feature contributes to their prominent role in neurocognitive.

As shown in Figure 6.6, the linkage between the synapse and the intraneuronal matrix is as follows: (1) Sensory inputs relay information to synapses located on spines activating AMPA and NMDA receptors. (2) Attention systems release acetylcholine or other neuromodulators on dendrite shafts near to spines. (3) Actin filaments transmit signals to microtubules in the subsynaptic zone that are undergoing polymerization/depolymerization cycles. These microtubules are in turn linked to other microtubules by MAP2

[3] A counter point of view is that volition and other seemingly self-initiated acts are illusory.

connecting various synapses of the neuron together (as well as connecting to the cell nucleus and axon terminal).

Experience (e.g., with learning) would be expected to facilitate the restructuring of microtubules, which under the appropriate conditions would become stabilized (i.e., polymerized). Many means are available to stabilize information storage in microtubules. Cross-linking with MAP2 or tau markedly increases the structural stability of microtubules [138]-[140], as do post-translational modifications to tubulin such as detyrosination, polyglutamylation or acetylation – all of which are known to enhance microtubule stability by increasing their binding to MAPs and neurofilaments [141]. Detyrosinated microtubules are so exceptionally resistant to polymerization-depolymerization cycles that they must be fragmented in the presence of calcium before significant polymerization will occur [142]. Normally, calcium ion is compartmentalized in the spine away from microtubules such that stabilization can be maintained. Moreover, microtubules in certain parts of the neuron are more stable than in others, mostly due to variations in post-translational modifications. Highly stable, acetylated microtubules are particularly concentrated in the subsynaptic region of the neuromuscular junction, for example [143]. In the brain, post-translational modifications of microtubules afford an experimental probe into long-term storage of events related to neural plasticity. The abrupt changes in tyrosinated tubulins and acetylated tubulins have been described as giving a "snapshot" of a microtubule's growth history [144]. It is conceivable that in the brain such snapshots correspond to memory.

As previously discussed in earlier sections of this chapter and in Chapter 4, stabilized microtubules may store histories of past synaptic activity as patterns of tubulin conformations representative of transport that occurred along that microtubule. It is also possible that transport routes are stored in the aperiodic patterns of bound tau and MAP2, since these stabilizing MAPs affect motor protein processivity. Kinesin attaches more rapidly to microtubules having bare stretches lacking stabilizing MAPs [121]. Acetylation of tubulin also increases kinesin binding [145]. The faster the attachment, the more cargo (i.e., synaptic proteins and mRNAs) will arrive at the subsynaptic zone. Thus, histories of previous transport rates (and shortest transport routes) might be embedded in patterns of these cross-linking proteins-tau and MAP2.

To the extent that the intraneuronal matrix provides a storage mechanism for familiar patterns of input, this matrix would presumably be activated by that same input pattern, similar input patterns, partial input patterns, or the neuron could autonomously activate that information spontaneously. As signals repeatedly circulated in the intraneuronal matrix (in accordance with re-entrant signaling extending to other microtubule segments located in other dendrites or neurons), these extended patterns would be expected to become stored at multiple sites in the intraneuronal matrix at multiple brain sites. The entire dendrite arbor and cell body contain microtubules capable of storing information; nonetheless, a wealth of information relevant to an individual synapse is likely stored in the subsynaptic zone just beneath the synapse or the spine. Spine necks range 100 - 300 nm in width (calculated from data in [47]). If each computational segment of a microtubule is the approximate length of the width of a spine neck, that would correspond to approximately 160 - 500 tubulins having 2^{100} possible computational states being able to calculate and store information about that individual synapse.

Storage of information directly beneath a spine would provide a means to restructure that spine to its original form if it were transiently altered. Cooling of brain tissue results in the retraction of spines and loss of cytoskeletal proteins, which subsequently reappear in the same location upon warming the tissue [146]. The intraneuronal matrix of microtubules located in the subsynaptic zone beneath the spine could be the neural compartment that stores the essential information upon which that reappearance and restructuring is based – enabling that newly restructured synapse to respond to inputs as it had in the past, or to incorporate new information with information about past inputs.

6.4.2 The Intraneuronal Matrix in Perception, Cognition, and Consciousness

Intrinsic microtubule properties (e.g., properties that culminate in self-directed processes such as dynamic polymerization/depolymerization cycles) endow the intraneuronal matrix with mechanisms for self-organization and autonomy. This provides the matrix with the incredible potential for arriving upon mental abstractions that go beyond being representations based solely on inputs arriving from stimuli (i.e., external energies impinging on sensory organs or internal states of the body). It is arguably impossible to account for the biomolecular basis of mind without including a biomolecular mechanism for autonomous behavior.

How does the mind make that giant leap whereupon it no longer simply responds to sensory inputs based on some topographic sensory map? Each synapse in primary visual cortex, for example, represents a point in visuotopic space, nothing more (assuming color and motion are processed in higher visual areas where similar arguments would apply). A point in visuotopic space is not any kind of meaningful representation. Only the relationship between multiple points in visual space can form a meaningful abstraction (such as a vertically oriented line, a horizontally oriented line, a curve, etc.). Thus, the storage of any abstract representation must be of a particular pattern of synaptic input distributed across a neuron or set of neurons. Hubel and Wiesel, Nobel Prize laureates working in vision discovered that complex cells in the visual cortex, which happen to be large pyramidal neurons, were first in the visual pathway to respond to the abstract idea of a line of a particular orientation not strictly bound to a precise location in the visual field [147]. Something that is not derived strictly from the visuotopic map drives these complex cell responses. Is it conceivable that these abstracted responses reflect a coherent spatio-temporal pattern of activity and transport in the intraneuronal matrix of microtubules located across a collective of neighboring pyramidal neurons? Pyramidal cells have basilar dendrites extending in all directions parallel to the surface of the cortex, similar to the fine branches arising from the apical dendrite. Since visual stimuli are processed in the plane parallel to the surface of the cortex [148], visual stimuli consisting of specifically oriented lines would be expected to maximally excite similarly oriented dendrite branches. According to the intraneuronal matrix model, microtubule segments would store that information as a spatio-temporal pattern of synaptic activation in that dendrite segment. But other dendrites would likely be stimulated given that the visual field is seldom stationary. Thus, as a stimulus such as an oriented line moved across the visual field, it would be expected to create multiple storage sites in multiple segments of microtubules among similarly oriented dendrites. Based on the close temporal relationships between these activations, coherent activation of a large ensemble of those microtubule segments would be expected. The largest pyramidal cells, having the highest concentration of microtubules and MAPs are the best candidates for complex cell responses, consistent with what has been experimentally established.

The intraneuronal matrix also offers novel ways to integrate top-down influences into the basic mechanisms of perception and cognition – circumventing the need to have these influences derive solely from outside cortical regions (such as the prefrontal areas of the cerebral cortex). When it comes to object recognition and working memory, recent studies have revealed evidence that top-down influences may not derive exclusively from prefrontal cortex [149], and that associations between objects processed in other brain regions are also critical [150]. Because of increased computational power, collectives of interactive intraneuronal matrices of microtubules are, in principle, capable of storing high-level, top-down influences relevant to perceptual processes executed by primary and secondary sensory cortical circuits. Given the extensive nature of the intraneuronal matrix within each neuron, a functional hierarchy of the type hypothesized for the entire cerebral cortex, could be operative in each individual neuron. Ideally, these two hierarchies would have a means to interact.

The intraneuronal matrix – made up of nanowires capable of controlling the efficacy of ensembles of synapses through coherent activity and transport – provides a way for conscious recollections of the past to be compared to perceptions occurring in the present. According to Jeffrey Gray, once a conscious entity is stored in memory, it can then be compared to present inputs; this is a comparator function built into consciousness [151]. Gray's notion fits perfectly with the intraneuronal matrix model. Any pattern of inputs resembling a previously learned pattern, whole or in part, will result in that same essential "quality" of conscious content (often termed "qualia" by neurophilosophers). When only some of the synapses related to the abstract idea are reactivated (such as when a familiar, yet partially occluded stimuli is presented), the intraneuronal matrix of nanowires would be expected to still fit the best match and connect to the appropriate set of synapses. This is because the intraneuronal matrix of microtubules links different synapses on a neuron together. Rather than each neuron acting as a simple processing unit in a large assembly possessing great computational power, each neuron by virtue of its intraneuronal matrix possesses the computational capability similar to that of a supercomputer with links to many other such supercomputer-like entities.

It is a long-held notion that activity in both localized and distributed assemblies of interconnected neurons is critical to specific functions mediated by the brain [152]. While the intraneuronal matrix model endows each neuron with much more computing power than do conventional neural information processing models, it is nonetheless compatible with some such models. The theory of neural group selection advanced by Gerald Edelman, for example, specifies that alliances of neurons fire in concert based on reentrant interactions within a dynamic core, consisting of thalamic-cortical-thalamic loops [153]. Activity in the dynamic core is reflected in the electroencephalogram (EEG), which detects synchronous brain activity and coherent oscillations. During wakefulness, inputs to the cortex trigger stationary EEG components spanning hundreds of milliseconds, which are necessary for consciousness. A person in slow-wave sleep lacks stationary thalamic-cortical-thalamic responses lasting more than a hundred milliseconds [154], suggesting such stationary responses are necessary to build up coherent signaling in increasingly larger numbers of neurons (and presumably in the intraneuronal matrix of microtubules). Although the relationship between nanoscale (and below) quantum coherence and macroscopic coherent states is far from being understood, macroscopic coherent states, such as those measured in the EEG during sleep and wakefulness would be expected to impact the activity of intraneuronal matrices of cytoskeletal proteins.

According to E. Roy John, consciousness derives from coherent oscillations among distributed local field potentials [155]. Field theories of consciousness, including those that rely on quantum fields, account for conscious wholes or "gestalts" as spatio-temporal patterns of amplitude and phase coherent EEG activity that can extend "from a few hypercolumns to an entire hemisphere" [156]. The relationship between coherent EEG states and coherence among microtubule computations remains to be determined.

Besides being consistent with a number of conventional neural computational models, incorporating the intraneuronal matrix into conventional neural information processing models adds: (1) a vastly enhanced capacity for storing spatio-temporal patterns of neural activity, (2) a source for spontaneous neural activation that is intimately connected to previous experience, and (3) a possible mechanism for the subjective nature of consciousness (e.g., qualia). These potential advantages are speculative in nature, but are consistent with experimental evidence that electric signals are propagated along microtubules and actin filaments, and that these cytoskeletal proteins are dipolar, electromagnetic, ferroelectric, and potentially governed by quantum mechanical forces, which depending on the existence of possible shielding mechanisms, could exert meaningful effects [17, 18, 22, 23, 113].

6.4.3 Quantum Models of Perception, Cognition, and Consciousness

Intense interest in quantum computers and the intrigue of quantum computations in brain have motivated quantum computational models of neurocognitive functions ranging from associative memory and pattern recognition to attention, arousal, and consciousness. These models illustrate some advantages afforded by quantum computing; however, most of these models do not specify any particular biological mechanism.

Khrennikov has proposed a quantum-like model of mind, which differs substantively from "quantum mind" models, and can be implemented using quantum algorithms [157]. According to his model, cognitive quantum-like behavior arises as a consequence of interference of probabilities. The human brain is assumed to contain millions of separate minds or mental states, yet at any given instant can only realize conscious awareness of one mental state, represented as Ψ_C. This quantum-like cognitive model is able to derive a numerical measure of the contextual incompatibility for decision-making tasks having high uncertainty [158]. In another model, called counterfactual quantum computation, the computer is set in a superposition state of "running" and "not running" [159]. Surprisingly, it is sometimes possible to infer parts of the solution counterfactually (i.e., from the as-yet-unknown solution). This approach has been used to implement the Grover's search algorithm and appears to be able to eliminate errors that arise due to decoherence.

Ventura and Martinez propose an associative memory model in which quantum computations are combined with classical neural network theory to produce results exponentially faster than can be obtained with classical neural network models [160]. In addition to being potentially faster, quantum computing-based models are also adept at filling in missing information, as demonstrated by a quantum associative memory algorithm executed by a small number of qubits that reproduces entire patterns when presented with partial input [161]. Part of the advantage of these quantum computing-based models is that they rely on different search modes than do conventional neural computing models. Rather than being limited to content-addressable retrieval processes, quantum memory models can implement probabilistic searches [162]. Moreover, it is not uncommon for models of quantum associative memory to incorporate principles typical of conventional neural computing models to their advantage. As an example, qubits have been modeled as having long-range interactions that operate according to the Hebbian learning rule (i.e., that increased activity strengthens connectivity) [163].

Quantum computing-based models have also yielded positive results relevant to pattern recognition tasks and bistable perception of ambiguous images. A quantum wave implementation applied to image reconstruction enabled high-resolution image retrieval and increased capacity to an almost infinite size [164]. The bistable perception of the Necker cube has been successfully modeled according to "weak quantum theory" and the quantum Zeno effect (i.e., a strong coupling to the environment which decreases decoherence). The Necker cube is an image that can be viewed in one of two ways – with either of the two faces appearing as the front of the cube (see Figure 6.7). Implementing "weak quantum theory" and the quantum Zeno effect predicts the measured lower-bound threshold of 30 msec, oscillation period for switching of 3 sec, and cognitive time scale (time it takes to reach conscious levels) of 300 msec for perception of the Necker cube [165]. This cognitive time scale is roughly equivalent to the 0.5 sec delay to consciousness described by Libet [166].

In yet another model, object recognition has been compared to a series of quantum observations with a mental state being equivalent to a "perpetually evolving fitness landscape" [167]. There are more models that relate conscious visual perception to quantum mechanical computations. Gamma-band oscillations, which have long been suggested to correlate with visual perception, for example, can be fitted to the Schrödinger equation and may thereby enable quantum mechanics to account for the binding of different aspects of a visual stimulus represented in different cortical areas by near-zero phase lag synchrony [168]. Finally, one model suggests quantum computations in microtubules might account for the conscious aspects of visual perception [101].

In addition to associative memory and perception, quantum models have been applied to general neurocognitive states such as attention, arousal, and consciousness. Pop-Jordanov and Pop-Jordanova have proposed a field-dipole quantum interaction model for attention and arousal [169]. These authors argue that background consciousness, as measured by the EEG, can be analytically expressed and numerically modeled as the probability of collective transitions of quantum dipoles in the cortical electric field. For their calculations, they assume that a neuron has $N = 10^{12}$ dipole molecules. This number is several orders of magnitude larger than the estimated 10^8 tubulins per neuron; however, assemblies of 10^4 neurons would contain that many dipoles. This is approximately the number of neurons in functionally organized slabs of cortex, sometimes referred to as cortical columns [170].

A number of additional quantum mechanical models of consciousness have been proposed; those dealing directly or indirectly with microtubules have already been discussed in this chapter. A number of models related to consciousness deal with electromagnetic fields, which are governed by both classical and quantum mechanical principles. Electromagnetic fields are of particular interest, and because of their penetrability, are uniquely capable of affecting the atom, molecule, cell, and organism as a whole [171]. Hallucinogenic drugs, for example, may induce electromagnetic fields responsible for the altered states of consciousness they produce [172].

Among the key advantages that quantum computing models of neurocognition have over conventional neural computing models are: (1) exponentially increased computational power, (2) a built-in mechanism for autonomy (in the case of models that employ microtubules), and (3) the possibility of further uncovering the biophysical underpinnings of higher cognitive functions. If, as suggested here and elsewhere in this book, microtubules and actin filaments connected in intraneuronal matrices contribute in a major way to these computations, their biophysical properties are highly relevant to way in which these computations are carried out.

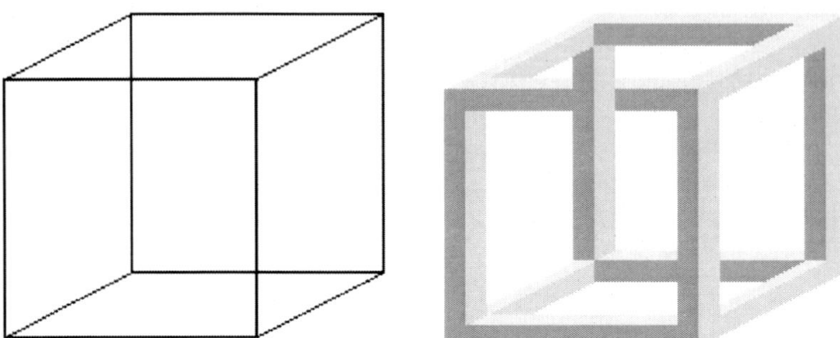

Fig. 6.7. The Necker cube, named after Swiss crystallographer Louis Necker, produces a bistable image that can be perceived as either of the faces being the front of the figure.

6.5 Future Directions and Conclusions

Nanoneuroscience methods stand to increase our understanding of higher brain function in dramatically new ways. This chapter has focused on the application of this perspective to novel modes of neural computation by cytoskeletal proteins, but the same perspective could be applied to matrices that further include neuroreceptors, ion channels, and membrane components. That cytoskeletal polymers resemble nanowires more than other subcellular components makes these biomolecular networks a good starting point in the search for intrinsic control mechanisms relevant to neural information processing underlying neurocognition.

Molecular modeling is one approach that can be used to test ideas about how microtubules and actin filaments compute specific outcomes relevant to neural activity, transport, and neural plasticity – especially concerning how these processes relate to higher neurocognition [98]. Methods enabling direct experimentation include those in which tubulin is labeled by luminescent nanorods (a method which has already been used to study self-assembly of microtubules) [173]. Studies relating self-assembly of microtubules to transport in living neurons could employ this technique. Self-assembly and transport could also be studied following specific training paradigms administered to laboratory animals. Other ways to label tubulin include microinjection of fluorescent-labeled tubulin into neurons or transfection of cells with tubulins fused with green fluorescent proteins (GFPs) [174]. Such methods could be used to analyze rates of polymerization and depolymerization among microtubules in different compartments of the neuron in relation to neural plasticity and higher neurocognitive function. There are also immunocytochemical methods that enable direct visualization of acetylated and tyrosinated tubulins, which could be used to trace histories of microtubule growth. Single-molecule fluorescence spectroscopy and microscopy have been applied to study kinesin transport on microtubules [175], as has the high-resolution, metal-shadowing, cryoelectron microscopy technique [110]. These methods could be employed in studies that directly assess neural plasticity and function.

Focusing on the most computationally relevant biophysical properties, there exist patch clamp methods capable of measuring currents passed along microtubules and actin filaments, which enable the testing of effects of electrical stimulation, increased ionic concentration, and electromagnetic fields [4, 5, 176]. Questions that could be addressed in future experiments using this methodology include: Do transverse or longitudinal electromagnetic fields produce conformational changes in microtubules or actin filaments? Do electromagnetic fields alter MAP binding or kinesin binding? Are adjacent microtubules aggregated into bundles capable of influencing each other through electrostatic interactions or electromagnetic fields?

There have been major inroads recently in designing neural prosthetics that may someday be capable of restoring higher neurocognitive function. Theodore Berger and colleagues have constructed potentially implantable hippocampal circuit boards capable of receiving complex inputs, performing non-linear computations, and delivering complex outputs [177, 178]. Although these designs currently focus on neural spiking activity as described by conventional neural network models, prosthetics that enable higher neurocognitive functions may benefit from including properties and capabilities of intraneuronal matrices composed of cytoskeletal filaments or other biomolecules found in neurons.

References

1. Koch C. Biophysics of Computation: Information Processing in Single Neurons, 2004, Oxford University Press, USA.
2. Häusser M, Mel B. Dendrites: bug or feature? Curr Opin Neurobiol. 2003 Jun;13(3):372-83.
3. US Department of Energy http://www.oe.energy.gov/smartgrid.htm (retrieved Aug. 23, 2008).
4. Tuszyński JA, Portet S, Dixon JM, Luxford C, Cantiello HF. Ionic wave propagation along actin filaments. Biophys J. 2004 Apr;86(4):1890-903.
5. Priel A, Ramos AJ, Tuszyński JA, Cantiello HF. A biopolymer transistor: electrical amplification by microtubules. Biophys J. 2006 Jun 15;90(12):4639-43.
6. Woolf, NJ., Young, S. L., Johnson, G. V., and Fanselow, M. S. (1994) Pavlovian conditioning alters cortical microtubule-associated protein-2. Neuroreport, 5, 1045-1048.
7. Woolf, NJ., Zinnerman, M. D., and Johnson, G. V. (1999) Hippocampal microtubule-associated protein-2 alterations with contextual memory. Brain Research, 821, 241-249.
8. Woolf NJ. A structural basis for memory storage in mammals. Prog Neurobiol. 1998 May;55(1):59-77.
9. Mershin, A., Pavlopoulos, E., Fitch, O., Braden, B. C., Nanopoulos, D.V., and Skoulakis, E.M. (2004b) Learning and memory deficits upon TAU accumulation in Drosophila mushroom body neurons. Learning and Memory, 11, 277-287.
10. Khuchua, Z., Wozniak, D. F., Bardgett, M. E., Yue, Z., McDonald, M., Boero, J., Hartman, R. E., Sims, H., and Strauss, A. W. (2003) Deletion of the N-terminus of murine MAP2 by gene targeting disrupts hippocampal ca1 neuron architecture and alters contextual memory. Neuroscience, 119, 101-111.
11. Kaech S, Brinkhaus H, Matus A. Volatile anesthetics block actin-based motility in dendritic spines. Proc Natl Acad Sci U S A. 1999 Aug 31;96(18):10433-7.
12. Turina D, Loitto VM, Björnström K, Sundqvist T, Eintrei C. Propofol causes neurite retraction in neurones. Br J Anaesth. 2008 Sep;101(3):374-9.
13. Matus A. Actin-based plasticity in dendritic spines. Science. 2000 Oct 27;290(5492):754-8.
14. Yuste, R., and Bonhoeffer, T. (2001) Morphological changes in dendritic spines associated with long-term synaptic plasticity. Annual Review of Neuroscience, 24, 1071-1089.

15. Lisman J. Actin's actions in LTP-induced synapse growth. Neuron. 2003 May 8;38(3):361-2.
16. Matsuzaki M, Honkura N, Ellis-Davies GC, Kasai H. Structural basis of long-term potentiation in single dendritic spines. Nature. 2004 Jun 17;429(6993):761-613.
17. Sataric MV, Tuszyński JA, Zakula RB. Kinklike excitations as an energy-transfer mechanism in microtubules. Phys Rev E Stat Phys Plasmas Fluids Relat Interdiscip Topics. 1993 Jul;48(1):589-597.
18. Tuszyński, J., S. Hameroff, M.V. Sataric, B. Trpisova and M.L.A. Nip. 1995. Ferroelectric behavior in microtubule dipole lattices: implications for information processing, signaling and assembly/disassembly. J. Theor. Biol. 174, 371 (1995).
19. Hameroff, S. R. (1987) Ultimate computing: biomolecular consciousness and nanotechnology. Elsevier Science Publishers.
20. Hameroff, S. (1998) Quantum computation in brain microtubules? The Penrose-Hameroff "Orch OR" model of consciousness. Philosophical Transactions: Mathematical, Physical and Engineering Sciences, 356, 1869-1896.
21. Hameroff, S.R. and Penrose, R. (1996) Conscious events as orchestrated space-time selections. Journal of Consciousness Studies, 3(1), 36-53.
22. Mavromatos N.E. and Nanopoulos, D.V., 1998. On quantum mechanical aspects of microtubules. Int. J. Mod. Phys. B B12, pp. 517-527
23. Mershin A, Kolomenski AA, Schuessler HA, Nanopoulos DV. Tubulin dipole moment, dielectric constant and quantum behavior: computer simulations, experimental results and suggestions. Biosystems. 2004 Nov;77(1-3):73-85.
24. Faber J, Portugal R, Rosa LP. Information processing in brain microtubules. Biosystems. 2006 Jan;83(1):1-9.
25. Rosa LP, Faber J. Quantum models of the mind: are they compatible with environment decoherence? Phys Rev E Stat Nonlin Soft Matter Phys. 2004 Sep;70(3 Pt 1):031902.
26. Sporns O, Chialvo DR, Kaiser M, Hilgetag CC. Organization, development and function of complex brain networks.
27. Casey BJ, Giedd JN, Thomas KM. Structural and functional brain development and its relation to cognitive development. Biol Psychol. 2000 Oct;54(1-3):241-57.
28. Orpwood RD. A possible neural mechanism underlying consciousness based on the pattern processing capabilities of pyramidal neurons in the cerebral cortex. J Theor Biol. 1994 Aug 21;169(4):403-18.
29. LaBerge D. Apical dendrite activity in cognition and consciousness. Conscious Cogn. 2006 Jun;15(2):235-57.
30. Petanjek Z, Judas M, Kostović I, Uylings HB. Lifespan alterations of basal dendritic trees of pyramidal neurons in the human prefrontal cortex: a layer-specific pattern. Cereb Cortex. 2008 Apr;18(4):915-29.
31. Woolf NJ Microtubules in the cerebral cortex: role in memory and consciousness. In: The Emerging Physics of Consciousness, (Ed.) Tuszyński,J., 2006.
32. Olmsted, J. B., and Borisy, G. G. (1975) Ionic and nucleotide requirements for microtubule polymerization in vitro. Biochemistry, 14, 2996-3005.
33. Wolff, J., Sackett, D. L., and Knipling, L. (1996) Cation selective promotion of tubulin polymerization by alkali metal chlorides. Protein Science, 5, 2020-2028.

34. Feria-Velasco, A., Castillo-Medina, S., Verdugo-Diaz, L., Castellanos, E., Orozco-Suarez, S., Sanchez-Gomez, C., and Drucker-Colin, R. (1998) Neural differentiation of chromaffin cells in vitro, induced by extremely low frequency magnetic fields or nerve growth factor: a histological and ultrastructural comparative study. Journal of Neuroscience Research, 53, 569-582.
35. Sanchez, C., Diaz-Nido, J., and Avila, J. (2000) Phosphorylation of microtubule-associated protein 2 (MAP2) and its relevance for the regulation of the neural cytoskeleton function. Progress in Neurobiology, 61, 133-168.
36. Shea TB. Selective stabilization of microtubules within the proximal region of developing axonal neurites. Brain Res Bull. 1999 Feb;48(3):255-61.
37. Kurachi M, Komiya Y, Tashiro T. Biophysical properties of stable microtubules in neurites revealed by optical techniques. Cell Struct Funct. 1999 Oct;24(5):405-12.
38. Nakayama, T., and Sawada, T. (2002) Involvement of microtubule integrity in memory impairment caused by colchicine. Pharmacology Biochemistry and Behavior, 71, 119-138.
39. Wong, R. W., Setou, M., Teng, J., Takei, Y., and Hirokawa N. (2002) Overexpression of motor protein KIF17 enhances spatial and working memory in transgenic mice. Proceedings of the National Academy of Sciences of the United States of America, 99, 14500-14505.
40. Washbourne, P., Bennett, J. E., and McAllister, A. K. (2002) Rapid recruitment of NMDA receptor transport packets to nascent synapses. Nature Neuroscience 5, 751-759.
41. Kakegawa, W., Tsuzuki, K., Yoshida, Y., Kameyama, K., Ozawa, S. (2004) Input- and subunit-specific AMPA receptor trafficking underlying long-term potentiation at hippocampal CA3 synapses. European Journal of Neuroscience, 20, 101-110.
42. Kim, C. H., and Lisman, J. E. (2001) A labile component of AMPA receptor-mediated synaptic transmission is dependent on microtubule motors, actin, and N-ethylmaleimide-sensitive factor. The Journal of Neuroscience, 21, 4188-4194.
43. Yuen EY, Jiang Q, Feng J, Yan Z. Microtubule regulation of N-methyl-D-aspartate receptor channels in neurons. J Biol Chem. 2005 Aug 19;280(33):29420-7.
44. Ouyang Y, Wong M, Capani F, Rensing N, Lee CS, Liu Q, Neusch C, Martone ME, Wu JY, Yamada K, Ellisman MH, Choi DW. Transient decrease in F-actin may be necessary for translocation of proteins into dendritic spines. Eur J Neurosci. 2005 Dec;22(12):2995-3005.
45. Ruiz-Canada, C., Ashley, J., Moeckel-Cole, S., Drier, E., Yin, J., and Budnik V. (2004) New synaptic bouton formation is disrupted by misregulation of microtubule stability in aPKC mutants. Neuron, 27, 567-580.
46. Sarmiere, P. D., and Bamburg, J. R. (2002) Head, neck, and spines: a role for LIMK-1 in the hippocampus. Neuron, 35, 3-5.
47. Meng, Y., Zhang, Y., Tregoubov, V., Janus, C., Cruz, L., Jackson, M., Lu, W. Y., MacDonald, J. F., Wang, J. Y., Falls, D. L., Jia Z. (2002) Abnormal spine morphology and enhanced LTP in LIMK-1 knockout mice. Neuron, 35, 121-133.
48. Adamatzky A. Collision-based computing in biopolymers and their automata models. International Journal of Modern Physics, 2000 Vol:11 No: 7 pp.1321-1346.
49. Glade, N. Existence and persistence of microtubule chemical trails - a step toward microtubule collision-based computing In: From Utopian to Genuine

Unconventional Computers - Splendeurs et misčres du calcul peu usual, Part of the 5th International Conference on Unconventional Computation, UC 2006, York: Royaume-Uni.

50. Scott JF. Applications of modern ferroelectrics. Science. 2007 Feb 16;315(5814):954-9.
51. Craddock, T.J.A., Beauchemin, C., Tuszyński, J.A. Cellular automata modeling of information processing mechanisms in microtubules at biological temperature. Masters Thesis, University of Alberta, Edmonton, Canada.
52. Lupacchini R. Finite machines, mental procedures, and modern physics. Acta Biomed. 2007;78 Suppl 1:39-46.
53. Tegmark M. Importance of quantum decoherence in brain processes. Phys Rev E Stat Phys Plasmas Fluids Relat Interdiscip Topics. 2000 Apr;61(4 Pt B):4194-206.
54. Smith CU. The 'hard problem' and the quantum physicists. Part 2: Modern times. Brain Cogn. 2007
55. Koch C, Hepp K. Quantum mechanics in the brain. Nature. 2006 Mar 30;440(7084):611.
56. Eccles JC. Do mental events cause neural events analogously to the probability fields of quantum mechanics? Proc R Soc Lond B Biol Sci. 1986 May 22;227(1249):411-28.
57. Wiesendanger M. Eccles' perspective of the forebrain, its role in skilled movements, and the mind-brain problem. Prog Neurobiol. 2006 Feb-Apr;78(3-5):304-21.
58. Scott A. Stairway to the Mind: The Controversial New Science of Consciousness, 1995, Springer-Verlag, New York.
59. Vedral V. Quantifying entanglement in macroscopic systems. Nature. 2008 Jun 19;453(7198):1004-7.
60. Mancini S, Giovannetti V, Vitali D, Tombesi P. Entangling macroscopic oscillators exploiting radiation pressure. Phys Rev Lett. 2002 Mar 25;88(12):120401
61. Brukner, C., Vedral, V. and Zeilinger, A. Crucial role of entanglement in bulk properties of solids. Phys. Rev. A 73, 012110 (2006).
62. Davies PC. Does quantum mechanics play a non-trivial role in life? Biosystems. 2004 Dec;78(1-3):69-79.
63. Nagy A, Prokhorenko V, Miller RJ. Do we live in a quantum world? Advances in multidimensional coherent spectroscopies refine our understanding of quantum coherences and structural dynamics of biological systems. Curr Opin Struct Biol. 2006 Oct;16(5):654-63.
64. Numata J, Wan M, Knapp EW. Conformational entropy of biomolecules: beyond the quasi-harmonic approximation. Genome Inform. 2007;18:192-205.
65. Abdel-Halim H, Hanrahan JR, Hibbs DE, Johnston GA, Chebib M. A molecular basis for agonist and antagonist actions at GABA(C) receptors. Chem Biol Drug Des. 2008 Apr;71(4):306-27.
66. Peters MB, Raha K, Merz KM Jr. Quantum mechanics in structure-based drug design. Curr Opin Drug Discov Devel. 2006 May;9(3):370-9.
67. Guzel, Y., and Sivritas, K. (2004) Three-dimensional model of nonsteroidal estrogen receptor ligand binding/electron topological method. Arzneimittelforschung, 54, 348-354.
68. Abbott, J. Gea-Banacloche, PCW Davies, S. Hameroff, A. Zeilinger, J. Eisert, H. Wiseman, SM Bezrukov, and H. Frauenfelder. Plenary debate: quantum effects in biology: trivial or not? 2008, 8: C5-C26.

69. Cavalcanti EG, Reid MD. Signatures for generalized macroscopic superpositions. Phys Rev Lett. 2006 Oct 27;97(17):170405.
70. Matsuno K. Cell motility as an entangled quantum coherence. Biosystems. 1999 Jul;51(1):15-9.
71. Freeman WJ. The wave packet: an action potential for the 21st century. J Integr Neurosci. 2003 Jun;2(1):3-30.
72. Beck F, Eccles JC. Quantum aspects of brain activity and the role of consciousness. Proc Natl Acad Sci U S A. 1992 Dec 1;89(23):11357-61.
73. Schwartz JM, Stapp HP, Beauregard M. Quantum physics in neuroscience and psychology: a neurophysical model of mind-brain interaction. Philos Trans R Soc Lond B Biol Sci. 2005 Jun 29;360(1458):1309-27.
74. Ibarretxe G, Perrais D, Jaskolski F, Vimeney A, Mulle C. Fast regulation of axonal growth cone motility by electrical activity. J Neurosci. 2007 Jul 18;27(29):7684-95.
75. Hilfiker S, Benfenati F, Doussau F, Nairn AC, Czernik AJ, Augustine GJ, Greengard P. Structural domains involved in the regulation of transmitter release by synapsins. J Neurosci. 2005 Mar 9;25(10):2658-69.
76. Tuszyński, J. A., Trpisova, B., Sept, D., and Brown, J. A. (1997) Selected physical issues in the structure and function of microtubules. Journal of Structural Biology, 118, 94-106.
77. Neven H, Aertsen A. Rate coherence and event coherence in the visual cortex: a neuronal model of object recognition. Biol Cybern. 1992;67(4):309-22.
78. Perus M, Bischof H, Caulfield HJ, Loo CK. Quantum-implementable selective reconstruction of high-resolution images. Appl Opt. 2004 Nov 20;43(33):6134-8.
79. Kassal I, Jordan SP, Love PJ, Mohseni M, Aspuru-Guzik A. Polynomial-time quantum algorithm for the simulation of chemical dynamics. Proc Natl Acad Sci U S A. 2008 Dec 2;105(48):18681-6.
80. Dearne, K. The Shock of things to come. Austalian IT, November 11, 2008, http://www.australianit.news.com.au/story/0,24897,24631944-15302,00.html, retrieved 11/30/08.
81. Zhao, Z., Chen, Y. A., Zhang, A. N., Yang T., Briegel, H. J., and Pan, J. W. (2004) Experimental demonstration of five-photon entanglement and open-destination teleportation. Nature, 430, 54-58.
82. Aaronson S. The limits of quantum computers. Sci Am. 2008 Mar;298(3):50-7.
83. Negrevergne C, Mahesh TS, Ryan CA, Ditty M, Cyr-Racine F, Power W, Boulant N, Havel T, Cory DG, Laflamme R. Benchmarking quantum control methods on a 12-qubit system. Phys Rev Lett. 2006 May 5;96(17):170501.
84. Robledo L, Elzerman J, Jundt G, Atatüre M, Högele A, Fält S, Imamoglu A. Conditional dynamics of interacting quantum dots. Science. 2008 May 9;320(5877):772-5.
85. Galland C, Imamoğlu A. All-optical manipulation of electron spins in carbon-nanotube quantum dots. Phys Rev Lett. 2008 Oct 10;101(15):157404.
86. Ardavan, A., Austwick, M., Benjamin, S. C, Briggs, G. A., Dennis, T. J., Ferguson, A., Hasko, D. G., Kanai, M., Khlobystov, A. N., Lovett, B. W., Morley, G. W., Oliver, R. A., Pettifor, D. G., Porfyrakis, K., Reina, J. H., Rice, J.H., Smith, J. D., Taylor, R. A., Williams, D.A., Adelmann, C., Mariette, H., and Hamers, R. J. (2003) Nanoscale solid-state quantum computing. Philosophical Transactions: Mathematical, Physical and Engineering Sciences, 361, 1473-1485.

87. Khlobystov, A. N., Porfyrakis, K., Kanai, M., Britz, DA., Ardavan, A., Shinohara, H., Dennis, T. J., Briggs, G. A. (2004) Molecular motion of endohedral fullerenes in single-walled carbon nanotubes. Angewandte Chemie International Edition, 116, 1410-1413.
88. Monroe CR, Wineland DJ. Quantum computing with ions. Sci Am. 2008 Aug;299(2):64-71.
89. Vinh NQ, Greenland PT, Litvinenko K, Redlich B, van der Meer AF, Lynch SA, Warner M, Stoneham AM, Aeppli G, Paul DJ, Pidgeon CR, Murdin BN. Silicon as a model ion trap: time domain measurements of donor Rydberg states. Proc Natl Acad Sci U S A. 2008 Aug 5;105(31):10649-53.
90. Xu X, Sun B, Berman PR, Steel DG, Bracker AS, Gammon D, Sham LJ. Coherent optical spectroscopy of a strongly driven quantum dot. Science. 2007 Aug 17;317(5840):929-32.
91. Morton, JJL, Tyryshkin AM, Brown RM, Shankar S, Lovett BW, Ardavan A, Schenkel T, Haller EE, Ager JW, Lyon SA. Solid-state quantum memory using the 31P nuclear spin. Nature Lett. 2008, 455: 1085-1088.
92. Fu P. Biomolecular computing: is it ready to take off? Biotechnol J. 2007 Jan;2(1):91-101.
93. Willner I, Shlyahovsky B, Zayats M, Willner B. DNAzymes for sensing, nanobiotechnology and logic gate applications. Chem Soc Rev. 2008 Jun;37(6):1153-65.
94. Karafyllidis IG. Quantum mechanical model for information transfer from DNA to protein. Biosystems. 2008 Sep;93(3):191-8.
95. Pampaloni F, Florin EL. Microtubule architecture: inspiration for novel carbon nanotube-based biomimetic materials. Trends Biotechnol. 2008 Jun;26(6):302-10.
96. Rasmussen S, Karampurwala H, Vaidyanath R, Jensen KJ, Hameroff S. Computational connectionism within neurons: a model of cytoskeletal automata subserving neural networks. Proceedings of the ninth annual international conference of the Center for Nonlinear Studies on Self-organizing, Collective, and Cooperative Phenomena in Natural and Artificial Computing Networks on Emergent computation, 1990, 428-449.
97. Hameroff S, Nip A, Porter M, Tuszyński J. Conduction pathways in microtubules, biological quantum computation, and consciousness. Biosystems. 2002 Jan;64(1-3):149-68.
98. Priel A, Tuszyński JA, Woolf NJ. Transitions in microtubule C-termini conformations as a possible dendritic signaling phenomenon. Eur Biophys J. 2005 Dec;35(1):40-52.
99. Moehring DL, Maunz P, Olmschenk S, Younge KC, Matsukevich DN, Duan LM, Monroe C. Entanglement of single-atom quantum bits at a distance. Nature. 2007 Sep 6;449(7158):68-71.
100. Jibu M, Hagan S, Hameroff SR, Pribram KH, Yasue K. Quantum optical coherence in cytoskeletal microtubules: implications for brain function. Biosystems. 1994;32(3):195-209.
101. Woolf, N. J., and Hameroff, S. R. (2001) A quantum approach to visual consciousness. Trends in Cognitive Science, 5, 472-478.
102. Woolf, N. J. (1996) Global and serial neurons form a hierarchically-arranged interface proposed to underlie learning and cognition. Neuroscience, 74, 625-651.

103. Van der Zee, E. A., Douma, B. R., Bohus, B., and Luiten, P. G. (1994) Passive avoidance training induces enhanced levels of immunoreactivity for muscarinic acetylcholine receptor and coexpressed PKC gamma and MAP-2 in rat cortical neurons. Cerebral Cortex, 4, 376-90.
104. Pan JZ, Xi J, Tobias JW, Eckenhoff MF, Eckenhoff RG. Halothane binding proteome in human brain cortex. J Proteome Res. 2007 Feb;6(2):582-92.
105. Xi J, Liu R, Asbury GR, Eckenhoff MF, Eckenhoff RG. Inhalational anesthetic-binding proteins in rat neuronal membranes. J Biol Chem. 2004 May 7;279(19):19628-33.
106. Pan JZ, Xi J, Eckenhoff MF, Eckenhoff RG. Inhaled anesthetics elicit region-specific changes in protein expression in mammalian brain. Proteomics. 2008 Jul;8(14):2983-92.
107. Eckenhoff RG. Promiscuous ligands and attractive cavities: how do the inhaled anesthetics work? Mol Interv. 2001 Dec;1(5):258-68.
108. Nogales, E. (2001) Structural insight into microtubule function. Annual Review of Biophysics and Biomolecular Structure, 30, 397-420.
109. Keskin, O., Durell, S. R., Bahar, I., Jernigan, R.L., and Covell, D. G. (2002) Relating molecular flexibility to function: a case study of tubulin. Biophysical Journal, 83, 663-680.
110. Santarella, R.A., Skiniotis, G., Goldie, K. N., Tittmann, P., Gross, H., Mandelkow, E. M., Mandelkow, E., and Hoenger, A. (2004) Surface-decoration of microtubules by human tau. Journal of Molecular Biology, 339, 539-553.
111. Kielpinski D, Meyer V, Rowe MA, Sackett CA, Itano WM, Monroe C, Wineland DJ. A decoherence-free quantum memory using trapped ions. Science. 2001 Feb 9;291(5506):1013-5.
112. Lidar DA, Whaley KB. Decoherence-free subspaces and subsystems. In: Irreversible Quantum Dynamics, F. Benatti and R. Floreanini (Eds.), pp. 83-120 (Springer Lecture Notes in Physics vol. 622, Berlin, 2003)
113. Hagan, S., Hameroff, S. R., and Tuszyński, J. A. (2002) Quantum computation in brain microtubules: decoherence and biological feasibility. Physical Review E, 65, 061901.
114. Meunier V, Sumpter BG. Amphoteric doping of carbon nanotubes by encapsulation of organic molecules: electronic properties and quantum conductance. J Chem Phys. 2005 Jul 8;123(2):24705.
115. Krenner HJ, Pryor CE, He J, Petroff PM. A semiconductor exciton memory cell based on a single quantum nanostructure. Nano Lett. 2008 Jun;8(6):1750-5.
116. Ouyang M, Awschalom DD. Coherent spin transfer between molecularly bridged quantum dots. Science. 2003 Aug 22;301(5636):1074-8.
117. Hu H, Wu M. Spin-mediated consciousness theory: possible roles of neural membrane nuclear spin ensembles and paramagnetic oxygen. Med Hypotheses. 2004;63(4):633-46.
118. Fahle M. Figure-ground discrimination from temporal information. Proc Biol Sci. 1993 Dec 22;254(1341):199-203.
119. Morley GW, McCamey DR, Seipel HA, Brunel L-C, van Tol J, Boehme C. Long-Lived Spin Coherence in Silicon with an Electrical Spin Trap Readout. Physical Rev Lett. 2008, 101:207602.
120. Bodovitz S. Consciousness is discontinuous: the perception of continuity requires conscious vectors and needs to be balanced with creativity. Med Hypotheses. 2004;62(6):1003-5.

121. Seitz A, Kojima H, Oiwa K, Mandelkow EM, Song YH, Mandelkow E. Single-molecule investigation of the interference between kinesin, tau and MAP2c. EMBO J. 2002 Sep 16;21(18):4896-905.
122. Bressloff PC. Stochastic model of protein receptor trafficking prior to synaptogenesis. Phys Rev E Stat Nonlin Soft Matter Phys. 2006 Sep;74(3 Pt 1):031910.
123. Bressloff PC, Earnshaw BA. Diffusion-trapping model of receptor trafficking in dendrites. Phys Rev E Stat Nonlin Soft Matter Phys. 2007 Apr;75(4 Pt 1):041915.
124. Eriksson M, Samuelsson H, Samuelsson EB, Liu L, McKeehan WL, Benedikz E, Sundström E. The NMDAR subunit NR3A interacts with microtubule-associated protein 1S in the brain. Biochem Biophys Res Commun. 2007 Sep 14;361(1):127-32.
125. Williams JM, Guévremont D, Mason-Parker SE, Luxmanan C, Tate WP, Abraham WC. Differential trafficking of AMPA and NMDA receptors during long-term potentiation in awake adult animals. J Neurosci. 2007 Dec 19;27(51):14171-8.
126. Mok, H., Shin, H., Kim, S., Lee, J. R., Yoon, J., and Kim, E. (2002) Association of the kinesin superfamily motor protein KIF1Balpha with postsynaptic density-95 (PSD-95), synapse-associated protein-97, and synaptic scaffolding molecule PSD-95/discs large/zona occludens-1 proteins. The Journal of Neuroscience, 22, 5253-5258.
127. Steward, O., and Schuman, E. M. (2003) Compartmentalized synthesis and degradation of proteins in neurons. Neuron, 40, 347-359.
128. Ostroff, L. E., Fiala, J. C., Allwardt, B., and Harris, K. M. (2002) Polyribosomes redistribute from dendritic shafts into spines with enlarged synapses during LTP in developing rat hippocampal slices. Neuron, 35, 535-545.
129. Craddock TJA, Tuszyński JA. Role of the microtubules in cognitive functions NeuroQuantology, 2007 , 5:32-57.
130. Tuszyński J, Luchko T, Carpenter E, Crawford E. Results of molecular dynamics computations of the structural and electrostatic properties of tubulin and their consequences for microtubules. Journal of Computational and Theoretical Nanoscience, 2005, 1:1-6.
131. Brunel N. Dynamics of sparsely connected networks of excitatory and inhibitory spiking neurons, J. Comput. Neurosci. 8 (2000), pp. 183-208.
132. Rioult-Pedotti MS, Friedman D, Donoghue JP. Learning-induced LTP in neocortex. Science. 2000 Oct 20;290(5491):533-6.
133. Monfils MH, Teskey GC. Skilled-learning-induced potentiation in rat sensorimotor cortex: a transient form of behavioural long-term potentiation. Neuroscience. 2004;125(2):329-36.
134. Rogan MT, Stäubli, U.V. and LeDoux, J.E., 1997. Fear conditioning induces associative long-term potentiation in the amygdala. Nature 390, pp. 604-607
135. Withers GS and Greenough WT, Reach training selectively alters dendritic branching in subpopulations of layer II-III pyramids in rat motor-somatosensory forelimb cortex. Neuropsychologia 1989. 27, 61-69.
136. Escobar G, Fares T, Stepanyants A. Structural plasticity of circuits in cortical neuropil. J Neurosci. 2008 Aug 20;28(34):8477-88.
137. Benavides-Piccione R, Ballesteros-Yáñez I, DeFelipe J, Yuste R. Cortical area and species differences in dendritic spine morphology. J Neurocytol. 2002 Mar-Jun;31(3-5):337-46.

138. Dehmelt, L., Smart, F. M., Ozer, R. S., and Halpain, S. (2003) The role of microtubule-associated protein 2c in the reorganization of microtubules and lamellipodia during neurite initiation. The Journal of Neuroscience, 23, 9479-9490.
139. Harada, A., Teng, J., Takei, Y., Oguchi, K., Hirokawa, N. (2002) MAP2 is required for dendrite elongation, PKA anchoring in dendrites, and proper PKA signal transduction. The Journal of Cell Biology, 158, 541-549.
140. Matus, A. (1994) Stiff microtubules and neural morphology. Trends in Neurosciences, 17, 19-22.
141. Westermann, S., and Weber, K. (2003) Post-translational modifications regulate microtubule function. Nature Reviews Molecular Cell Biology, 4, 938-947.
142. Infante, A. S., Stein, M. S., Zhai, Y., Borisy, G. G., and Gundersen, G. G. (2000) Detyrosinated (Glu) microtubules are stabilized by an ATP-sensitive plus-end cap. Journal of Cell Science, 113, 3907-3919.
143. Jasmin, B. J., Changeux, J.P., and Cartaud, J. (1990) Compartmentalization of cold-stable and acetylated microtubules in the subsynaptic domain of chick skeletal muscle fibre. Nature, 344, 673-675.
144. Brown, A., Li, Y., Slaughter,T., and Black, M. M. (1993) Composite microtubules of the axon: quantitative analysis of tyrosinated and acetylated tubulin along individual axonal microtubules. Journal of Cell Science, 104, 339-352.
145. Reed NA, Cai D, Blasius TL, Jih GT, Meyhofer E, Gaertig J, Verhey KJ. Microtubule acetylation promotes kinesin-1 binding and transport. Curr Biol. 2006 Nov 7;16(21):2166-72.
146. Cheng HH, Huang ZH, Lin WH, Chow WY, Chang YC. Cold-induced exodus of postsynaptic proteins from dendritic spines. J Neurosci Res. 2008 Aug 28.
147. Hubel DH, Wiesel TN. Receptive fields, binocular interaction and functional architecture in the cat's visual cortex. J Physiol. 1962 Jan;160:106-54.
148. Tootell RB, Switkes E, Silverman MS, Hamilton SL. Functional anatomy of macaque striate cortex. II. Retinotopic organization. J Neurosci. 1988 May;8(5):1531-68.
149. Fenske MJ, Aminoff E, Gronau N, Bar M. Top-down facilitation of visual object recognition: object-based and context-based contributions. Prog Brain Res. 2006;155:3-21.
150. Miller BT, D'Esposito M. Searching for "the top" in top-down control. Neuron. 2005 Nov 23;48(4):535-8.
151. Gray, J. A. (2004) Consciousness: creeping up on the hard problem. Oxford University Press.
152. Gage N, Hickok G. Multiregional cell assemblies, temporal binding and the representation of conceptual knowledge in cortex: a modern theory by a "classical" neurologist, Carl Wernicke. Cortex. 2005 Dec;41(6):823-32.
153. Edelman, G. M. (2003) Naturalizing consciousness: a theoretical framework. Proceedings of the National Academy of Sciences of the United States of America, 100, 5520-5524.
154. Massimini M, Rosanova M, and Mariotti M. EEG slow (approximately 1 Hz) waves are associated with nonstationarity of thalamo-cortical sensory processing in the sleeping human. Journal of Neurophysiology 2003. 89, 1205-1213.
155. John ER. From synchronous neuronal discharges to subjective awareness? Prog Brain Res. 2005;150:143-71.
156. Freeman WJ. A field-theoretic approach to understanding scale-free neocortical dynamics. Biol Cybern. 2005 Jun;92(6):350-9.

157. Khrennikov A. Quantum-like brain: "Interference of minds". Biosystems. 2006 Jun;84(3):225-41.
158. Khrennikov A. Quantum-like model of cognitive decision making and information processing. Biosystems. 2008 Nov 1.
159. Hosten O, Rakher MT, Barreiro JT, Peters NA, Kwiat PG. Counterfactual quantum computation through quantum interrogation. Nature. 2006 Feb 23;439(7079):949-52.
160. Ventura D, and Martinez T. Quantum associative memory. Information Science 2000. 124, 273-296.
161. Howell J, Yeazell J, and Ventura D. Optically simulating a quantum associative memory, Physical Review A 2000. 62, 042303.
162. Trugenberger C. Probablistic quantum memories. Physical Review Letters 2001, 87, 067901.
163. Diamantini MC, Trugenberger CA. Quantum pattern retrieval by qubit networks with Hebb interactions. Phys Rev Lett. 2006 Sep 29;97(13):130503.
164. Perus̃ M, Bischof H, and Kiong LC. Quantum-implemented selective reconstruction of high-resolution images. Applied Optics 2004. 43, in press.
165. Atmanspacher H, Filk T, Römer H. Quantum Zeno features of bistable perception. Biol Cybern. 2004 Jan;90(1):33-40.
166. Libet B. Reflections on the interaction of the mind and brain. Prog Neurobiol. 2006 Feb-Apr;78(3-5):322-6.
167. Igamberdiev AU. Objective patterns in the evolving network of non-equivalent observers. Biosystems. 2008 May;92(2):122-31.
168. Robinson PA. Visual gamma oscillations: waves, correlations, and other phenomena, including comparison with experimental data. Biol Cybern. 2007 Oct;97(4):317-35.
169. Pop-Jordanov J, Pop-Jordanova N. Neurophysical substrates of arousal and attention. Cogn Process. 2008 Oct 31.
170. Voigt BC, Brecht M, Houweling AR. Behavioral detectability of single-cell stimulation in the ventral posterior medial nucleus of the thalamus. J Neurosci. 2008 Nov 19;28(47):12362-12367.
171. Persinger MA, Koren SA. A theory of neurophysics and quantum neuroscience: implications for brain function and the limits of consciousness. Int J Neurosci. 2007 Feb;117(2):157-75.
172. Steele RH. Harmonic oscillators: the quantization of simple systems in the old quantum theory and their functional roles in biology. Mol Cell Biochem. 2008 Mar;310(1-2):19-42.
173. Riegler J, Nick P, Kielmann U, and Nann T. Visualizing the self-assembly of tubulin with luminescent nanorods. Journal of Nanoscience and Nanotechnology 2003, 3, 380-385.
174. Freitag M, Hickey PC, Raju NB, Selker EU, and Read ND. GFP as a tool to analyze the organization, dynamics and function of nuclei and microtubules in Neurospora crassa. Genetics and Biology 2004, 41, 897-910.
175. Peterman EJ, Sosa H, and Moerner WE. Single-molecule fluorescence spectroscopy and microscopy of biomolecular motors. Annual Review of Physical Chemistry 2004. 55, 79-96.
176. Lin EC, Cantiello HF. A novel method to study the electrodynamic behavior of actin filaments. Evidence for cable-like properties of actin. Biophys J. 1993 Oct;65(4):1371-8.

177. Zanos TP, Courellis SH, Berger TW, Hampson RE, Deadwyler SA, Marmarelis VZ. Nonlinear modeling of causal interrelationships in neuronal ensembles. IEEE Trans Neural Syst Rehabil Eng. 2008 Aug;16(4):336-52.
178. Song D, Chan RH, Marmarelis VZ, Hampson RE, Deadwyler SA, Berger TW. Nonlinear dynamic modeling of spike train transformations for hippocampal-cortical prostheses. IEEE Trans Biomed Eng. 2007 Jun;54(6 Pt 1):1053-66.

Index